Nucleic Acids in Chemistry and Biology

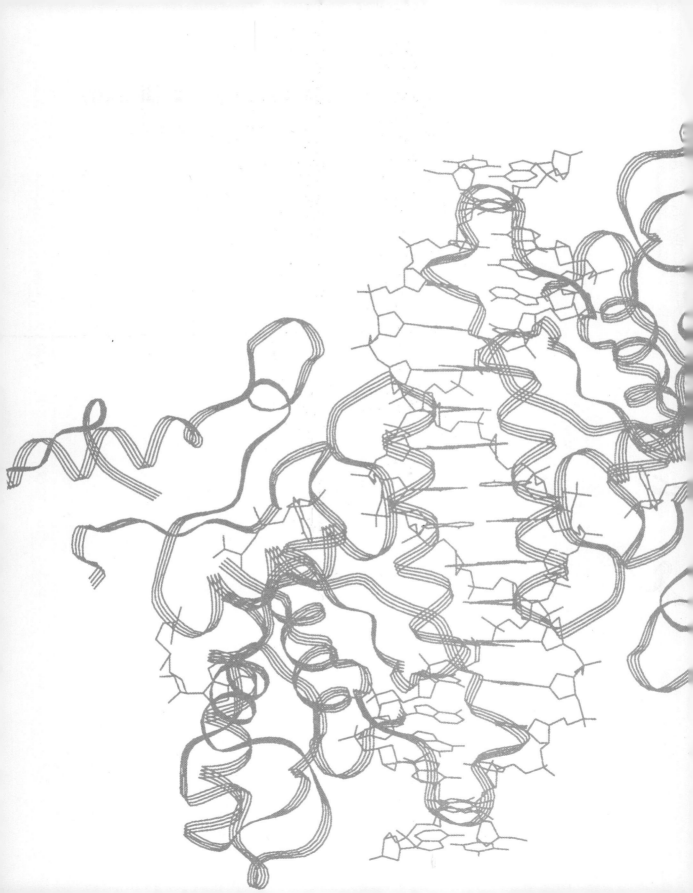

Nucleic Acids in Chemistry and Biology

Edited by

G. MICHAEL BLACKBURN
Department of Chemistry, University of Sheffield

and

MICHAEL J. GAIT
MRC Laboratory of Molecular Biology, Cambridge

IRL PRESS
—at—
OXFORD UNIVERSITY PRESS
Oxford New York Tokyo

1990

Oxford University Press, Walton Street, Oxford OX2 6DP

Oxford   New York   Toronto
Delhi   Bombay   Calcutta   Madras   Karachi
Petaling Jaya   Singapore   Hong Kong   Tokyo
Nairobi   Dar es Salaam   Cape Town
Melbourne   Auckland
and associated companies in
Berlin   Ibadan

Oxford is a trade mark of Oxford University Press

Published in the United States
by Oxford University Press, New York

British Library Cataloguing in Publication Data
Blackburn, G. M.
Nucleic acids in chemistry and biology.
1. Nucleic acids
I. Title   II. Gait, M. J. (Michael J.)
547.79
ISBN 0–19–963120–4
ISBN 0–19–963121–2 pbk

Library of Congress Cataloging-in-Publication Data
Nucleic acids in chemistry and biology/edited by G. Michael Blackburn and Michael J. Gait.
Includes bibliographical references and index.
1. Nucleic acids.   I. Blackburn, G. Michael.   II. Gait, M. J. (Michael J.)
QD433.N83 1990   547.7'9–dc20   90–14187
ISBN 0–19–963120–4 (hardback)
ISBN 0–19–963121–2 (paperback)

Set by Promenade Graphics, Cheltenham
Printed in Great Britain by
Butler & Tanner Ltd, Frome, Somerset

Preface

Nucleic acids dominate modern molecular science. They have vital roles that are fundamental for the storage and transmission of genetic information within cells. It follows that an accurate and detailed knowledge of their structure and function is of prime importance for molecular scientists of all descriptions. Just as significantly, the genius of biologists and chemists working together has made contemporary research into nucleic acids a rich source of discovery and invention that is dramatically transforming and improving the human condition.

Our own teaching and research experience, shared in discussions with colleagues around the world, has convinced us of the need to fill a significant gap in the modern science library by creating a broad-based yet concise and readable book on nucleic acids. Our single volume is designed to provide a compact, molecular perspective of this great subject. To that end, we have used it unashamedly to emphasize chemical and structural aspects of nucleic acids at all points. In it, we have surveyed a very broad field—up to the point where the frontiers of current studies in nucleic acids are only attainable by reading the latest issues of key journals! In particular, we have strengthened its production by drawing on the talents of an international group of co-authors whose expertise has extended the authority of this book from cover to cover. At the same time, we have tried to keep it selective and simple so as to make it widely accessible to students. This has meant that, of necessity, some sections have focused more on key concepts rather than on fine detail.

We have tried to provide a radically fresh and unified approach. This book builds on a general introduction to the chemistry and biology of the nucleic acids in order to reach out to some of the most significant modern developments of this subject. It is couched in an easily readable style and in a language which, while technically accurate, can yet be grasped quickly by those with a basic scientific background.

We begin with a brief historical perspective designed to point out the significance of later progress. We next provide an outline of the essential features of DNA and RNA structure, highlighting the new subtle insights which have been obtained by detailed analysis of crystals of synthetic oligonucleotides of defined sequence. The next four chapters are the core of the book. The first concentrates on modern chemistry applied to the synthesis of biologically important nucleosides, nucleotides, and oligonucleotides. Then comes a discussion of the biosynthesis of nucleotides, which is given a fresh presentation to emphasize how anti-cancer and anti-viral agents interfere with biosynthetic processes. The core is completed by two chapters which deal with the basic molecular biology of DNA and of RNA, showing how information stored in the form of nucleotide sequence is transmitted into cellular activity. Recent exciting developments in the auto-catalysis of RNA, ribozymes, are especially featured.

Three rather more specialized chapters then focus on the covalent and physical interactions of nucleic acids with small molecules, especially with mutagens and carcinogens and the relevant repair processes, and on their physical interactions with proteins. These important topics are at the forefront of much present research and typify the success of creative symbioses between chemistry and biology. The final chapter contrasts the *in vivo* rearrangements which DNA experiences with the *in vitro* techniques of manipulation of DNA sequences that are the essence of experimentation in cloning and mutagenesis.

This is not a textbook on the molecular biology of nucleic acids. From the outset, we have aimed this book especially at the needs of students and new research workers with a chemical or biochemical background. We hope that molecular biologists and more senior chemists and biochemists alike will find their knowledge of nucleic acids broadened through the special perspectives this book offers.

Sheffield and Cambridge
November 1989

G.M.B.
M.J.G.

Acknowledgements

Both Mikes express their sincere appreciation of the efforts of all who have supported the production of this book. Principally, our unqualified thanks go to those eight expert and understanding co-authors without whose contributions this book would not have been possible. We have taken considerable liberties with their manuscripts, perhaps more than is the norm with such a multi-author book, because we wanted to blend all of their contributions into a homogeneous final product. They have responded to these efforts with equanimity and understanding and have co-operated superbly in the numerous revision processes required for the production of the finished work.

We are very grateful to the following of our colleagues and fellow-scientists who have read and commented on portions of the text during its formative stages, namely Tom Brown, Chris Christodoulou, Bernard Connolly, David Hornby, Paul Kong, Christian Lehmann, Mick McLean, Daniella Rhodes, Ian Willis, and also to Joachim Engels for supplying most of the definitions in the glossary. We particularly wish to thank Lord Todd, Gobind Khorana, and Dan Brown for suggestions and comments on the early part of Chapter 1 which have, we believe, given us a genuine feeling for the key events in those seminal early years of nucleic acid studies.

The final production of this book has been supported by many able individuals. We are particularly indebted to Colin Yeomans for a substantial amount of the original artwork, to Peter Artymiuk (Sheffield University), Simon Phillips (Leeds University), and Tom Steitz (Yale University) for access to original graphics for nucleic acids and protein structures, and above all to the staff of OUP for establishing new standards for the style and presentation of this volume.

Finally, it is inevitable in a book of this breadth that omissions, occasional errors, and lapses in the accuracy of interpretation will have escaped the detection of even the most assiduous proof-readers. We hope that any such mistakes are both minor and minimal and we accept full and exclusive responsibility for them. We shall be grateful to receive your help in their identification for future rectification.

Contents

Contributors

G. MICHAEL BLACKBURN,
Department of Chemistry, University of Sheffield, Sheffield, S3 7HF, UK

MICHAEL J. GAIT,
MRC Laboratory of Molecular Biology, Hills Road, Cambridge, CB2 2QH, UK

GORDON C. BARR,
Department of Biochemistry, Medical Sciences Institute, University of Dundee, Dundee, DD1 4HN, UK

ANDREW J. FLAVELL,
Department of Biochemistry, Medical Sciences Institute, University of Dundee, Dundee, DD1 4HN, UK

ADRIAN GOLDMAN,
Waksman Institute of Microbiology, State University of New Jersey, Rutgers, P.O. Box 759, Piscataway, New Jersey, 08854–0759, USA

DAVID L. OLLIS,
Departments of Biochemistry, Molecular Biology and Cell Biology, Northwestern University, 2153 Sheridan Drive, Evanston, Illinois, 60201, USA

UTTAM L. RAJBHANDARY,
Departments of Biology and Chemistry, Massachusetts Institute of Technology, 77 Massachusetts Avenue, Cambridge, Massachusetts, 02138, USA

DIETER SÖLL,
Department of Molecular Biophysics and Biochemistry, Yale University, P.O. Box 6666, New Haven, Connecticut, 06511, USA

RICHARD T. WALKER,
Department of Chemistry, University of Birmingham, P.O. Box 363, Birmingham, B15 2TT, UK

W. DAVID WILSON,
Department of Chemistry, State University of Georgia, Atlanta, Georgia, 30303, USA

Nomenclature

The nomenclature for nucleic acids and their constituents used in this book is derived from the following recommendations:

IUPAC–IUB Joint Commission on Biochemical Nomenclature. Abbreviations and symbols for nucleic acids, polynucleotides and their constituents. Recommendations 1970. (1970) *Biochemistry* **9**, 4022–7.

IUPAC–IUB Joint Commission on Biochemical Nomenclature. Abbreviations and symbols for the description of conformations of polynucleotide chains. Recommendations 1982. (1983) *Eur. J. Biochem.*, **131**, 9–15.

Definitions and nomenclature of nucleic acid structure parameters. (1989) *EMBO J.*, **8**, 1–4.

Stereodiagrams

The stereo-pair figures used in this book are for parallel viewing, i.e., left diagram to left eye and right to right. They can be viewed either unaided (a little practice helps) or with the help of a simple convex-lens viewer.

Introduction and overview

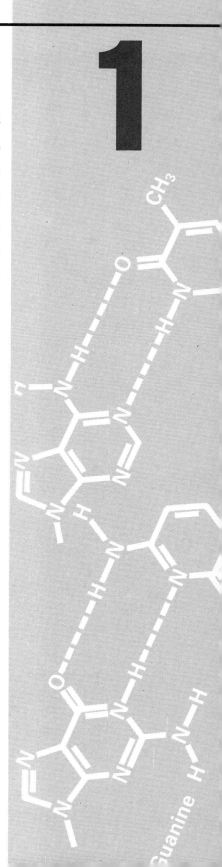

1.1 The biological importance of DNA

From the beginning, the study of nucleic acids has drawn together, as though by a powerful unseen force, a galaxy of scientists of the highest ability. Striving to tease apart its secrets, these talented individuals have brought with them a broad range of skills from other disciplines while many of the problems they have encountered have proved to be soluble only by new inventions. Looking at their work, one is constantly made aware that scientists in this field appear to have enjoyed a greater sense of excitement in their work than is given to most. Why?

For over 60 years, such men and women have been fascinated and stimulated by their awareness that the study of nucleic acids is central to a knowledge of life. Let us start by looking at Fred Griffith, who was employed as a scientific civil servant in the British Ministry of Health investigating the nature of epidemics. In 1923, he was able to identify the difference between a virulent, S, and a non-virulent, R, form of the pneumonia bacterium. Griffith went on to show that this bacterium could be made to undergo a permanent, hereditable change from non-virulent to virulent type. This discovery was a bombshell in bacterial genetics.

Oswald Avery and his group at the Rockefeller Institute in New York set out to identify the molecular mechanism responsible for the change Griffith had discovered, now technically called **bacterial transformation**. They achieved a breakthrough in 1940 when they found that non-virulent R pneumococci could be transformed *irreversibly* into a virulent species by treatment with a pure sample of high molecular weight DNA. Avery had purified this DNA from heat-killed bacteria of a virulent strain and showed that it was active at a dilution of 1 part in 10^9.

Avery concluded that '**DNA is responsible for the transforming activity**' and published that analysis in 1944, just three years after Griffith had died in a London air-raid. The staggering implications of Avery's work turned a searchlight on the molecular nature of nucleic acids and it soon became evident that ideas on the chemistry of nucleic acid structure at that time were wholly inadequate to explain such a momentous discovery. As a result, a new wave of scientists directed their attention to DNA and discovered that large parts of the accepted tenets of nucleic acid chemistry had to be set aside before real progress was possible. We need to examine some of the earliest features of that chemistry to appreciate fully the significance of later progress.

1.2 The origins of nucleic acids research

Friedrich Miescher started his research career in Tübingen by looking into the physiology of human lymph cells. In 1868, seeking a more readily available material, he began to study human pus cells which he obtained in abundant supply from the bandages discarded from the local hospital. After defatting the cells with alcohol, he incubated them with a crude preparation of pepsin from pig stomach and so obtained a grey precipitate of pure cell nuclei. Treatment of this with alkali followed by acid gave Miescher a precipitate of a phosphorus-containing substance which he named '**nuclein**'. He later found this material to be a common constituent of yeast, kidney, liver, testicular, and nucleated red blood cells.

After Miescher moved to Basel in 1872, he found the sperm of Rhine salmon to be a more plentiful source of nuclein. The pure nuclein was a strongly acidic substance which existed in a salt-like combination with a nitrogenous base which Miescher crystallized and called protamine. In fact, his nuclein was really a nucleoprotein and it fell subsequently to Richard Altman in 1889 to obtain the first protein-free material, to which he gave the name '**nucleic acid**'.

Following William Perkin's invention of mauveine in 1856, the development of aniline dyes had stimulated a systematic study of the colour-staining of biological specimens. Cell nuclei were characteristically stained by basic dyes, and around 1880 Walter Flemming applied that property in his study of the rod-like segments of chromatin (so called because of their colour-staining characteristic) which became visible within the cell nucleus only at certain stages of cell division. Flemming's speculation that the chemical composition of these **chromosomes** was identical with that of Miescher's nuclein was confirmed in 1900 by E. B. Wilson who wrote:

Now chromatin is known to be closely similar to, if not identical with, a substance known as nuclein which analysis shows to be a tolerably definite chemical compound of nucleic acid and albumin. And thus we reach the remarkable conclusion that inheritance may, perhaps, be affected by the physical transmission of a particular compound from parent to offspring.

While this insight was later to be realized in Griffith's 1928 experiments, all of this work was really far ahead of its time. We have to recognize that, at the turn of the century, tests for the purity and identity of substances were relatively primitive. Emil Fischer's classic studies on the chemistry of high molecular weight, polymeric organic molecules was under question until well into the twentieth century. Even in 1920, it was possible to argue that there were only two species of nucleic acids in nature: animal cells were believed to provide **thymus nucleic acid** (DNA), whilst nuclei of plant cells were thought to give **pentose nucleic acid** (RNA).

1.3 Early structural studies on nucleic acids

Accurate molecular studies on nucleic acids essentially date from 1909 when Levene and Jacobs began a reinvestigation of the structure of **nucleotides** at the Rockefeller Institute. Inosinic acid, which Liebig had isolated from beef muscle in 1847, proved to be hypoxanthine-riboside 5'-phosphate. Guanylic acid, isolated from the nucleoprotein of pancreas glands, was identified as guanine-riboside 5'-phosphate (Fig. 1.1). Each of these nucleotides was cleaved by alkaline hydrolysis to give phosphate and the corresponding **nucleosides**, inosine and guanosine respectively. Since then, all nucleosides are characterized as the condensation products of a pentose and a nitrogenous base while nucleotides are the phosphate esters of one of the hydroxyl groups of the pentose.

Pentose nucleic acid was available in plentiful supply from yeast and on mild hydrolysis with aqueous ammonia it gave the four pentose-nucleosides adenosine, cytidine, guanosine, and uridine. These were identified as derivatives of the four bases adenine, cytosine, guanine, and uracil (Fig. 1.1).

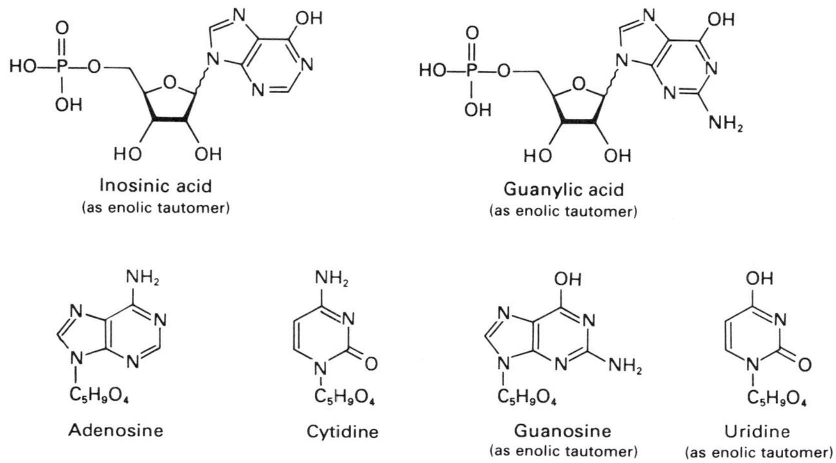

Fig. 1.1 Early nucleoside and nucleotide structures (using the enolic tautomers originally employed).

Thymus nucleic acid, which was readily available from calf tissue, was found to be resistant to alkaline hydrolysis. It was only successfully degraded into deoxy-nucleosides in 1929 when Levene adopted enzymes to hydrolyse the deoxyribo-nucleic acid followed by mild acidic hydrolysis of the deoxynucleotides. He identified its pentose as the hitherto unknown 2-deoxy-D-ribose. These deoxy-nucleosides involved the four heterocyclic bases, adenine, cytosine, guanine, and thymine, with the latter corresponding to uracil in ribonucleic acid.

Up to 1940, most groups of workers were convinced that hydrolysis of nucleic acids gave the appropriate four bases in **equal relative proportions**. This erroneous conclusion probably resulted from the use of impure nucleic acid or from the use of analytical methods of inadequate accuracy and reliability. It led, naturally enough, to the general acceptance of a '**Tetranucleotide hypothesis**' for the structure of both thymus and yeast nucleic acids, which materially retarded further progress on the molecular structure of nucleic acids.

Several of these tetranucleotide structures were proposed. They all had four nucleosides (one for each of the bases) with an arbitrary location of the two purines and two pyrimidines. They were joined together by four phosphate residues in a variety of ways, among which there was a strong preference for phosphodiester linkages. In 1932, Takahashi showed that yeast nucleic acid contained neither pyrophosphate nor phosphomonoester functions and so disposed of earlier proposals in preference for a neat, cyclic structure which joined the pentoses exclusively using phosphodiester units (Fig. 1.2). It was generally accepted that these bonded 5'- to 3'-positions of adjacent deoxyribonucleosides, but the linkage positions in ribonucleic acid were not known.

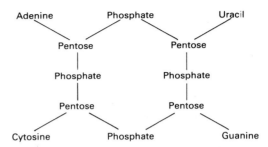

Fig. 1.2 The tetranucleotide structure proposed for nucleic acids by Takahashi (1932).

One property stuck out like a sore thumb from this picture: the molecular weight of nucleic acids was greatly in excess of that calculated for a tetranucleotide. The best DNA samples were produced by Einar Hammarsten in Stockholm and one of his students, Torjbörn Caspersson, showed that this material was greater in size than protein molecules. Hammarsten's DNA was examined by Rudolf Signer in Bern whose flow-birefringence studies revealed rod-like molecules with a molecular weight of $0.5–1.0 \times 10^6$ Daltons (Da). The same material provided Astbury in Leeds with X-ray fibre diffraction measurements that supported Signer's conclusion. Finally, Levene estimated the molecular weight of native DNA at between 200 000 and one million based on ultracentrifugation studies.

Scientists compromised. In his Tilden Lecture of 1943, Masson Gulland suggested that the concept of nucleic acid structures of polymerized, uniform tetranucleotides was limited, but he allowed that they could 'form a practical working hypothesis'.

This then was the position in 1944 when Avery published his great work on the transforming activity of bacterial DNA. One can sympathize with Avery's hesitance to press home his case. Levene, in the same Institute, and others were strongly persuaded that the tetranucleotide hypothesis imposed an invariance on the structure of nucleic acids which denied them any role in biological diversity. By contrast, Avery's work showed that DNA was responsible for completely transforming the behaviour of bacteria. It demanded a fresh look at the structure of nucleic acids.

1.4 The discovery of the structure of DNA

From the outset, it was evident that DNA exhibited greater resistance to selective chemical hydrolysis than did RNA. So, the discovery in 1935 that DNA could be cut into **mononucleotides** by an enzyme doped with arsenate was invaluable. Using this procedure, Klein and Thannhauser obtained the four crystalline deoxyribonucleotides whose structures (Fig. 1.3) were later put beyond doubt by total chemical synthesis by Alexander Todd and the Cambridge school he founded in 1944. Todd established the β-configuration of the glycosidic linkage for ribonucleosides in 1951, but found the chemical synthesis of the 2′-deoxyribonucleosides more taxing. The key to success for the Cambridge group was the development of methods of phosphorylation, illustrated in Fig. 1.4 for the preparation of the 3′- and 5′-phosphates of deoxyadenosine (Fig. 1.4).

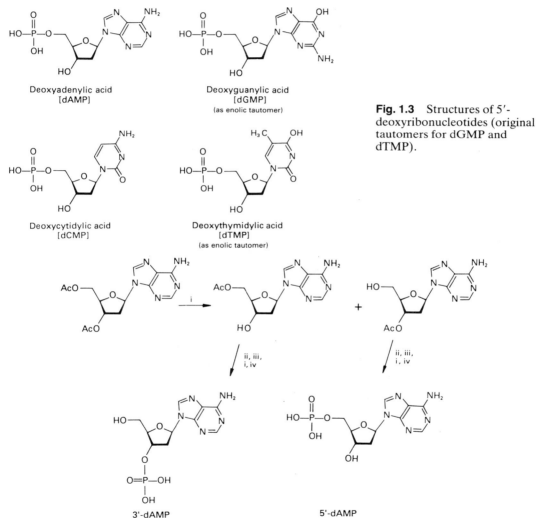

Deoxyadenylic acid
[dAMP]

Deoxyguanylic acid
[dGMP]
(as enolic tautomer)

Fig. 1.3 Structures of 5′-deoxyribonucleotides (original tautomers for dGMP and dTMP).

Deoxycytidylic acid
[dCMP]

Deoxythymidylic acid
[dTMP]
(as enolic tautomer)

ii, iii,
i, iv

ii, iii,
i, iv

3′-dAMP

5′-dAMP

Fig. 1.4 Todd's syntheses of deoxyadenosine 3′- and 5′-phosphates (Hayes, D. H., Michelson, A. M., and Todd, A. R. (1955). *J. Chem. Soc.*, 808–15).
Reagents: (i) MeOH, NH₃ (ii) (PhO)₂P(O)OP(H)(O)OCH₂Ph (iii) *N*-chlorosuccinimide (iv) H₂/PdC

All the facts were now available to establish the primary structure of DNA as a **linear polynucleotide** in which each deoxyribonucleoside is linked to the next by means of a 3′- to 5′-phosphodiester (Chapter 2, Fig. 2.15). The presence of only diester linkages was essential to explain the stability of DNA to chemical hydrolysis, since phosphate triesters and monoesters, not to mention pyrophosphates, are more labile. The measured molecular weights for DNA of about one million meant that a single strand of DNA would have some 3000 nucleotides. Such a size was much greater than that of enzyme molecules, but entirely compatible with Staudinger's established ideas on macromolecular structure for synthetic and natural polymers. But by the mid-point of the twentieth century, chemists could advance no further with the primary structure of DNA. Neither of the key requirements for sequence determination was to hand: there were no methods for obtaining pure samples of DNA with homogeneous base sequence nor were methods available for the cleavage of DNA strands at a specific base residue. Consequently, all attention came to focus on the secondary structure of DNA.

Two independent experiments in biophysics showed that DNA possesses an ordered secondary structure. Using a sample of DNA obtained from Hammarsten in 1938, Astbury obtained an X-ray diffraction pattern from stretched, dry fibres of DNA. From the rather obscure data he deduced ' . . . A spacing of 3.34 Å along the fibre axis corresponds to that of a close succession of flat or flattish nucleotides standing **out** perpendicularly to the long axis of the molecule to form a relatively rigid structure.' These conclusions roundly contradicted the tetranucleotide hypothesis.

Some years later, Gulland studied the viscosity and flow-birefringence of calf thymus DNA and thence postulated the presence of hydrogen bonds linking the purine–pyrimidine **hydroxyl** groups and some of the amino groups. He suggested that these hydrogen bonds could involve nucleotides either in adjacent chains or within a single chain, but he somewhat hedged his bets between these alternatives.

Sadly, Astbury returned to the investigation of proteins and Gulland died prematurely in a train derailment in 1947. Both of them left work that was vital for their successors to follow, but each contribution contained a misconception that was to prove a stumbling block for the next half-a-dozen years. Thus, Linus Pauling's attempt to create a helical model for DNA located the pentose-phosphate backbone in its core and the **bases pointing outwards**—as Astbury had decided. Gulland had subscribed to the wrong tautomeric forms for the heterocyclic bases thymine and guanine, believing them to be **enolic** and having hydroxyl groups. The importance of the true **keto forms** was only appreciated in 1952.

Erwin Chargaff began to investigate a very different type of order in DNA structure. He studied the base composition of DNA from a variety of sources using the new technique of paper chromatography to separate the products of hydrolysis of DNA and employing one of the first commercial ultraviolet spectrophotometers to quantify their relative abundance. His data showed that there is a variation in base composition of DNA between species which is overridden by a universal 1:1 ratio of adenine with thymine and of guanine with cytosine. This meant that the proportion of purines, (A+G), is always equal to the proportion of pyrimidines, (C+T). Although the ratio (G+C) / (A+T) varies from species to species, different tissues from a single species give DNA of the same composition. Chargaff's results finally discredited the tetranucleotide hypothesis, because it called for equal proportions of all four bases in DNA.

In 1951, Francis Crick and Jim Watson joined forces in the Cavendish Laboratory in Cambridge to tackle the problem of DNA structure. Both of them were persuaded that the model-building approach that had led Pauling and Corey to the α-helix structure for peptides should work just as well for DNA. Almost incredibly, they attempted no other line of direct experimentation but drew on the published and unpublished results of other research teams in order to construct a variety of models, each to be discarded in favour of the next until they created one which satisfied all the facts.

The best X-ray diffraction results were to be found in King's College, London. There, Maurice Wilkins had observed the importance of keeping DNA fibres in a moist state and Rosalind Franklin had found that the X-ray diffraction pattern obtained from such fibres showed the existence of an A-form of DNA at low humidity which changed into a B-form at high humidity. Both forms of DNA were highly crystalline and clearly helical in structure. Consequently, Franklin decided that this behaviour required the phosphate groups to be exposed to water on the **outside** of the helix, with the corollary that the bases were on the **inside** of the helix.

Watson decided that the number of nucleotides in the unit crystallographic cell favoured a double-stranded helix. Crick's physics-trained mind recognized the symmetry implications of the space-group of the A-form diffraction pattern, monoclinic C^2. There had to be local twofold symmetry axes normal to the helix, a feature which called for a double-stranded helix whose two chains must run in opposite directions.

Crick and Watson thus needed merely to solve the final problem: how to construct the core of the helix by packing the bases together in a regular structure. Watson knew about Gulland's conclusions regarding hydrogen-bonds joining the DNA bases. This convinced him that the crux of the matter had to be a rule governing hydrogen-bonding between bases. Accordingly, Watson experimented with models using the **enolic** tautomeric forms of the bases (Fig. 1.3) and pairing like-with-like. This structure was quickly rejected by Crick because it had the wrong symmetry for B-DNA. **Self-pairing** had to be rejected because it could not explain Chargaff's 1:1 base ratios, which Crick had perceived were bound to result if you had **complementary base-pairing**.

Based on advice from Jerry Donohue in the Cavendish Laboratory, Watson turned to manipulating models of the bases in their **keto forms** and paired adenine with thymine and guanine with cytosine. Almost at once he found a compellingly simple relationship involving two hydrogen bonds for an adenine–thymine pair and two or three hydrogen bonds for a guanine–cytosine pair. The special feature of this base-pairing scheme is that the relative geometry of the bonds joining the bases to the pentoses is virtually identical for the A:T and G:C pairs (Fig. 1.5). It follows that if a purine always pairs with a pyrimidine then an irregular sequence of bases in a single strand of DNA could nonetheless be paired regularly in the centre of a double helix and without loss of symmetry.

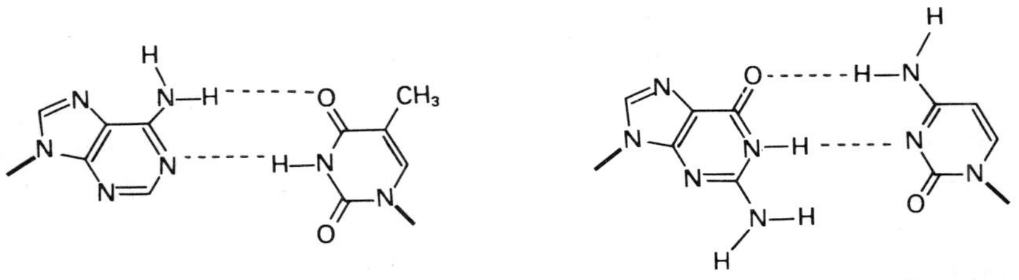

Fig. 1.5 Complementary hydrogen-bonded base-pairs as proposed by Watson and Crick (thymine and guanine in the revised ketoforms). *NB* The guanine–cytosine structure was later altered to include three hydrogen bonds.

Chargaff's 'Rules' were straightaway revealed as an obligatory consequence of a double-helical structure for DNA. Above all, since the base sequence of one chain automatically determines that of its partner, Crick and Watson could easily visualize how one single chain might be the template for creation of a second chain of complementary base sequence.

The structure of the core of DNA had been solved and the whole enterprise fittingly received the ultimate accolade of the scientific establishment when Crick, Watson, and Wilkins shared the Nobel prize for Chemistry in 1962, just four years after Rosalind Franklin's early death.

1.5 The advent of molecular biology

It is common to ascribe the publication of Watson and Crick's paper in *Nature* in April 1953 as the end of the 'classical' period in the study of nucleic acids, up to which time basic discoveries were made by a few gifted academics in an otherwise relatively unexplored field. The excitement aroused by the model of the double helix certainly drew the attention of a much wider scientific audience to the importance of nucleic acids than hitherto, particulary because of the biological implications of the model rather than because of the structure itself. It was immediately apparent that locked into the irregular sequence of nucleotide bases in the DNA of a cell was all the information required to specify the diversity of biological molecules needed to carry out the functions of that cell. The important question now was what was the key, the **genetic code**, through which the sequence of DNA could be translated into protein.

The solution to the coding problem is often attributed to the laboratories in the USA of Marshall Nirenberg and of Severo Ochoa who devised an elegant cell-free system for translating enzymatically synthesized polynucleotides into polypeptides and who by the mid 1960s had established the genetic code for a number of amino acids. In reality, the story of the elucidation of the code involves numerous strands of knowledge obtained from a variety of workers in different laboratories. An essential contribution came from Alexander Dounce in Rochester, New York, who in the early 1950s postulated that RNA, and not DNA, served as a template to direct the synthesis of cellular proteins and that a sequence of three nucleotides might specify a single amino acid. Sydney Brenner and Leslie Barnett in Cambridge later (1961) confirmed the code to be both triplet and non-overlapping. From Robert Holley in Cornell University, New York, and Hans Zachau in Cologne, came the isolation and determination of the sequence of three **transfer RNAs** (tRNA) 'adapter' molecules that each carry an individual amino acid ready for incorporation into protein and which are also responsible for recognizing the triplet code on the **messenger RNA** (mRNA). The mRNA species contain the sequences of individual genes copied from DNA (Chapter 6). Gobind Khorana and his group in Madison, Wisconsin, chemically synthesized all 64 ribotrinucleoside diphosphates and, with a combination of chemistry and enzymology, synthesized a number of polyribonucleotides with repeating di-, tri-, and tetranucleotide sequences. These were used as synthetic mRNAs to help identify each triplet in the code. This work was recognized by the award of the Nobel prize for Medicine in 1968 jointly to Holley, Khorana, and Nirenberg.

Nucleic acid research in the 1950s and 1960s was preoccupied by the solution to the coding problem and the establishment of the biological roles of tRNA and mRNA. This was not surprising bearing in mind that at that time the smaller size and attainable homogeneity made isolation and purification of RNA a much easier task than for DNA. It was clear that in order to approach the fundamental question of what constituted a **gene**—a single heritable element of DNA that up to then could be defined genetically but not chemically—it was going to be necessary to break down DNA into smaller, more tractable pieces in a specific and predictable way.

The breakthrough came in 1968 when Meselson and Yuan reported the isolation of a **restriction enzyme** from the bacterium *Escherichia coli*. Here at last was an enzyme, a nuclease, that could recognize a defined sequence in a DNA and cut it specifically (Section 9.3). The bacterium used this activity to break down and hence inactivate invading (e.g. phage) DNA. It was soon realized that this was a general property of bacteria and the isolation of other restriction enzymes with different specificities soon followed. But it was not until 1973 that the importance of these enzymes became apparent. At this time, Chang and Cohen at Stanford and Helling and Boyer at the University of California were able to construct in a test tube a biologically functional DNA that combined genetic information from two different sources. This **chimera** was created by cleaving DNA from one source with a restriction enzyme to give a fragment that could then be joined to a carrier DNA, a **plasmid**. The resultant **recombinant DNA** was shown to be able to replicate and express itself in *E. coli*.

This remarkable demonstration of genetic manipulation was to revolutionize biology. It soon became possible to dissect out an individual gene from its source DNA, to amplify it in a bacterium or other organism (**cloning**, Section 10.2), and to study its expression by the synthesis first of RNA and then of protein (Chapters 5 and 6). This single advance by the groups of Cohen and Boyer truly marked the dawn of modern molecular biology.

1.6 The partnership of chemistry and biology

In the 1940s and 1950s the disciplines of chemistry and biology were so separate that it was a rare occurrence for an individual to embrace both. Two young scientists who were just setting out on their careers at that time were exceptional in recognizing the potential of chemistry in the solution of biological problems and both, in their different ways, were to have a substantial and lasting effect in the field of nucleic acids.

One was Frederick Sanger, a product of the Cambridge Biochemistry School, who in the early 1950s set out to determine the sequence of a protein, insulin. This feat had been thought unattainable, since it was widely supposed that proteins were not discrete species with defined primary sequence. Even more remarkably, he went on to develop methods for sequence determination first of RNA and then of DNA (Section 5.2). These methods involved a subtle blend of enzymology and chemistry that few would have thought possible to combine. The results of his efforts have transformed DNA sequencing in only a few years into such a routine procedure that the determination of the sequence of the complete human genome is now regarded as a serious proposition. The award of two Nobel prizes to Sanger (1958 and 1980) hardly seems recognition enough!

The other scientist has already been mentioned in connection with the elucidation of the genetic code. From not long after his post-doctoral studies under George Kenner and Alexander Todd in Cambridge, Gobind Khorana was convinced that chemical synthesis of polynucleotides could make an important contribution to the study of the fundamental process of information flow from DNA to RNA to protein. Having completed the work on the genetic code in the mid 1960s and aware of Holley's recently determined (1965) sequence for an alanine tRNA, he then established a new goal of total synthesis of the corresponding DNA duplex, the gene specifying the tRNA. Like Sanger, he ingeniously devised a combination of nucleic acid chemistry and enzymology to form a general strategy of gene synthesis, which in principle remains unaltered to this day (Section 3.5). Knowledge became available by the early 1970s about the signals required for gene expression and the newly emerging recombinant DNA methods of Cohen and Boyer allowed a second synthetic gene, this time specifying the precursor of a tyrosine suppressor tRNA (Fig. 1.6), to be cloned and shown to be fully functional.

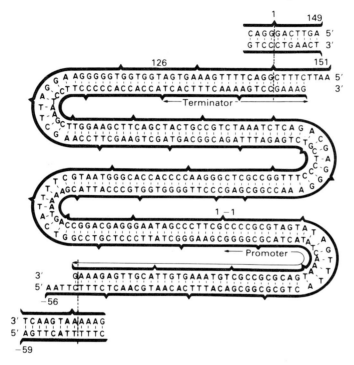

Fig. 1.6 Khorana's totally synthetic DNA corresponding to the tyrosine suppressor transfer RNA gene (from Belagaje, R. *et al.* (1978)). *Chemistry and biology of nucleosides and nucleotides* (ed. R. E. Harmon, R. K. Robins, and L. B. Townsend). Academic Press, New York.

It is ironic that even up to the early 1970s Khorana's gene syntheses were regarded by many biologists to be unlikely to have practical value. Today, synthetic genes are used routinely in the production of proteins. More than this, oligodeoxyribonucleotides, the short pieces of single-stranded DNA for which Khorana developed the first chemical syntheses, have become invaluable general tools in the manipulation of DNA. They are used as primers in DNA sequencing (Section 5.2), as probes in gene detection and isolation (Section 10.2.3), as mutagenic agents to alter the sequence of DNA (Section 10.3), and more recently as potential therapeutic agents (Section 3.4.6).

The availability of synthetic DNA also provided new impetus in the study of DNA structure. In the early 1970s, new X-ray crystallographic techniques had been developed and applied to solve the structure of ApU (Rich and co-workers in MIT, USA). This was followed by the complete structure of yeast phenylalanine tRNA (independently by Rich and by Klug and colleagues in Cambridge). For the first time, the complementary base-pairing between two strands could be seen in greater detail than was previously possible from studies of DNA and RNA fibres. ApU formed a double-helix by end-to-end packing of molecules, with Watson–Crick pairing clearly in evidence between each strand. The tRNA showed not only Watson–Crick pairs, but also a variety of alternative base-pairs and base-triples, many of which were entirely novel (Chapter 6).

Then in 1978, the structure of synthetic d(pATAT) was solved by Kennard and her group in Cambridge. This tetramer also formed an extended double-helix, but excitingly revealed that there was a substantial sequence dependence in its conformation. The angles between neighbouring dA and dT residues were quite different between the A–T sequence and the T–A sequence elements. Soon after, the discovery by Rich that synthetic d(CGCGCG) adopted a totally unpredicted, left-handed Z-conformation and then the demonstration of both a B-DNA helix in a synthetic dodecamer by Dickerson in California and an A-DNA helix in an octamer by Kennard finally put paid to the concept that DNA had a rigid, rod-like structure. Clearly, DNA could adopt different conformations dependent on sequence and also on its external environment (Chapter 2). More importantly, an immediate inference could be drawn that conformational differences in DNA (or the potential for their formation) might be recognized by other molecules. Thus, it was not long before synthetic DNA was also being used in the study of DNA binding to carcinogens and drugs (Chapters 7 and 8) and to proteins (Chapter 9).

These spectacular advances were only possible because of the equally dramatic improvements in methods of oligonucleotide synthesis that took place in the late 1970s and early 1980s. The laborious manual work of the early gene synthesis days was replaced by reliable automated DNA synthesis machines, which within hours could assemble sequences well in excess of 100 residues (Section 3.4). Khorana's vision of the importance of synthetic DNA has been fully realized.

1.7 Future developments

Predicting the future is a pastime more wisely left to fortune-tellers than to scientists. However, there might be some value in the identification of a few fascinating areas of emerging development and the indulgence in brief extrapolation into the future. A particularly clear case is that of synthetic RNA. Both chemical and enzymatic methods of RNA synthesis (Section 3.6) are improving rapidly such that supply of specific sequences should shortly become routine. This will have a profound effect on the study of RNA structure (Section 2.4) and of catalytic RNA (Section 6.5).

Major recent initiatives aimed at conquering the AIDS-causing viruses (HIV I and II), perhaps the greatest health threat of our generation, are resulting in a strong revival in synthesis of nucleoside analogues. AZT (Section 4.7.2) may be just the first in a line of new therapeutic agents aimed at selective interference with viral metabolic pathways.

Another major study area is the understanding of the processes involved in human development. Whereas we have learnt how to detect and isolate genes, to sequence them (Section 5.2), and to clone them (Section 10.2), our knowledge about the control of their *in vivo* expression (Section 5.5), particularly as a function of the various stages of development from embryo through fetus to baby, is still very limited. It is very likely that once again chemistry will play an important role. Sensitive probes of whether a gene is switched on or off at a particular time will need to be developed. Already synthetic oligonucleotides which are complementary to important control regions have been used to turn off gene expression, but more subtle methods are probably desirable. Detailed knowledge of gene control mechanisms will be essential if we are to stand any realistic chance of implementing gene therapy strategies for the major genetic diseases (Section 10.5) or to understand why cells become cancerous. Part of this knowledge will undoubtedly be gained by determination of the structures of control proteins bound to their target nucleic acids. So far, relatively few complexes of protein and nucleic acids have been studied at the atomic level (Chapter 9) because of difficulties in obtaining good co-crystals for X-ray analysis. More powerful 2-D and 3-D NMR techniques may provide us with the tools to probe protein–nucleic acid interactions in a much wider context.

The heady days of the discovery of the double-helix and the elucidation of the genetic code are long gone, but in their place have come even more exciting times when many more of us now have the opportunity to answer fundamental questions about genetic structure and function. 'You ain't heard nothin' yet, folks' (Al Jolson, *The Jazz Singer*, July 1927).

Further reading

1.1–1.4

Avery, O. T., MacLeod, C. M., and McCarty, M. (1944). Studies on the chemical nature of the substance inducing transformation of pneumococcal types. *J. Exp. Med.*, **79**, 137–58.

Chargaff, E. (1950). Chemical specificity of nucleic acids and mechanism of their enzymatic degradation. *Experientia*, **6**, 201–9.

Fruton, J. S. (1972). *Molecules and life*, pp. 180–224. Wiley-Interscience, New York.

Judson, H. F. (1979). *The eighth day of creation.* Jonathan Cape, London.

Miescher, F. (1897). *Die Histochemischen und Physiologischen Arbeiten.* Vogel, Leipzig.

Olby, R. (1973). *The path to the double helix.* Macmillan, London.

Portugal, F. H. and Cohen, J. S. (1977). *A century of DNA.* MIT Press, Cambridge, MA.

Watson, J. D. and Crick, F. H. C. (1953). A structure for deoxyribose nucleic acid. *Nature*, **171**, 737–8.

Watson, J. D. (1968). *The double helix.* Athenaeum Press, New York.

1.5

Cohen, S. N. (1975). The manipulation of genes. *Sci. Amer.*, **233**(1), 24–33.

Khorana, H. G. (1965). Polynucleotide synthesis and the genetic code. *Fed. Proc.*, **24** 1473–87.

Nirenberg, M. W., Matthai, J. H., Jones, O. W., Martin, R. G., and Barondes, S. H. (1963). Approximation of genetic code *via* cell-free protein synthesis directed by template RNA. *Fed. Proc.*, **22,** 55–61.

Ochoa, S. (1963). Synthetic polynucleotides and the genetic code. *Fed. Proc.*, **22**, 62–74.

1.6

Khorana, H. G. (1979). Total synthesis of a gene. *Science*, **203**, 614–25.

Sanger, F. (1988). Sequences, sequences, and sequences. *Ann. Rev. Biochem.*, **57**, 1–28.

DNA and RNA structure

2

2.1 Structures of components

Nucleic acids are very long, thread-like polymers, made up of a linear array of monomers called **nucleotides**. Different nucleic acids can have from around 80 nucleotides, as in tRNA, to over 10^8 nucleotide-pairs in a single eukaryotic chromosome. The unit of size of a nucleic acid is the base-pair (for double-stranded species) or base (for single-stranded species) with the units Mb (million base-pairs) and kb (thousand base-pairs). The chromosome in *E. coli* has 4 million base-pairs, 4 Mb, which gives it a molecular weight of 3×10^9 Da and a length of 1.5 mm. The size of the (haploid) fruit fly genome is 180 Mb which, shared between four chromosomes, gives a total length of 56 mm. How are these extraordinarily long molecules constructed?

2.1.1 Nucleosides and nucleotides

Nucleotides are the phosphate esters of nucleosides and these are components of both ribonucleic acid (RNA) and deoxyribonucleic acid (DNA). RNA is made up of ribonucleotides while the monomers of DNA are 2′-deoxyribonucleotides.

All nucleotides are constructed from three components: a nitrogen heterocyclic **base**, a pentose **sugar**, and a **phosphate** residue. The major bases are monocyclic **pyrimidines** or bicyclic **purines** (some species of tRNA have tricyclic minor bases such as the Wye (Chapter 3, Fig. 3.17)). The major purines are **adenine (A)** and **guanine (G)** and are found in both DNA and RNA. The major pyrimidines are **cytosine (C)**, **thymine (T)**, and **uracil (U)** (Fig. 2.1).

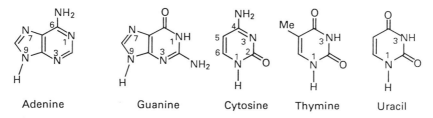

Fig. 2.1 Structures of the five major purine and pyrimidine bases of nucleic acids in their dominant tautomeric forms and with the IUPAC numbering systems for purines and pyrimidines.

In **nucleosides**, the purine or pyrimidine base is joined from a ring nitrogen to carbon-1 of a pentose sugar. In ribonucleic acid, the pentose is **D-ribose** which is locked into a five-membered **furanose** ring by the bond from C-1 of the sugar to N-1 of C or U or to N-9 of A or G. This bond is on the same side of the sugar ring as the C-5 hydroxymethyl group and is defined as a β-glycosidic linkage (Fig. 2.2). In DNA, the pentose is 2-deoxy-D-ribose and the four nucleosides are **deoxyadenosine**, **deoxyguanosine**, **deoxycytidine**, and **deoxythymidine** (Fig. 2.3). In DNA, the methylated pyrimidine base thymine takes the place of uracil in RNA,

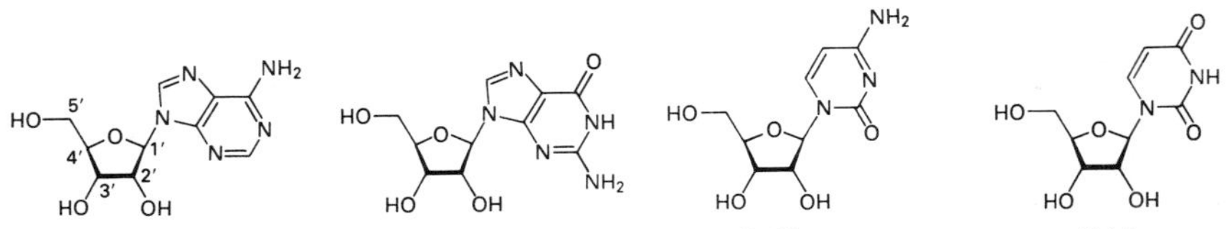

Fig. 2.2 Structures of the four ribonucleosides. The bases retain the same numbering system and the pentose carbons are numbered 1′ through 5′. By convention, the furanose ring is drawn with its ring oxygen at the back and C-2′ and C-3′ at the front. Hydrogen atoms are usually omitted for clarity.

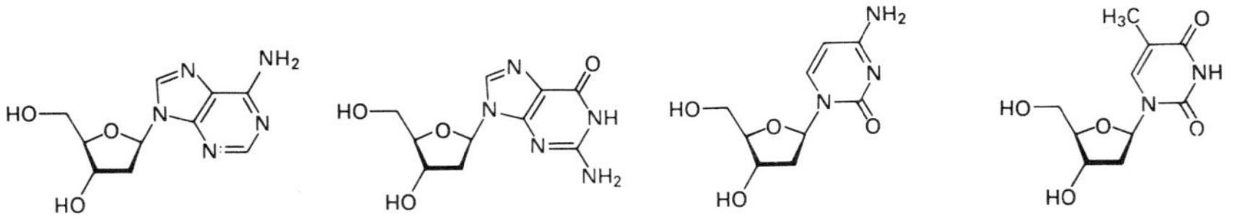

Fig. 2.3 Structures of the four major deoxynucleosides. By convention, only hydrogens bonded to oxygen or nitrogen are depicted.

and its nucleoside with deoxyribose is still commonly called thymidine. However, since the discovery of **ribothymidine** as a regular component of tRNA species (Section 6.4.1), it has been preferable to use the name deoxythymidine rather than thymidine. Unless indicated otherwise, it is assumed that nucleosides, nucleotides, and oligonucleotides are derived from D-pentofuranose sugars.

The **phosphate** esters of nucleosides are **nucleotides**, and the simplest of them have one of the hydroxyl groups of the pentose esterified by a single phosphate monoester function. Adenosine 5′-phosphate is a **5′-ribonucleotide** also called adenylic acid and abbreviated to AMP (Fig. 2.4). Similarly, deoxycytidine 3′-phosphate is a **3′-deoxynucleotide**, identified as 3′-dCMP. Nucleotides which have two phosphate monoesters on the same sugar are called **nucleoside bisphosphates** while nucleoside monoesters of pyrophosphoric acid are **nucleoside diphosphates**. By extension, nucleoside esters of tripolyphosphoric acid are **nucleoside triphosphates**, of which the classic example is adenosine 5′-triphosphate (ATP) (Section 3.3.2). Finally, cyclic nucleotides are nucleosides which have two neigh-

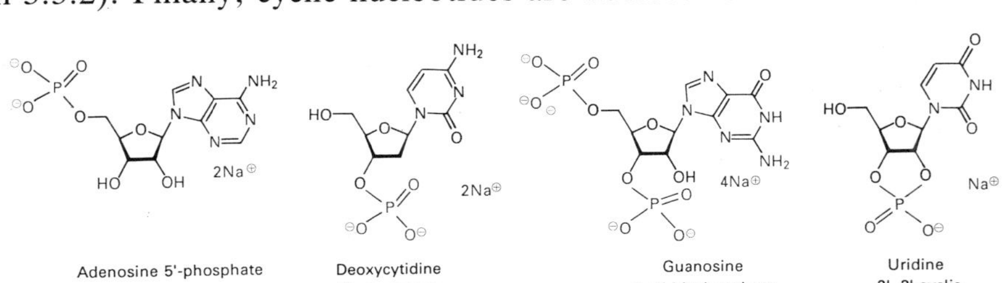

Fig. 2.4 Structures of some common nucleotides. All are presented as their sodium salts in the state of ionization observed at neutral pH.

bouring hydroxyl groups on the same pentose esterified by a single phosphate as a diester. The most important of these is adenosine 3′,5′-cyclic phosphate (cAMP).

In the most abbreviated nomenclature currently employed, **pN** stands for a 5′-nucleotide, **Np** for a 3′-nucleotide, and **dNp** for a 3′-deoxynucleotide (to be precise, a 2′-deoxynucleoside 3′-phosphate). This shorthand notation is based on the convention that an oligonucleotide chain is drawn horizontally with its 5′-hydroxyl group at the left and its 3′-hydroxyl group at the right-hand end. Thus, pppGpp is the shorthand representation for the 'magic spot' nucleotide, guanosine 3′-diphosphate 5′-triphosphate, while ApG is short for adenylyl-(3′→5′)-guanosine, whose 3′→5′ internucleotide linkage runs from the nucleoside on the left to that on the right of the phosphate.

2.1.2 Physical properties of nucleosides and nucleotides

Because of their polyionic character, nucleic acids are soluble in water up to about 1 per cent w/v according to size and are precipitated by the addition of alcohol. Their solutions are quite viscous and the long nucleic acid molecules are easily sheared by stirring or by passage through a fine nozzle (such as a hypodermic needle or a fine pipette).

Ionization

The acid–base behaviour of a nucleotide is its most important physical characteristic. It determines its charge, its tautomeric structure, and thus its ability to donate and accept hydrogen bonds, which is the key feature of base:base recognition. The pK_a values for the five bases in the major nucleosides and nucleotides are listed in Table 2.1.

Table 2.1. pK_a values for bases in nucleosides and nucleotides

Base (site of protonation)		Nucleoside	3′-Nucleotide	5′-Nucleotide
Adenine	(N–1)	3.52	3.70	3.88
Cytosine	(N–3)	4.17	4.43	4.56
Guanine	(N–7)	3.3	(3.5)	(3.6)
Guanine	(N–1)	9.42	9.84	10.00
Thymine	(N–3)	9.93	—	10.47
Uracil	(N–3)	9.38	9.96	10.06

These data approximate to 20°C and zero salt concentration. They correspond to *loss* of a proton for $pK_a > 9$ and *capture* of a proton for $pK_a < 5$

It is clear that all of the bases are uncharged in the physiological range 5 < pH < 9. The same is true for the pentoses, where the ribose 2′,3′-diol only loses a proton above pH 12 while isolated hydroxyl groups ionize only above pH 15. The nucleotide phosphates lose one proton at pH 1 and a second proton (in the case of monoesters) at pH 7. This pattern of proton equilibria is shown for AMP across the whole pH range (Fig. 2.5).

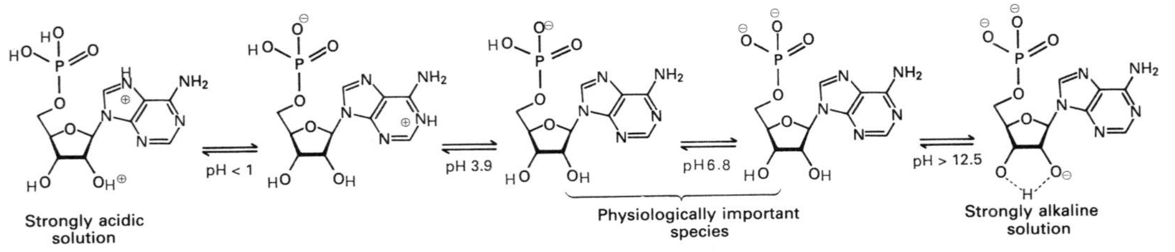

Fig. 2.5 States of protonation of adenosine 5'-phosphate (AMP) from strongly acidic solution (left) to strongly alkaline solution (right).

The three amino bases, A, C, and G, each become protonated on one of the ring nitrogens rather than on the exocyclic amino group since this does not interfere with delocalization of the NH_2 electron lone-pair into the aromatic system. The C–NH_2 bonds of A, C, and G are about 1.34 Å long which means that they have 40–50 per cent double bond order, while the C=O bonds of C, G, T, and U have some 85–90 per cent double bond order. It is also noteworthy that the proximity of negative charge on the phosphate residues has a secondary effect, making the ring nitrogens more basic ($\Delta pK_a \approx +0.4$) and the amine N–Hs less acidic ($\Delta pK_a \approx +0.6$).

Tautomerism

A tautomeric equilibrium involves alternative structures which differ only in the location of hydrogen atoms. The choices available to nucleic acid bases are illustrated by the **keto–enol** equilibrium between 2-pyridone and 2-hydroxypyridine and the **amine–imine** equilibrium for 2-aminopyridine (Fig. 2.6). UV, NMR, and IR spectroscopies have established that the five major bases exist overwhelmingly (>99.99 per cent) in the **amino-** and **keto**-tautomeric forms at physiological pH (see Fig. 2.1) and not in the benzene-like **enol** tautomers in common use before 1950 (Chapter 1, Fig. 1.5).

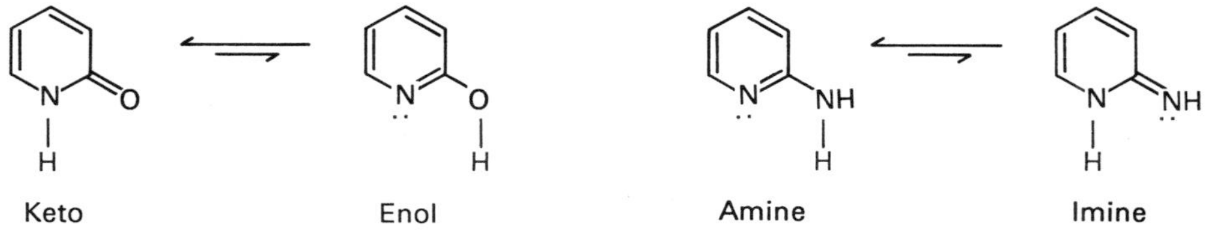

Fig. 2.6 Keto-enol tautomers for 2-pyridone:2-hydroxypyridine (left) and amine–imine tautomerism for 2-aminopyridine (right).

Hydrogen-bonding

The mutual recognition of A by T and of C by G uses hydrogen-bonds to establish the fidelity of DNA transcription and translation. The N–H groups of the bases are good hydrogen-bond donors (**d**), while the sp^2-hybridized electron pairs on the oxygens of the base C=O groups and on the ring nitrogens are much better hydrogen-bond acceptors (**a**) than are the oxygens of either the phosphate or the pentose. The **a·d** hydrogen-bonds so formed are largely ionic in character, with a charge of about +0.2e on the hydrogens and about -0.2e on the oxygens and nitrogens, and they seem to have an average strength of 8–12 kJ mol^{-1}

The predominant amino–keto tautomer for cytosine has a pattern of hydrogen-bond acceptor and donor sites which for $O^2.N^3.N^4$ can be expressed as **a·a·d** (Fig. 2.7). Its minor tautomer has a very different pattern: **a·d·a**. In the same way we can establish that the corresponding pattern for the dominant tautomer of dT is **a·d·a** while the pattern for $N^2.N^1.O^6$ of dG is **d·d·a** (Fig. 2.7) and that for dA is **(−)·a·d**.

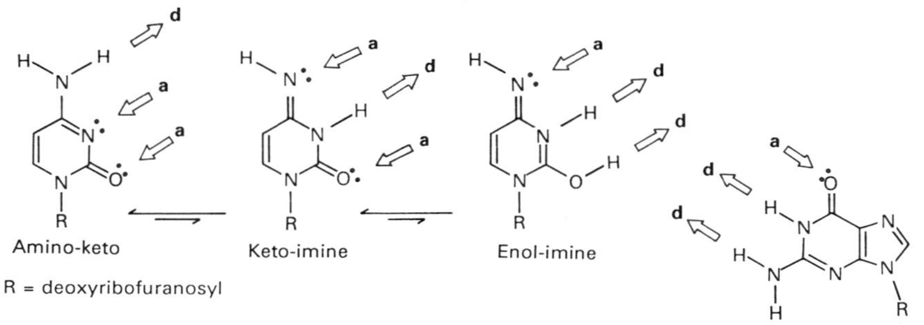

Amino-keto   Keto-imine   Enol-imine

R = deoxyribofuranosyl

Fig. 2.7 Tautomeric equilibria for deoxycytidine showing hydrogen-bond acceptor **a** and donor **d** sites as used in nucleic acid base-pairing. The major tautomer for deoxyguanosine is drawn to show its characteristic **d·d·a** hydrogen-bond donor–acceptor capacity.

When Jim Watson was engaged in DNA model-building studies in 1952 (Section 1.4), he recognized that the hydrogen-bonding capability of an A:T base-pair uses complementarity of **(−)·a·d** to **a·d·a**, while a C:G pair uses the complementarity of **a·a·d** to **d·d·a**. This base-pairing pattern rapidly became known as **Watson–Crick pairing** (Fig. 2.8). There are two hydrogen-bonds in an A:T pair and three in a C:G pair (cf. Chapter 1, Fig. 1.5 for the original pairing scheme). The geometry of these pairs has been fully analysed in many structures from dinucleoside phosphates through oligonucleotides to tRNA species, both by the use of X-ray crystallography and, more recently, by NMR spectroscopy.

In planar base-pairs, the hydrogen-bonds join nitrogen and oxygen atoms that are 2.8 Å to 2.95 Å apart. This geometry gives a C-1′:C-1′ distance of 10.60 ± 0.15 Å with an angle of 68 ± 2° between the two glycosidic bonds for both the A:T and the C:G base-pairs. As a result of this **isomorphous geometry**, the four

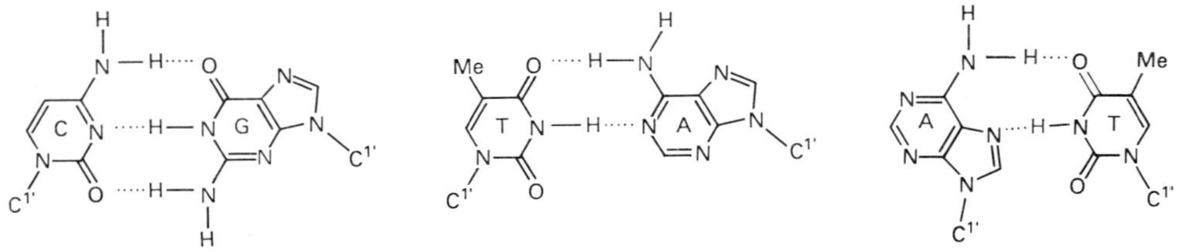

Fig. 2.8 Watson–Crick base-pairing for C:G (left) and T:A (centre). Hoogsteen base-pairing for A:T (right).

base-pair combinations, A:T, T:A, C:G, and G:C, can all be built into the same regular framework of the DNA duplex.

While Watson–Crick base-pairing is the dominant pattern, other pairings have been suggested of which the most significant to have been identified so far are **Hoogsteen pairs** and Crick **'wobble' pairs**. Hoogsteen pairs, illustrated for A:T, are not isomorphous with Watson–Crick pairs because they have an 80° angle between the glycosidic bonds and an 8.6 Å separation of the anomeric carbons (Fig. 2.8). In the case of Reversed Hoogsteen pairs and Reversed Watson–Crick pairs (not shown), one base is rotated through 180° relative to the other.

. Francis Crick proposed the existence of 'wobble' base-pairings to explain the degeneracy of the genetic code (Section 6.6.8). This phenomenon calls for a single base in the 5'-anticodon position of tRNA to be able to recognize either of the pyrimidines or, alternatively, either of the purines as its 3'-codon base partner. Thus a G:U 'wobble' pair has two hydrogen-bonds, $G-N^1-H \cdots O^2-U$ and $G-O^6 \cdots H-N^3-U$ and this requires a sideways shift of one base relative to its position in the regular Watson–Crick geometry (Fig. 2.9). The resulting loss of a hydrogen-bond leads to reduced stability which may be offset in part by the improved base-stacking (Section 2.3.1) that results from such sideways base displacement.

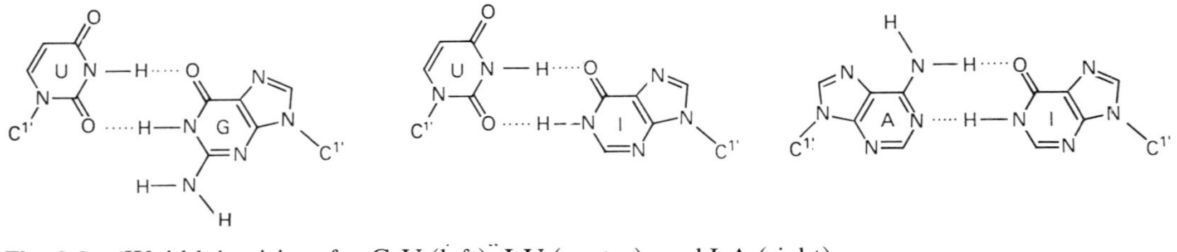

Fig. 2.9 'Wobble' pairings for G:U (left), I:U (centre), and I:A (right).

Base-pairing of these and other non-Watson–Crick patterns is significant in three structural situations. First, the compact structures of tRNAs maximize both base-pairing and base-stacking wherever possible. This has led to the identification of a considerable variety of Reversed Hoogsteen and 'wobble' base-pairs as well as of tertiary base-pairs (or base-triplets) (Section 6.3.2). Secondly, in triple-stranded helices for DNA and RNA, such as (poly(dA).2poly(dT)) and (poly(rG).2poly (rC)), the second pyrimidine chain binds to the purine in the major groove by Hoogsteen hydrogen-bonds and runs parallel to the purine chain (Section 2.4.5). Thirdly, mismatched base-pairs are necessarily identified with anomalous hydrogen-bonding and many such patterns have been revealed by X-ray studies on synthetic oligodeoxyribonucleotides (Section 2.3.2).

2.1.3 Spectroscopic properties of nucleosides and nucleotides

Neither the pentose nor the phosphate components of nucleotides show any significant UV absorption above 230 nm. This means that both nucleosides and nucleotides have UV absorption profiles rather similar to those of their constituent bases and absorb strongly with λ_{max} close to 260 nm and molar extinction coefficients of around 10^4 (Table 2.2).

Table 2.2. Some light absorption characteristics for nucleotides

Compound	$[\alpha_D]^*$	pH 1–2		pH > 11	
		λ_{max} /nm	$10^{-4} \times \varepsilon$	λ_{max} /nm	$10^{-4} \times \varepsilon$
Ado 5'-P	−26°	257	1.5	259	1.54
Guo 3'-P	−57°	257	1.22	257	1.13
Cyd 3'-P	+27°	279	1.3	272	0.89
Urd 2'-P	+22°	262	0.99	261	0.73
Thd 5'-P	+7.3°	267	1.0	267[a]	1.0
3',5'-cAMP	−51.3°	256	1.45	260[b]	1.5

[a] pH 7.0; [b] pH 6.0; * specific molar rotation

There are two important applications of UV absorption. First, it is used in the determination of temperature-dependent changes in base-stacking. Secondly, it permits the monitoring of changes in the asymmetric environment of the bases by circular dichroism (CD), or by optical rotatory dispersion (ORD) effects. Both of these techniques are especially valuable for studying helix-coil transitions (Section 2.5.1).

IR analysis of nucleic acid components has been less widely used but the availability of laser Raman and Fourier transform IR methods is making a growing contribution.

NMR has had a dramatic effect on studies of oligonucleotides, largely as a result of a variety of complex spin techniques such as NOESY and COSY for proton spectra, the use of ^{17}O, ^{18}O, and sulphur substituent effects for ^{31}P NMR, and the analysis of nuclear Overhauser effects (NOE). These provide a useful measure of internuclear distances and with computational analysis can provide solution conformations of oligonucleotides (Section 2.2). Nucleosides, nucleotides, and their analogues have relatively simple ^{1}H NMR spectra. The aromatic protons of the pyrimidines and purines resonate at low field ($\delta 7.6$ to $\delta 8.3$ with C^5-H close to $\delta 5.9$). The anomeric hydrogen is a doublet for ribonucleosides and a double-doublet for 2′-deoxynucleosides at $\delta 5.8$–6.4. The pentoses provide a multi-spin system which generally moves from low to high field in the series: H-2′, H-3′, H-4′, H-5′, and H-5″ in the region $\delta 4.3$ to 3.7. Lastly, 2′-deoxynucleosides have H-2′ and H-2″ as an ABMX system near $\delta 2.5$. The 400 MHz spectrum of a simple nucleoside, cytidine (Fig. 2.10), shows how essential 2-D spin techniques are in making possible the complete analysis of the spectrum of a large oligomer, which may be equivalent to a dozen such monomer spectra superimposed!

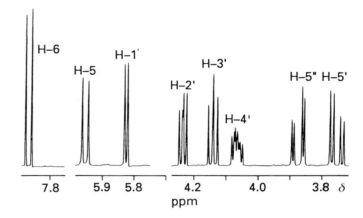

Fig. 2.10 Proton NMR spectrum for cytidine (run in D$_2$O at 400 MHz).

2.1.4 Shapes of nucleotides

Nucleotides have rather compact shapes with several interactions between non-bonded atoms. Their molecular geometry is so closely related to that of the corresponding nucleotide units in oligomers and nucleic acid helices that it has been argued that helix structure is a consequence of the conformational preferences of individual nucleotides. The details of conformational structure are accurately defined by the torsion angles α, β, γ, δ, ϵ, and ζ in the phosphate backbone, θ_0 to θ_4 in the furanose ring, and χ for the glycosidic bond (Fig. 2.11). Because many of these torsional angles are interdependent, we can more simply describe the shapes of nucleotides in terms of four parameters: the sugar pucker, the *syn–anti* conformation of the glycosidic bond, the orientation of $C^{4'}$–$C^{5'}$, and the shape of the phosphate ester bonds.

Sugar pucker

The furanose rings are twisted out of plane in order to minimize non-bonded interactions between their substituents. This 'puckering' is described by identifying the major displacement of carbons-2′ and -3′ from the median plane of $C^{1'}$–$O^{4'}$–$C^{4'}$. Thus, if the *endo*-displacement of C–2′ is greater than the *exo*-displacement of C–3′ the conformation is called $C^{2'}$-*endo* and so on (Fig. 2.11). (The *endo*-face of the furanose is on the same side as $C^{5'}$ and the base; the *exo*-face is on the opposite face to the base.) These sugar puckers are located in the north (**N**) and south (**S**) domains of the pseudorotation cycle of the furanose ring and so spectroscopists frequently use **N** and **S** designations, which also fortuitously reflect the relative shapes of the C–C–C–C bonds in the $C^{2'}$-*endo*- and -*exo*- forms respectively.

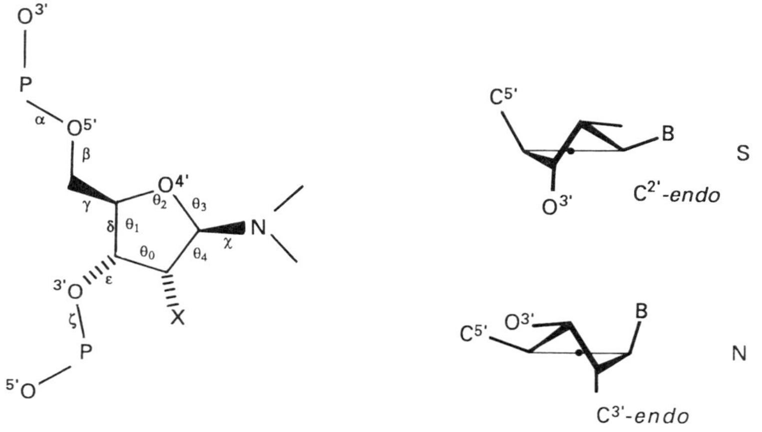

Fig. 2.11 Torsion angle notation (IUPAC) for polynucleotide chains and structures for the C2′-*endo* (**S**) and C3′-*endo* (**N**) preferred sugar puckers.

In solution, the **N** and **S** conformations are in rapid equilibrium and are separated by an energy barrier of less than 20 kJ mol^{-1}. The average position of the equilibrium can be estimated from the magnitudes of the ^3J NMR coupling constants linking H$^{1'}$–H$^{2'}$ and H$^{3'}$–H$^{4'}$. This is influenced by (i) the preference of electronegative substituents at $C^{2'}$ and $C^{3'}$ for axial orientation, (ii) the orientation of the base (*syn* goes with $C^{2'}$-*endo*), and (iii) the formation of an intra-strand hydrogen-bond from O$^{2'}$ in one RNA residue to O$^{4'}$ in the next, which favours $C^{3'}$-*endo*-pucker.

Syn–anti conformation

The plane of the bases is almost perpendicular to that of the sugars and approximately bisects the O$^{4'}$–C$^{1'}$–C$^{2'}$ angle. This allows the bases to occupy either of two principal orientations. The *anti*-conformer has the smaller H-6 (Py) or H-8 (Pu) atom above the sugar ring while the *syn*-conformer has the larger O-2 (Py) or N-3 (Pu) in that position. Pyrimidines occupy a narrow range of *anti*-conformations (Fig. 2.12) while purines are found in a wider range of *anti*-conformations which can even extend into the high-*anti* range for 8-azapurine nucleosides such as formycin.

Fig. 2.12 *Anti-* and *syn-*conformational ranges for glycosidic bonds in pyrimidine (left) and purine (right) nucleosides and drawings of the *anti*-conformation for deoxycytidine (lower left) and the *syn*-conformation for deoxyguanosine 5'-phosphate (lower right).

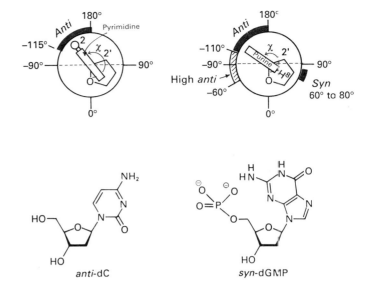

anti-dC   syn-dGMP

There is one important exception to the general preference for *anti*-forms. NMR, CD, and X-ray analyses all show that guanine prefers the *syn*-glycoside in mononucleotides, in alternating oligomers like d(CpGpCpG), and in Z-DNA. Theoretical calculations suggest that this effect comes from a favourable electrostatic attraction between the phosphate anion and the C^2-amino group in guanine nucleotides. It results from polarization of one of the nitrogen non-bonding electrons towards the ring. Most unusually, this *syn*-conformation can only be built into left-handed helices!

$C^{4'}$–$C^{5'}$ orientation

The conformation of the exocyclic $C^{4'}$–$C^{5'}$ bond determines the position of the 5'-phosphate relative to the sugar ring. The three favoured conformers for this bond (Fig. 2.11) are the classical **synclinal (sc)** and **antiperiplanar (ap)** rotamers. For pyrimidine nucleosides, +**sc** is preferred while for purine nucleosides +**sc** and **ap** are equally populated. However, in the nucleotides, the 5'-phosphate reduces the conformational freedom and the dominant conformer for this γ-bond is +**sc** (Figure 2.13). Once again, the demands of Z-DNA have a major effect and the **ap** conformer is found for the *syn*-guanine deoxynucleotides.

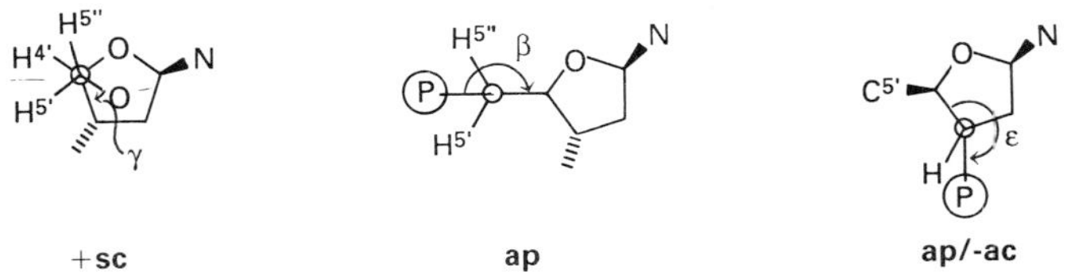

Fig. 2.13 Preferred nucleotide conformations: +**sc** for $C^{4'}$–$C^{5'}$ (left); **ap** for $C^{5'}$–$O^{5'}$ (centre); and **ap/–ac** for $C^{3'}$–$O^{3'}$ (right).

+sc   ap   ap/-ac

C–O and P–O ester bonds

Phosphate diesters are tetrahedral and show antiperiplanar conformations for the $C^{5'}$–$O^{5'}$ bond. Similarly, the $C^{3'}$–$O^{3'}$ bond lies in the **antiperiplanar** to **anticlinal** sector. This conformational uniformity has led to the use of the **virtual bond concept** in which the chains $P^{5'}$–$O^{5'}$–$C^{5'}$–$C^{4'}$ and $P^{3'}$–$O^{3'}$–$C^{3'}$–$C^{4'}$ can be analysed as rigid, planar units linked at phosphorus and at $C^{4'}$. Such a simplification has been used to speed up initial calculations of some complex polymeric structures.

Our knowledge of P–O bond conformations comes largely from X-ray structures of tRNA and DNA oligomers. In general, $H^{4'}$–$C^{4'}$–$C^{5'}$–$O^{5'}$–P adopts an extended W-conformation in these structures. A skewed conformation for the C–O–P–O–C system has been observed in structures of simple phosphate diesters such as dimethyl phosphate and also for polynucleotides. This has been described as a *gauche* effect and attributed to the favourable interactions of a non-bonding electron pair on $O^{5'}$ with the P–$O^{3'}$ bond, and *vice versa* for the P–$O^{5'}$ bond (Fig. 2.14). (This may arise from interaction of the electron lone pair with either phosphorus d-orbitals or, more likely, with the P–O antibonding σ orbital. The interaction has been calculated at 30 kJ mol^{-1} more favourable than the extended W-conformation for the C–O–P–O–C system). Other non-bonded interactions dictate that α and ζ

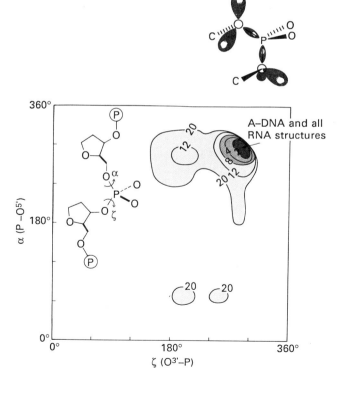

Fig. 2.14 (Upper) Gauche conformation for phosphate diesters showing the antiperiplanar alignment of an occupied non-bonding oxygen orbital with the adjacent P–O bond. (Lower) Contour map for P–O bond rotations calculated for diribose triphosphate (energies in kJ mol^{-1}) (adapted from Govil, G. (1976). *Biopolymers*, **15**, 2303–7. Copyright (1976) John Wiley and Sons, Inc.).

both have values close to +300° in helical structures though values of +60° are seen in some dinucleoside phosphate structures.

Other P–O conformations have been observed in non-helical nucleotides while left-handed helices also require changed P–O conformations. These changes take place largely in the rotamers for α. In Z-DNA these are +**sc** for guanines but broadly **antiperiplanar** for cytosines while ζ is +**sc** for cytosines but broadly **syn-periplanar** for guanines (Section 2.2.2).

Summary

The building blocks of nucleic acids are nucleotides, which are the phosphate esters of nucleosides. These are formed by condensation of a base and a pentose. In RNA, the pentose is D-ribose and is linked in its furanose form from C-1′ to N-9 of a purine, adenine, or guanine, or N-1 of a pyrimidine, cytosine, or uracil. In DNA, 2-deoxy-D-ribose is joined in the same way to the four bases, among which thymine takes the place of uracil. The phosphate esters are strong acids and exist as anions at neutral pH. The 'bases' are, in reality, only very weakly basic and A, C, and G become protonated only below pH 4. The amide N–Hs in G, T, and U are deprotonated at pHs above 9.

Hydrogen-bonds can be formed between the major *amino–keto* tautomers of the bases to link A with T and C with G in Watson–Crick base-pairing. Such hydrogen-bonds are largely ionic in character. 'Wobble' and Hoogsteen base-pairs offer minor variations to Watson–Crick pairing, and are seen in tRNA structures.

The nucleotides have well defined shapes with a general preference for *anti*-conformers of the glycosidic bond χ, for the $C^{4'}–C^{5'}$ bond γ, and for the two C–O(P) bonds β and ε. The furanose ring is puckered to relieve strain and adopts both the $C^{2'}$-*endo* and $C^{3'}$-*endo* conformations, which are in rapid equilibrium at room temperature.

Nucleotides which have this standard shape can be built into regular double-stranded helices using variable –O–P–O– torsion angles. These helices show a preference for a right-handed duplex, which has minimal internucleotide non-bonding interactions.

2.2 Regular DNA structures

Structural studies on DNA began with the nature of the primary structure of DNA. The classical analysis, completed in mid-century, is easily taken for granted today when we have machines for DNA oligomer synthesis that presuppose the integrity of the 3′-to-5′ phosphodiester linkage. Nonetheless, the classical analysis was the essential key that opened the door to later studies on the regular secondary structure of double-stranded DNA and thereby primed the modern revolution known as molecular biology. **Regular structures** for DNA have generally been determined on heterogeneous duplex material and are thus independent of sequence and apply only to Watson–Crick base-pairing.

2.2.1 Primary structure of DNA

Klein and Thannhauser's work (Section 1.4) established that the primary structure of DNA has each nucleoside joined by a phosphodiester from its 5'-hydroxyl group to the 3'-hydroxyl group of one neighbour and by a second phosphodiester from its 3'-hydroxyl group to the 5'-hydroxyl of its other neighbour. There are no 5'–5' or 3'–3' linkages in the regular DNA primary structure (Fig. 2.15). This means that the uniqueness of a given DNA primary structure resides solely in the sequence of its bases.

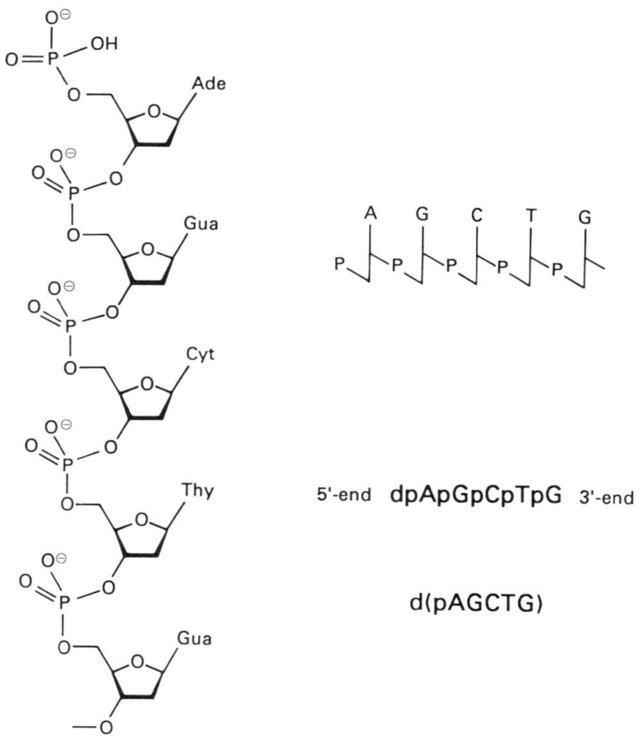

Fig. 2.15 The primary structure of DNA (left) and three of the common shorthand notations: 'Fischer' (upper right), linear alphabetic (centre right) and condensed alphabetic (lower right).

5'-end **dpApGpCpTpG** 3'-end

d(pAGCTG)

2.2.2 DNA secondary structure

In the first phase of investigation of DNA secondary structure, diffraction studies on heterogeneous DNA fibres identified two distinct conformations for the DNA double-helix. At low humidity (and high salt) the favoured form is the highly crystalline A-DNA while at high humidity (and low salt) the dominant structure is B-DNA. B-DNA has now grown into a family of structures encompassing B-, B'-, C-,

C′-, C″-, D-, E-, and T-DNAs. Rich's discovery in 1979 of a left-handed helical structure, named Z-DNA, was one of the first dramatic discoveries to result from the synthesis of oligonucleotides in sufficient quantity for crystallization and X-ray diffraction analysis. Such syntheses also enabled the features of A-DNA and B-DNA to be determined at atomic resolution.

As more structures have become available, the idea that these three families of DNA conformations are restricted to regular structures has been whittled away. We now accept that there are local, sequence-dependent modulations of structure which are primarily associated with changes in the orientation of bases. Such changes seek to minimize non-bonded interactions between adjacent bases and maximize base-stacking. They are generally tolerated by the relatively flexible sugar-phosphate backbone. Other studies have explored perturbations in regular helices which result from deliberate mismatching of base-pairs and of lesions caused by chemical modification of bases, such as base-methylation and thymine photodimers (Section 7.8.1). In all of these areas, the results derived from X-ray crystallography have been carried into the solution phase by high resolution NMR analysis.

Finally, our knowledge of higher order structures, which began with Vinograd's work on DNA supercoiling in 1965, has been extended to studies on DNA cruciform structures, to 'bent' DNA, and to other unusual features of DNA structures.

Regular DNA structures are described by a range of characteristic features. The global parameters of **average rise** D_z and **helix rotation** Ω per base-pair define the pitch of the helix. Sideways tilting of the base pairs through a **tilt angle** τ permits the separation of the bases along the helix axis D_z, to be smaller than the van der Waals distance, 3.4 Å, and so gives a shorter, fatter cylindrical envelope for DNA. The angle τ is positive for A-DNA (positive means a clockwise rotation of the base-pair when viewed end-on and towards the helix axis) but is smaller and negative for B-DNA helices. At the same time, the base-pairs are displaced laterally from the helix axis by a distance D_a. This parameter together with the groove width defines the depth of the major and minor grooves (Table 2.3).

2.2.3 A-DNA

Among the first synthetic oligonucleotides to be crystallized in the late 1970s were d(GGTATACC), an iodinated-d(CCGG), and d(GGCCGGCC). They all proved to have A-type DNA structures, similar to the classical A-DNA deduced from fibre analysis at low resolution. Several other oligomers, mostly octamers, also form crystals of the A-structure, but NMR studies suggest that some of these may have the B-form in solution. It is conceivable that crystal packing might especially favour A-DNA for octanucleotides.

The general anatomy of A-DNA follows the Watson–Crick model with anti-parallel, right-handed double-helices. The sugar rings are parallel to the helix axis

Table 2.3. Average helix parameters for the major DNA conformations

Structure Type	Helix sense	Residues per turn	Twist per bp $t/°$	Displacement bp $D/Å$	Rise per bp $/Å$	Base tilt $\tau°$	Sugar pucker	Groove Width/Å minor	major	Groove Depth/Å minor	major
A-DNA	R	11	32.7	4.5	2.56	20	C-3'-endo	11.0	2.7	2.8	13.5
dGGCCGGCC	R	11	32.6	3.6	3.03	12	C-3'-endo	9.6	7.9	—	—
B-DNA	R	10	36	-0.2 to -1.8	3.3–3.4	-6	C-2'-endo	5.7	11.7	7.5	8.8
dCGCGAATTCGCG	R	9.7	37.1	—	3.34	-1.2	C-2'-endo	3.8	11.7	—	—
C-DNA	R	9.33	38.5	-1.0	3.31	-8	C-3'-exo	4.8	10.5	7.9	7.5
D-DNA	R	8	45	-1.8	3.03	-16	C-3'-exo	1.3	8.9	6.7	5.8
T-DNA	R	8	45	-1.43	3.4	-6	C-2'-endo	narrow	wide	deep	shallow
Z-DNA	L	12	-9, -51	-2 to -3	3.7	-7	C-2'-endo(syn)	2.0	8.8	13.8	3.7
A-RNA	R	11	32.7	4.4	2.8	16–19	C-3'-endo				
A'-RNA	R	12	30	4.4	3.0	10	C-3'-endo				

and the phosphate backbone is on the outside of a cylinder of about 24 Å diameter (Fig. 2.16).

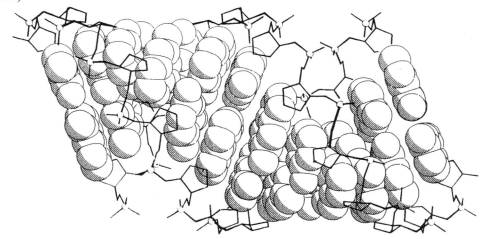

Fig. 2.16 3-D structure of one turn of A-DNA showing the bases (grey and pink) and sugar-phosphate backbone (black and red). The atoms of the base-pairs are drawn at 80 per cent of their van der Waals radii, phosphorus atoms at 0.3 per cent, and backbone structure in line form for clarity.

X-ray diffraction data at atomic resolution shows that the bases are displaced 4.5 Å away from the helix axis, creating a hollow core down the axis which is 3 Å in diameter. There are 11 bases in each turn of 28 Å, which gives a vertical rise of 2.56 Å per base-pair. In order to maintain the normal van der Waals separation of 3.4 Å, the stacked bases are tilted sideways through 20°. The sugar backbone has skewed phosphate ester P–O bonds and antiperiplanar conformations for the adjacent C–O ester bonds. Finally, the furanose ring has a $C^{3'}$-*endo*-pucker and the glycoside is in the *anti*-conformation (Table 2.3). This leaves a 5.4 Å P–P separation between adjacent intrastrand phosphates.

2.2.4 The B-DNA family

The general features of the B-type structure, obtained from DNA fibres at high relative humidity (95 per cent RH) have been put into sharper focus by X-ray studies on crystals of the dodecamer d(CGCGAATTCGCG) and of its bromo-derivative at cytosine-9. This sequence was designed to have a non-Z-core flanked by C–G-rich ends and to contain a recognition site for the *Eco R1* restriction enzyme (Section 9.3.3). The B-conformation has also been observed in crystals of a thiophosphate hexamer, d(Gp$_S$CpGp$_S$CpGp$_S$C), and for a mismatch decamer, d(CCAAGATTGG).

B-DNA has major and minor grooves of similar depth (Table 2.3). Its bases stack predominantly above their neighbours in the same strand and are perpendicular to the helix axis, which they straddle. The sugars have the $C^{2'}$-*endo*-pucker (with some $O^{4'}$-*endo*-pucker for the dodecamer), all the glycosides have the *anti*-conformation, and most of the other rotamers have normal populations (Table 2.4). Adjacent phosphates in the same chain are a little further apart, P–P = 6.7 Å, than in A-DNA (Fig. 2.17).

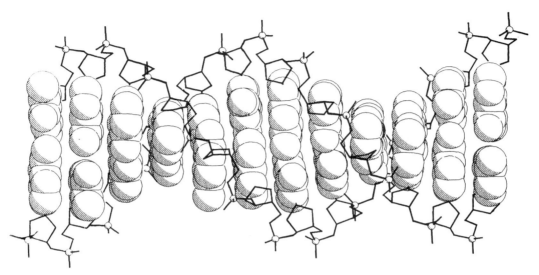

Fig. 2.17   3-D structure of one turn of B-DNA showing the bases (grey and pink) and the sugar-phosphate backbone (black and red). Atomic representations are as for Fig. 2.16.

The X-ray data not only provide accurate detail of the regular features of B-DNA, now fully supported by solution-phase NMR studies, but they serve two other purposes. On the one hand they have given us details of the hydration pattern of this high humidity conformation of DNA. On the other, they have established that B-DNA has systematic, sequence-dependent structural modulations—which we shall discuss later (Section 2.3.1).

X-ray structures identify many ordered water molecules in the major and minor grooves of B-DNA. The broad major groove is 'coated' by a unimolecular layer of water molecules which interact with the exposed C=O, N, and NH functions and also extensively solvate the phosphate backbone. More significantly, the narrow minor grooves contain well-ordered, zig-zag chains of two water molecules per base-pair. In the narrow minor groove of the A:T region of the dodecamer structure, these form a **spine of hydration** with alternating first shell oxygens on the floor of the groove and second shell oxygens above and between them (Fig. 2.18). In the broader minor groove of the mismatched decamer, the water is arranged in two **strings of hydration** down each side of the groove. It has been argued that the transition between B-DNA at 95 per cent relative humidity and A-DNA at 75 per

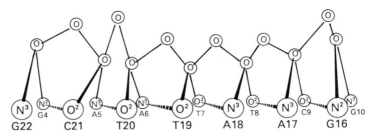

Fig. 2.18 Unrolled view of the spine of hydration as found in the minor groove of the A:T region of d(CGCGAATTCGCG). Broken lines link the base-pair atoms (bottom). Unbroken lines (colour) show hydrogen-bonds between water molecules (oxygens only shown) in the spine first hydration shell and the atoms in the bases (adapted from Drew, H. R. and Dickersen, R.E. (1981). *J. Mol. Biol.*, **151**, 535–56).

cent is governed by the integrity of such hydration spines or strings which stabilize the B-DNA helix.

Other B-type DNA structures have rather lesser significance. C-DNA is obtained from the lithium salt of natural DNA at rather low humidity. It has 28 bases in three full turns of the helix and in other respects is rather like B-DNA. D-DNA is observed for alternating A:T regions of DNA and has an overwound helix compared to B-DNA with 8 base-pairs per turn. In phage T2 DNA, where cytosine bases have been replaced by glucosylated 5-hydroxymethylcytosines, the B-conformation observed at high humidity changes into a T-DNA form at low humidity, <60 per cent relative humidity, which also has eightfold symmetry around the helix (see Table 2.3).

2.2.5 Z-DNA

Two of the earliest crystalline oligodeoxyribonucleotides, d(CGCGCG) and d(CGCG), provided structures of a new type of DNA conformer, the left-handed Z-DNA, which has also been found for d(CGCATGCG). Initially it was thought that left-handed DNA had a strict requirement for alternating Pur.Pyr sequences.

Table 2.4. Average torsion angles (°) for DNA helices

Structure type	α	β	γ	δ	ε	ζ	χ
A–DNA[a]	−50	172	41	79	−146	−78	−154
GGCCGGCC	−75	185	56	91	−166	−75	−149
B–DNA[a]	−41	136	38	139	−133	−157	−102
CGCGAATTCGCG	−63	171	54	123	−169	−108	−117
Z–DNA (C residues)	−137	−139	56	138	−95	80	−159
(G residues)	47	179	−169	99	−104	−69	68
DNA–RNA decamer	−69	175	55	82	−151	−75	−162
A–RNA	−68	178	54	82	−153	−71	−158

[a] Fibres

We now know that this condition is neither necessary nor sufficient since left-handed structures have been found for crystals of d(CGATCG) in which cytosines have been modified by C-5 bromination or methylation and have been identified for GTTTG and GACTG sequences by supercoil relaxation studies (Section 2.3.4).

The left-handedness of this antiparallel duplex is a result of a switch in the glycosidic torsion angle from the regular *anti-* to the unusual *syn*-conformation for one nucleoside in each base-pair, which is normally the purine. This *anti–syn* feature **alternates regularly along the DNA backbone**, causing the phosphorus atoms to follow a zig-zag course: hence the name Z-DNA. At the same time, the sugar pucker alternates from C$^{2'}$-*endo* for the *anti*-residues to C$^{3'}$-*endo* for the *syn*-nucleosides (Tables 2.3 and 2.4). These changes have a profound effect on base-stacking in Z-DNA. While this is normal for the GpC sequences ($\Omega = -45°$ to $-51°$), it is much reduced by slide, Dy, for CpG sequences ($\Omega = -9°$ to $-15°$). The net result of these changes is that the minor groove of Z-DNA is so deep that it actually contains the helix axis and the 'major groove' of Z-DNA is a convex surface on which cytosine-C^5, guanine-N^7, and -C^8 are exposed (Fig. 2.19).

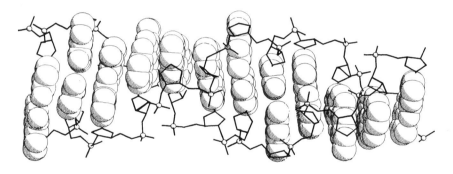

Fig. 2.19   3-D structure of one turn of Z-DNA showing the bases (grey and pink) and the sugar-phosphate backbone (black and red). Atomic representations are as for Fig. 2.16.

Solution studies on poly(dG–dC) have shown a salt-dependent transition between conformers that can be monitored either by circular dichroism or by ^{31}P NMR. In particular there is a near inversion in the CD spectrum above 4 M NaCl which has been identified with a change from B- to Z-DNA. It appears that a high salt concentration stabilizes the Z-conformation because it has a much smaller separation between the phosphate anions in opposite strands than for B-DNA, 8 Å as opposed to 11.7 Å. This analysis is supported by the fact that a high alcohol content in the solvent also has a stabilizing effect on Z-DNA, which results from its lower dielectric constant.

The scanning tunnelling microscope has the power to resolve the structure of biological molecules with atomic detail. Much progress has been made with dried samples of duplex DNA, in recording images of DNA under water, and in revealing details of single-stranded poly(dA). Such STM microscopy has provided images of poly(dG.me⁵dC).poly(dG-me⁵dC) in the Z-form. Both the general appearance of the fibres and measurements of helical parameters are in good agreement with models derived from X-ray diffraction data.

Summary

The primary structure of DNA has a string of nucleosides, each joined to its two neighbours through phosphodiester linkages. Each regular 5′-hydroxyl group is linked through a phosphate to a 3′-hydroxyl group, the uniqueness of any primary structure depending only on the sequence of bases in that chain.

A-DNA and B-DNA are the major regular DNA secondary structures with right-handed double-helices and Watson–Crick base-pairs.

A-DNA has 11 residues per turn, these bases are tilted 20° to enhance stacking, and they lie 4.5 Å away from the helix axis. This gives a fairly stiff helix which shows little sequence-dependent variation in structure. The major groove is deep and narrow, the minor groove is broad and shallow.

B-DNA has ten bases per turn with little tilting of the bases. The wide major groove and narrow minor groove are both of moderate depth and both grooves are well solvated by water molecules. The B-form structure is sufficiently flexible to permit a conformational response in the backbone to particular local base sequences.

Z-DNA is a left-handed double-helical structure stabilized by high concentrations of $MgCl_2$, NaCl, or ethanol. It is most favoured for alternating G-C sequences. Watson–Crick pairing is maintained but the purines adopt the *syn*-glycoside and the C³′-*endo* sugar pucker. The phosphate backbone has a zig-zag appearance. The minor groove is narrow and very deep but the major groove has become so shallow that normally inaccessible parts of the bases C and G are exposed.

2.3 Irregular DNA structures

2.3.1 Sequence-dependent modulation of DNA structure

So far we have emphasized the importance of hydrogen-bonds in base-pairing and DNA structure and have said rather less about base-stacking. We shall see later that both these two features are important for the energetics and dynamics of DNA helices (Section 2.5) but it is now time to look at the major part played by base-stacking in sequence-dependent DNA structure. Two particular hallmarks of B-DNA, in contrast to the A- and Z-forms, are its flexibility and its capacity to make small adjustments in local helix structure in response to particular base sequences.

Different base sequences have their own characteristic signature: they influence groove width, helical twist, curvature, mechanical rigidity, and resistance to bending. It seems probable that these features help proteins to read and recognize one base sequence in preference to another (Chapter 9) possibly only through changes in the positions of the phosphates in the backbone. What do we know about these sequence-dependent structural features?

Dickerson has contrasted the crystal structure of the dodecanucleotide d(CGCATATATGCG) with that of four other 'poly-A' sequences, d(CGCAAAAAAGCG), d(CGCAAATTTGCG), d(CGCGAATTCGCG), and d(CGCGAATT^{Br}CGC), all of which have B-DNA structures. While all five helical structures show a characteristically narrow minor groove in their A:T centres, the base-pairs show a **propeller twist** (see below) which is generally lower in the alternating poly(AT) helix than in the polyA helices. Three general principles emerge from structural analysis of this family of B-helices:

(a) the minor groove is wide in G:C and mixed sequence B-DNA, but narrow in hetero- or homo-polymer A:T sequences;

(b) propeller twist is low for G:C base-pairs and may (but need not) be high for A:T base-pairs; high propeller twist can be stabilized by cross-strand hydrogen-bonding in the major or minor groove; and

(c) stiffening of homopoly-A tracts can result from such cross-strand major groove hydrogen-bonds.

The major irregularities in the positions of the bases in the structures contrasts with only secondary conformational changes in the sugar–phosphate backbone. The main source of these modulations is **propeller twist**. This is caused by the bases rotating clockwise by some 5° to 25° relative to their hydrogen-bonded partner around the long axis through C-8 of the purine and C-6 of the pyrimidine (Fig. 2.20). Sections of these dodecanucleotides which have consecutive A residues show high propeller twist. The first of the dodecanucleotides has medium propeller twist in its alternating poly(AT) structure whereas mixed sequence B-DNAs have low propeller twist. Another feature of the A:T-rich sequences found in these and other oligonucleotides is a very narrow minor groove, which has particular significance in the case of groove-binding drugs (Section 8.6.2).

Why should the bases twist in this way? The advantage of propeller twist is that it gives improved face-to-face contact between adjacent bases in the same strand and this leads to increased stacking stability in the double-helix. However, there is a penalty! The larger purine bases occupy the centre of the helix so that in alternating purine–pyrimidine sequences they overlap with neighbouring purines in the opposite strand. Consequently, propeller twist causes a clash between such pairs of adjacent purines in opposite strands. For pyrimidine-(3→5′)-purine steps, these purine–purine clashes take place in the minor groove where they involve guanine-N^3 and -N^2, and adenine-N^3 atoms. For purine-(3′→5′)-pyrimidine steps, they take place in the major groove between guanine-O^6 and adenine-N^6 atoms (Fig. 2.20). There are no such clashes for purine–purine and pyrimidine–pyrimidine sequences.

One of the consequences of these effects is that bends may occur at junctions between polyA tracts and mixed-sequence DNA as a result of propeller twist, base-pair inclination, and base-stacking differences on two sides of the junction (see below).

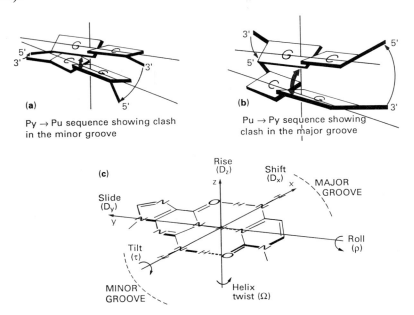

Fig. 2.20 Diagrams illustrating (a) clockwise propeller twist for a C–(3′→5′)–G sequence with clashes between guanines in the **minor** groove; (b) clockwise propeller twist for a G–(3′→5′)–C sequence showing purines clashing in the **major** groove; and (c) base-pair movements relative to the helix axis (vertical) and the major (C^8pu–C^6py) and minor grooves in the base-pair median plane.

Calladine's rules

B-DNA structures respond to minimize this problem in four ways, which were deduced by Calladine and articulated as follows:

(1) flatten the propeller twist locally for either or both base-pairs;

(2) roll the base-pairs away from their clashing edges;

(3) slide one or both of the base pairs along their axis to push the **purine** away from the helix axis;

(4) unwind the helix axis locally to diminish interstrand purine–purine overlap.

The relative motions required to achieve these effects are described by six parameters of which the most significant are ϱ for roll, D_y for slide, and Ω for helix twist. These motions are illustrated for neighbouring G:C base-pairs (Fig. 2.20c).

In practice, the structures of crystalline oligomers have exhibited the following six types of conformational modulation which are sequence-dependent and which support these rules:

(1) the B-DNA helix axis need not be straight but can curve with a radius of 112 Å;

(2) the twist angle, Ω, is not constant at 36° but can vary from 28° to 43°;

(3) propeller twist averages +11° for C:G pairs and +17° for A:T pairs;

(4) base-pairs 'roll' along their long axes to reduce clashing;

(5) sugar pucker varies from $C^{3'}$-*exo* to $O^{4'}$-*endo* to $C^{2'}$-*endo*;

(6) there can be local improved overlap of bases by slide, as in d(TCG) where C–2 moves towards the helix axis to increase stacking with G–3.

At the same time, these irregularities have revealed the incompleteness of the Calladine model because it ignores such important factors as electrostatic interactions, hydrogen-bonding, and hydration. For example, a major stabilizing influence proposed for the high propeller twist in sequences with consecutive adenines is the existence of cross-strand hydrogen-bonding between adenine N-6 in one strand with thymine O-4 of the next base-pair in the opposite strand.

Modulations of B-DNA structure which have been observed in the solid state have to some extent been mirrored by the results of solution studies for d(GCATGC) and d(CTGGATCCAG) obtained by a combination of NMR analysis and restrained molecular dynamics calculations. These oligomers have B-type structures which show clear, sequence-dependent variations in torsion angles and helix parameters. There is strong curvature to the helix axis of the hexamer which results from large positive roll angles at the Pyr→Pur steps. The decamer has a straight central core but there are bends in the helix axis at the second (TpG) and eighth (CpA) steps which result from positive roll angles and large slide values.

Taken together, these X-ray and NMR analyses give good support for the general conclusion that minor groove clashes at Pyr→Pur steps are twice as severe as major groove clashes at Pur→Pyr steps. As a result, it is possible to calculate the behaviour of the helix twist angle, Ω, using sequence data only.

The more rigid structure of A-DNA has yielded less information on sequence-dependent variations. However, a few examples have been observed of a reduction in the local helix twist angle and of the opening of the roll angle to alleviate purine–purine clashes. This picture supports the general view that bending in double-helical DNA is produced by local roll of base-pairs about their long axes.

Bending at helix junctions

Bent DNA was first identified as a result of modelling the junction between an A-type and a B-type helix. The best solution to this problem requires a bend of 26° in the helix axis in order to maintain full stacking of the bases. Bent DNA has gained support not only from NMR and CD studies on a DNA.RNA hybrid, [poly(dG).(rC)$_{11}$–(dC)$_{16}$], but from studies on regular homopolymers which contain $(dA)_5.(dT)_5$ sections occurring in phase in each turn of a 10-fold or 11-fold helix. Moreover, bent DNA containing such dA.dT repeats has been investigated from a variety of natural sources.

It appears that bending of this sort happens at junctions between the stiff [dA.dT] helix and the regular B-helix (see above). In situations where such junctions occur every five bases and in an alternating sense, the net result is a progression of bends which is equivalent to a continuous curve in the DNA.

2.3.2 Mismatch base-pairs

The fidelity of transmission of the genetic code rests on the specific pairings of A:T and C:G bases. Consequently, if changes in shape result from base mismatches, such as A:G, they must be recognized and be repaired by enzymes with high efficiency (Section 7.11). X-ray studies have given accurate and detailed analyses for a range of mismatched base-pairing introduced into central positions of otherwise self-complementary oligonucleotides, and high resolution NMR studies have extended the picture to solution conformations. The different types of base-pair mismatch can be grouped into **transition mismatches**, which pair a purine with the wrong pyrimidine, and **transversion mismatches**, which pair either two purines or two pyrimidines.

Transition mismatches

The G:T base-pair has been observed in crystal structures for A-, B-, and Z-conformations of oligonucleotides. In every case it has been found to be a typical 'wobble' pair having *anti–anti* glycosidic bonds. The structure of the dodecamer, d(CGC**G**AATT**T**GCG), which has two G:T^9 mismatches, can be superimposed on that of the regular dodecamer and shows excellent correspondence of backbone atomic positions.

The A:C pair has been examined in the dodecamer d(CGC**A**AATT**C**GCG) and, once again, the two A^4:C mismatches are typical 'wobble' pairs, achieved by the protonation of adenine-N^1 (Fig. 2.21). It is notable that there is no significant worsening of base-stacking and little perturbation of the helix conformation. However, it appears that no water molecules are bonded to these bases in the minor groove.

Transversion mismatches

The G:A mismatch is the most thoroughly studied in solution and in the solid state and two different patterns have been found. Crystals of the dodecamer, d(CGC**G**AATT**A**GCG) have an (*anti*)G:A(*syn*) mismatch with hydrogen-bonds from Ade-N^7 to Gua-N^1 and from Ade-N^6 to Gua-O^6 (Fig. 2.21). A similar (*anti*)I:A(*syn*) mismatch has been identified in a related dodecamer structure. Calculations on both of these mismatches suggest that they can be accommodated into a regular B-helix with minimal perturbation.

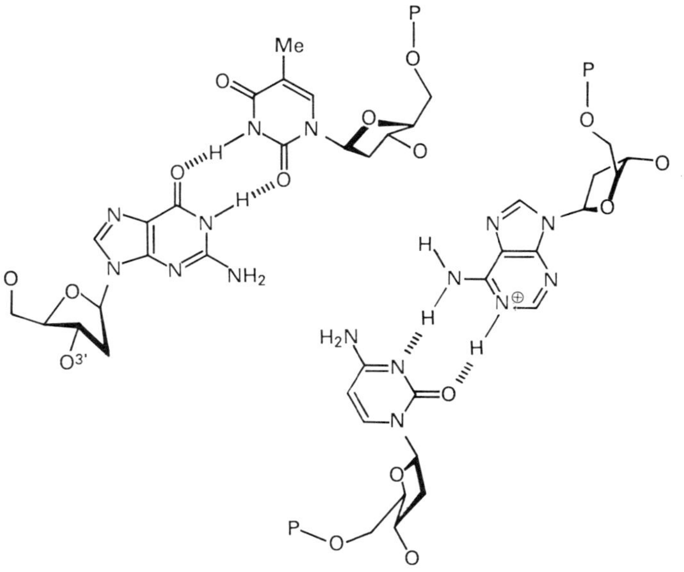

Fig. 2.21 'Wobble' pairs for transition mismatches G:T (left) and A:C (right).

This work contrasts with both NMR and X-ray studies on d(CCAA**GA**TTGG) and NMR work on d(CG**A**GAATTC**G**CG) which have identified (*anti*)G:A(*anti*) pairings with two hydrogen-bonds (Fig. 2.22). The X-ray analysis of the decamer (see above) shows a typical B-helix with a broader minor groove and a changed pattern of hydration. This arises in part because the two mismatched G:A pairs are 2.0 Å wider (from $C^{1'}$ to $C^{1'}$) than a conventional Watson–Crick pair.

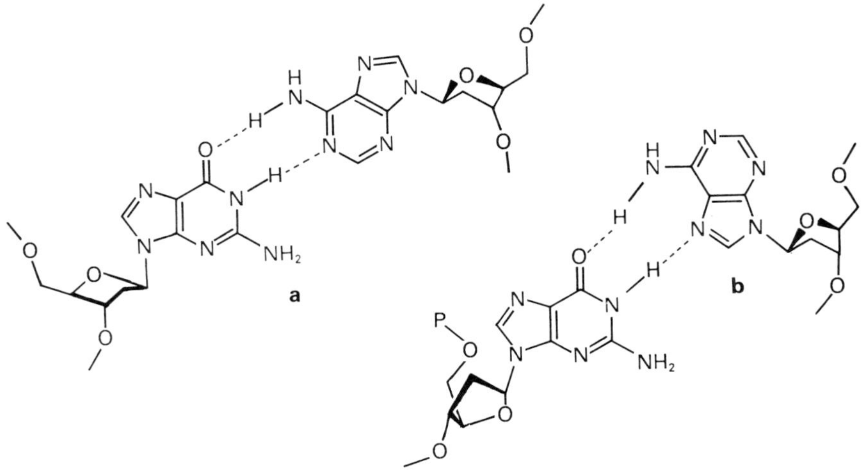

Fig. 2.22 Mismatched G:A pairings (**a**) for the decamer d(CCAAGATTGG) with (*anti*)G:A(*anti*) and (**b**) for the dodecamer d(CGAGAATTCGCG) with (*anti*)G:A(*syn*) conformation.

Insertion–deletion mispairs

When one DNA strand has one nucleotide more than the other, the extra residue can either be accommodated in an intrastrand position or be forced into an extra-strand location. Tridecanucleotides containing an extra adenine, cytosine, or thymine residue have been examined in the crystalline solid and solution states. In one case, an extra adenine has been accommodated into the helix stack while in others a cytosine or adenine is seen to be extruded into an extrahelical, unstacked location.

In addition to such work on mismatched base-pairs, related investigations have made good progress into structural changes caused by covalent modification of DNA. On the one hand, crystal structures of DNA adducts with *cis*platin have characterized its monofunctional linking to guanine sites in a B-DNA helix and, on the other, NMR studies of O^4-methylthymine residues and of thymine photodimers and psoralen:DNA photoproducts are advancing our understanding of the modifications to DNA structure that result from such lesions (Section 7.8.2). It seems likely that the range of patterns of recognition of structural abnormalities may be as wide as the range of enzymes available to repair them!

2.3.3 Unusual DNA structures

Since 1980, there has been a rapid expansion in our awareness of the heterogeneity of DNA structures which has resulted from a widening use of new analytical techniques, notably structure-dependent nuclease action, structure-dependent chemical modification, and physical analysis. Unusual structures are generally sequence-specific, as we have already described for the A–B helix junction (Section 2.3.1). Some of them are also dependent on DNA supercoiling which provides the necessary driving energy for their formation due to the release of torsional strain, as is particularly well defined for cruciform DNA. Consequently, much use has been made of synthetic DNA both in short oligonucleotides and cloned into circular DNA plasmids where the effect of DNA supercoiling can be explored.

Curved DNA

Kinetoplast DNA is a special form of DNA obtained from trypanosomatids. It provides a source of **open** DNA minicircles whose circularity is sequence-dependent and is not enforced by covalent closure of the circles. These circles have been seen in an electron micrograph of a 200 bp region of such DNA and have a 360° curve. The discovery that kinetoplast DNA has short adenine tracts spaced at 10 bp intervals has led to the use of synthetic oligonucleotides to study the origin of such strong intrinsic curvature. This can result in altered physical properties, such as anomalously slow electrophoretic migration through polyacrylamide gels. It has been identified in the centres of a set of four $CA_{5-6}T$ sequences, each of which is spaced from the next by 2 or 3 nucleotides (Fig. 2.23). There is a 20°–25° bend in each of these sequences which is more pronounced at the 3'-end of the adenine tracts than it is at the 5'-end. One can easily see that if such sequences are repeated in phase then a succession of bends will result in curvature.

Such intrinsic, sequence-dependent curvature must be distinguished from the bending of DNA which results from the application of an external force, such as the binding to DNA of a sequence-specific protein.

These bends also appear to have a sequence-defined 'inside' and 'outside'. Thus, A:T regions have a narrower minor groove which faces inwards while the broader, less compressible minor groove of the G:C regions faces outwards. As a result, it is apparently possible to use these features to design DNA sequences which will be naturally curved!

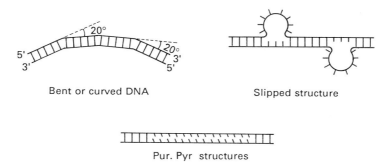

Bent or curved DNA   Slipped structure

Pur. Pyr structures

Fig. 2.23 Some unusual DNA structures. Bent (curved) DNA (left) with A:T base-pairs in colour and G:C base-pairs in black showing 20° bending at the 3'-ends of a run of 5(A:T) pairs. Slipped DNA (right) with compensating loops in alternate strands (all base-pairs in colour). Pur.pyr structure (centre below) showing region of anomalous stacking conformations.

Slipped structures have been postulated to occur at direct repeat sequences, and they have been found upstream of important regulatory sites. The structures described (Fig. 2.23) are consistent with the pattern of cleavage by single-strand nucleases but otherwise are not well characterized.

Anisomorphic DNA is the description given to DNA conformations associated with direct repeat, DR2, sequences at 'joint regions' in viral DNA, which are known to have unusual chemical and physical properties. The two complementary strands have different structures and this leads to structural aberrations at the centre of the tandem sequences that can be seen under conditions of torsional stress induced by negative supercoiling.

Cruciform structures are much the best explored species of unusual DNA structure. These 'paired stem-loop' formations were identified around 1980, some 20 years after they had been described theoretically. Such cruciforms have now been characterized for many inverted repeat sequences in plasmids and in phage where supercoiling plays a critical role (Fig. 2.24).

```
TGTGGATCCGGTACCAGAATTCTGGTACCGGATCCTCT
-------------><-------------
ACACCTAGGCCATGGTCTTAAGACCATGGCCTAGGAGA
```

Fig. 2.24 Plasmid **pIRbke8** (inverted repeat sequence).

When such DNA structures are torsionally underwound by negative supercoiling (Section 2.3.4), stem-loop structures are formed by intrastrand pairing (Fig. 2.25). These features can be detected by global changes in the supercoil structure or by chemical and enzymatic transformations which are specific for the loop and four-way junction regions. The loops are sensitive to single-strand nucleases, such as S1 and P1, and to reagents such as bromoacetaldehyde, osmium tetroxide, bisulphite, and glyoxal (Sections 7.4 and 7.5). In addition, the junctions are cleavage sites for yeast resolvase and for T4 endonuclease VII. The loop has an optimum size of four to six bases, which show some base-stacking, while both NMR and chemical analyses suggest that there is no loss of base-stacking in the junctions, which may be tetrahedrally oriented.

The formation of a single cruciform calls for a gain in free energy of some 75 kJ mol^{-1}. This energy is provided by relaxation of negative supercoiling whose magnitude depends on the length of the arms of the cruciform: a cruciform arm of 10.5 base-pairs unwinds the supercoil by one turn. There is also a kinetic barrier to cruciform formation and David Lilley has suggested two mechanisms which have clearly distinct physical parameters and may be sequence-dependent.

The faster process for cruciform formation, the S-pathway, has $\Delta G^{\ddagger}$ *ca.*100 kJ mol^{-1} with a small positive entropy of activation. This more common pathway is typified by the behaviour of plasmid **pIRbke8** (Fig. 2.24). Following the formation of a relatively small unpaired region, a proto-cruciform intermediate is produced which then grows to equilibrium size by branch migration through the four-way junction (Fig. 2.25). The slower mechanism, the C-pathway, involves the formation of a large bubble followed by its condensation to give the fully-developed cruciform. This behaviour explains the data for the **pColIR315** plasmid whose cruciform kinetics show $\Delta G^{\ddagger}$ *ca.* 180 kJ mol^{-1} with a large entropy of activation.

Such extrusion of cruciforms provides the most complete example of the characterization of unusual DNA structures by combined chemical, enzymatic, kinetic, and spectroscopic techniques. However, there has as yet been no identification of a significant biological role for these structures.

2.3.4 Circular DNA and supercoiling

The replicative form of bacteriophage φX174 DNA was found to be a double-stranded closed circle. It was later shown that bacterial DNA exists as closed circular duplexes, that DNA viruses have either single- or double-helical circular DNA, and that RNA viruses have circular single stranded RNA as their genomic material. Plasmid DNAs also exist as small, closed circular duplexes.

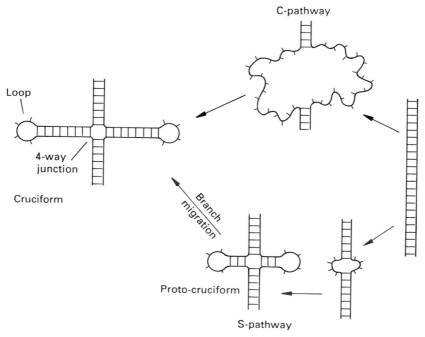

Fig. 2.25 Structures of a cruciform and alternative pathways for its formation (base-paired sections are helical throughout).

Topologically unconstrained dsDNA in its linear, relaxed state is either biologically inactive or displays reduced activity in key processes such as recombination, replication, or transcription. It follows that topological changes associated with the constraints of circularization of dsDNA have a profound biological significance. While such circularization can be achieved directly by covalent closure, the same effect can be achieved for eukaryotic DNA as a result of holding DNA loops together by means of a protein scaffold.

The molecular topology of closed circular DNA was described by Vinograd in 1965 and is especially associated with the phenomenon of superhelical DNA, which is also called supercoiled or supertwisted DNA. Vinograd's basic observation was that when a planar, relaxed circle of DNA is strained by changing the pitch of its helical turns, it relieves this torsional strain by winding around itself to form a superhelix whose axis is a diameter of the original circle.

This behaviour is most directly observed by following the sedimentation of negatively supercoiled DNA as the pitch of its helix is changed by intercalating a drug, typically ethidium bromide (Section 8.4.2). The DNA helix responds first by reducing the number of negative, right-handed supercoils until it is fully relaxed and then by increasing the number of positive, left-handed supercoils. As this happens, the sedimentation coefficient of the DNA first decreases, reaching a minimum when fully relaxed, and then increases as it becomes positively supercoiled. As a control process, the same circular DNA can be nicked in one strand to make it fully relaxed. The result is that it now shows a low sedimentation coeffiecient at all concentrations of the intercalator species (Fig. 2.26).

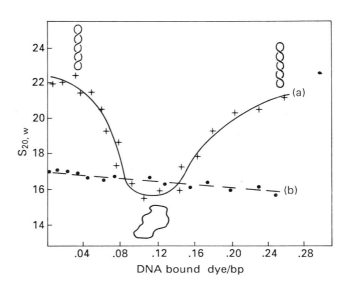

Fig. 2.26 Sedimentation velocity for SV40 DNA as a function of bound ethidium (**a**) for closed circular DNA ($-+-+-$) and (**b**) for nicked circular DNA ($-\cdot-\cdot-$) showing the transition from a negative supercoil (left) through a relaxed circle (centre) to a positive supercoil (right) (adapted from Bauer, W. and Vinograd, J. (1968). *J. Mol. Biol.*, **33**, 141).

Vinograd showed that the topological state of these covalently closed circles can be defined by three parameters and that the fundamental topological property is **linkage**. The topological winding number, $\textbf{\textit{T}}_w$ is the number of right-handed helical turns in the relaxed, planar DNA circle and the writhing number, $\textbf{\textit{W}}_r$, gives the number of left-handed crossovers in the supercoil. The sum of these two is the linking number, $\textbf{\textit{L}}_k$, a topological integer which cannot be changed unless either or both of the circles is opened. Thus, $\textbf{\textit{L}}_k$ is the number of times one strand of the helix winds around the other (clockwise is +ve) when the circle is constrained to lie in a plane. The simple equation is:

$$L_k = T_w + W_r$$

Such behaviour can be illustrated simply (Fig. 2.27) for a relaxed closed circle with 20 helical turns, $T_w = 20$, $L_k = 20$, $W_r = 0$. One strand is now cut, unwound two turns, and resealed to give $L_k = T_w = 18$. This circle is thus underwound by two turns. To restore fully the normal B-DNA base-pairing and base-stacking, the circle needs to gain two right-handed helical turns, $\Delta T_w = +2$, to give $T_w = 20$. Since the DNA circles have remained closed and the linking number stays at 18, the formation of the right-handed helical turns is balanced by the creation of one right-handed supercoil, making $W_r = -2$.

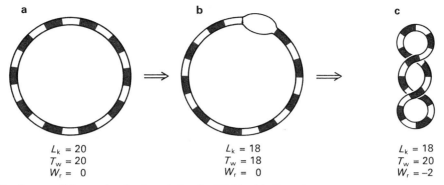

Fig. 2.27 Supercoil formation in closed circular DNA. (**a**) Closed circle of 20 duplex turns (alternate turns in colour); (**b**) circle nicked, underwound 2 turns, and resealed; (**c**) base-pairing and stacking forces result in the formation of B-helix with two new right-handed helix turns and one compensating right-handed supercoil.

The behaviour of a supercoil can be modelled using a length of rubber tubing. The ends are first held together to form a relaxed, closed circle. If the end in your right hand is given one turn clockwise (right-handed twist) and the other end is given one turn in the opposite sense, the tube will relieve this strain by forming one left-handed supercoil. This is equivalent to unwinding the DNA helix by two turns which generates one positive supercoil (four turns generate two supercoils, and so on). This model shows the relationship: two turns equals one supercoil.

In practice it is sometimes useful to describe the degree of supercoiling using the **superhelical density**, $\sigma = W_r/T_w$, which is close to the number of superhelical turns per 10 bp and is typically around 0.06 for superhelical DNA from cells and virions. The **energy of supercoiling** is a quadratic function of the density of supercoils as described by the equation:

$$\triangle G_s = 1050 \times \frac{RT}{N} . \triangle L_k{}^2 \qquad \text{kJ mol}^{-1}$$

where R is the gas constant, T is the absolute temperature, and N is the number of base-pairs.

B–Z transitions are especially important for supercoiling since the conversion of one right-handed B-turn into a left-handed Z-turn causes a change in T_w of -2. This must be complemented by $\Delta W_r = +2$ through the formation of one left-handed superturn.

Enzymology of DNA supercoiling

DNA topoisomers are circular molecules which have identical sequences and differ only in their linking number. A group of enzymes, discovered by Jim Wang, can change that linking number and fall into two classes: Class I topoisomerases effect integral changes in the linking number, $\Delta L_k = n$, while Class II enzymes interconvert topoisomers with a step rise of $\Delta L_k = \pm 2n$. Topoisomerase I enzymes use a 'nick–swivel–close' mechanism to operate on supercoiled DNA. They break a phosphate diester linkage, hold its ends, and reseal them after allowing exothermic (i.e. passive) free rotation of the other strand. Such enzymes from eukaryotes can operate on either left- or right-handed supercoils while prokaryotic enzymes only work on negative supercoils. The products of topoisomerase I action on plasmid DNA can be observed by gel electrophoresis, and show a ladder of bands, each corresponding to unit change in W_r as the supercoils are unwound, half-at-a-time (Fig. 2.28).

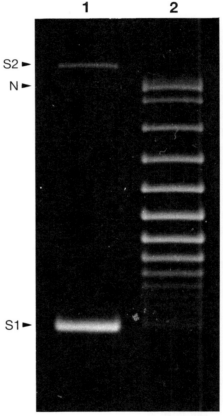

Fig. 2.28 Topoisomers of plasmid **pAT153** after incubation with topoisomerase I to produce partial relaxation. Electrophoresis in a 1 per cent agarose gel: Track 1 shows native supercoiled **pAT153** (S1), supercoiled dimer (S2), and nicked circular DNA (N); Track 2 shows products of topoisomerase I where ΔL_k up to 14 can be seen clearly (from Lilley, D. M. J. (1986). *Symp. Soc. Gen. Microbiol.*, **39**, 105–17).

By contrast, Class II topoisomerases use a 'double-strand passage' mechanism to effect unit change in the number of supercoils, $\Delta W_r = \pm 2$, and such prokaryotic enzymes can drive the endothermic supercoiling of DNA by coupling the reaction to hydrolysis of ATP. These topoisomerases cleave two phosphate esters to produce an enzyme-bridged gap in both strands. The other DNA duplex is passed through the gap (using energy provided by hydrolysis of ATP), and the gap is resealed. DNA gyrase from *E. coli* is a special example of the Class II enzyme. It is an **A₂B₂** tetramer with the energy-free topoisomerase activity of the **A** subunit being inhibited by quinolone antibiotics such as nalidixic acid. The energy transducing activity of the **B** subunit can be inhibited by novobiocin and other coumarin anti-biotics. We should point out that such topoisomerases also operate on linear DNA that is torsionally stressed by other processes, most notably at the replication fork in eukaryotic DNA.

Supercoiling is important for a growing range of enzymes as illustrated by two examples. RNA polymerase *in vitro* appears to work ten times faster on super-coiled DNA, σ = 0.06, than on relaxed DNA, and this phenomenon appears to be related to the enhanced binding of the polymerase to the promoter sequence. Secondly, the *tyrT* promoter in *E. coli* is expressed *in vitro* at least a hundred times stronger for supercoiled than for relaxed DNA and this behaviour seems linked to 'preactivation' of the DNA promoter region by negative supercoiling (Section 5.3.13).

Catenated and knotted DNA circles

While Type II topoisomerases usually only effect passage of a duplex from the same molecule through the separated double strands, they can also manipulate a duplex from a second molecule. As a result, two different DNA circles can be interlinked with the formation of a **catenane** (Fig. 2.29). Such catenanes have been identified by electron microscopy and can be artificially generated in high yield from mam-malian mitochondria. **Knotted DNA** circles are another unusual topoisomer species which are also formed by intramolecular double-strand passage from an incom-pletely unwound duplex (Fig. 2.29).

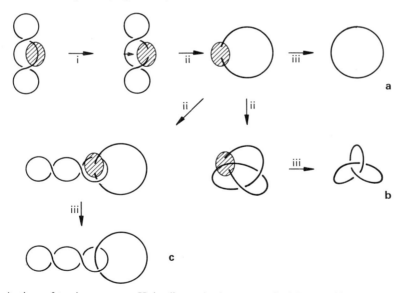

Fig. 2.29 Action of topiosomerase II (red) on singly supercoiled DNA: (i) double-strand opening; (ii) double-strand passage; (iii) resealing to give (**a**) relaxed circle, (**b**) knotted circle, and (**c**) cate-nated DNA circles.

Summary

Irregular structures for DNA range in scale from local conformational perturbations of the flexible B-form, through isomers interchanged by unpairing and re-pairing of bases, to topological isomers such as catenanes and knots which can only be inter-converted with single circles by the cleavage of both DNA strands.

Sequence-dependent modulation of DNA structure results largely from propeller twist of base-pairs which optimizes base-stacking between adjacent pairs. This results in purine clashes between neighbouring bases in opposite strands which are relieved by local conformational adjustments in the positions of the bases. The energetic requirements of stacking interactions are readily accommodated by changes in the backbone conformation, which are of relatively low energy.

Mismatched base-pairs also cause local structural irregularities which can be accommodated by 'wobble' base-pairing for A:C and G:T pairs or by more extensive changes for G:A pairs.

While DNA shows bends in the helix at junctions between A-form and B-form helices, curved DNA is the result of regular repeats of A_{5-6} tracts.

The topological behaviour of closed circles of DNA involves superhelices whose energy can be used, *inter alia*, to drive the formation of cruciform structures. Topological isomers have a clear biological role and can be interconverted by topoisomerase enzymes.

2.4 Structures of RNA species

As with DNA, studies on RNA structure began with its primary structure. This quest was pursued side-by-side with that of DNA, but had to deal with the extra complexity of the 2′-hydroxyl group in ribonucleosides. Today, we also recognize that RNA has greater structural versatility than DNA in the variety of its species, in its diversity of conformations, and in its chemical reactivity. Different natural RNAs can either form long, double-stranded structures or adopt a globular shape composed of short duplex domains connected by single-stranded segments. Watson–Crick base-pairing seems to be the norm, though tRNA structures have provided a rich source of unusual base-pairs and base-triplets (Section 6.3.2). In general, it is now possible to predict double-helical sections by computer analysis of primary sequence data, and this technique has been used extensively to identify secondary structural components of ribosomal RNA and viral RNA species. In this section, we shall focus attention mainly on regular RNA secondary structure.

2.4.1 Primary structure of RNA

The first degradation studies of RNA using mild alkaline hydrolysis gave a mixture of mononucleotides, originally thought to have only four components—one for each base A, C, G, and U. However, Waldo Cohn used ion-exchange chromatography to separate each of these four into pairs of isomers which were identified as the ribonucleoside 2'- and 3'-phosphates. This duplicity was overcome by use of a phosphodiesterase isolated from spleen tissue which digests RNA from its 5'-end (Section 9.4.3) to give the four 3'-phosphates, Ap, Cp, Gp, and Up, whilst an internal diesterase (snake venom phosphodiesterase was later used) cleaved RNA to the four 5'-phosphates, pA, pC, pG, and pU. It follows that RNA chains are made up of nucleotides which have 3'→5'-phosphodiester linkages just like DNA (Fig. 2.30).

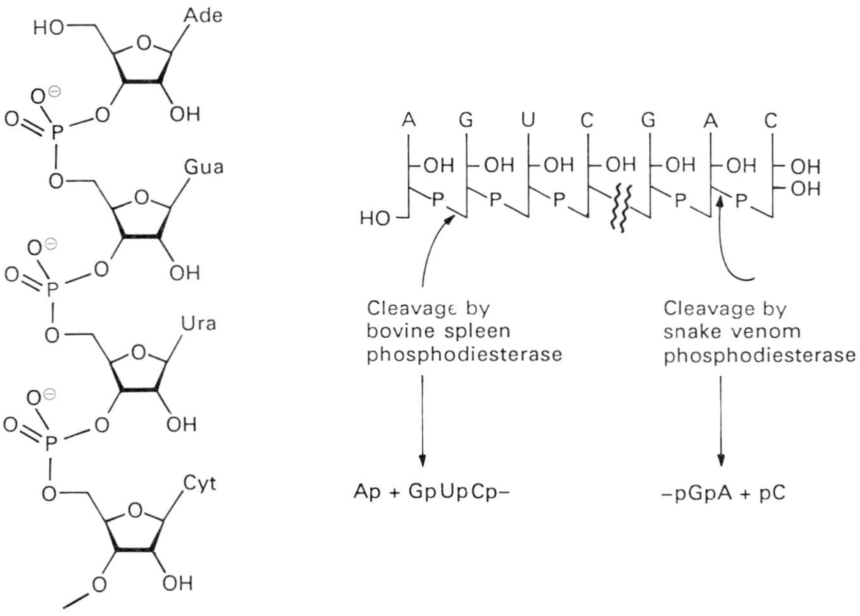

Fig. 2.30 The primary structure of RNA (left) and cleavage patterns with spleen (centre) and snake venom (right) phosphodiesterases.

The 3'→5' linkage in RNA is, in fact, thermodynamically less stable than the 'unnatural' 2'→5' linkage, which might therefore have had an evolutionary role. A rare example of such a polymer is produced in vertebrate cells in response to viral infection. Such cells make a glycoprotein called **interferon** which stimulates the production of an oligonucleotide synthetase. This polymerizes ATP to give oligo-adenylates with 2'→5' phosphodiester linkages and 3 to 8 nucleotides long. Such $(2'→5')(A)_n$ (Fig. 2.31) then activates an interferon-induced ribonuclease, RNase N, whose function seems to be to break down the viral messenger RNA. (Note also the 2'→5' ester linkage is a key feature of self-splicing RNA (Section 6.5).)

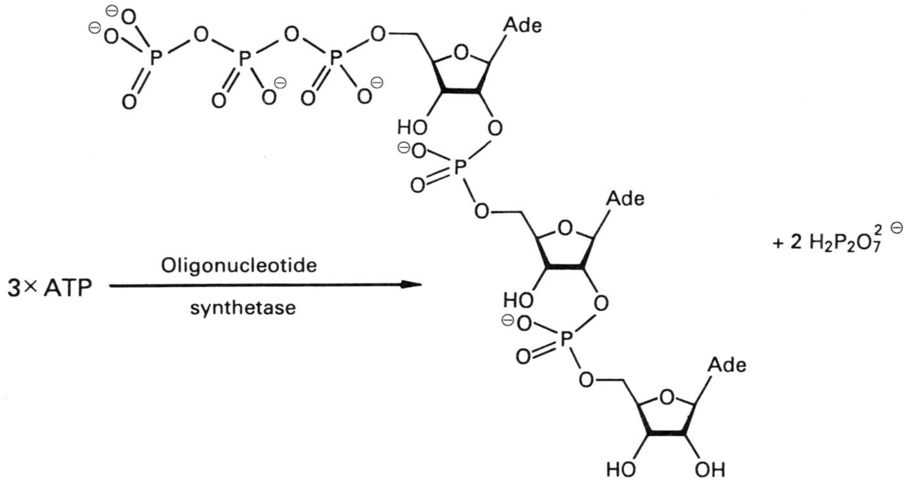

Fig. 2.31 Structure and formation of interferon-induced $(2'{\rightarrow}5')(A)_n$.

2.4.2 RNA secondary structures: A-RNA AND A′-RNA

The presence of the 2′-hydroxyl group in RNA hinders the formation of a B-type helix but can be accommodated within an A-type helix. Two varieties of A-type helices have been observed for fibres of RNA species such as poly(rA).poly(rU). At low ionic strength, A-RNA has 11 base-pairs per turn in a right-handed, anti-parallel double-helix. The sugars adopt a $C^{3'}$-*endo* pucker and the other geometric parameters are all very similar to those for A-DNA (see Tables 2.3 and 2.4). If the salt concentration is raised above 20 per cent, an A′-RNA form is observed which has 12 base-pairs per turn of the duplex. Both structures have typical Watson–Crick base-pairs, which are displaced 4.4 Å from the helix axis and so form a very deep major groove and a rather shallow minor groove.

In one of the first NMR studies of an RNA duplex, Gronenborn and Clore have combined 2-D NOE analysis with molecular dynamics to identify an A-RNA solution structure for the hexaribonucleotide, $5'r(GCAUGC)_2$. It shows sequence-dependent variations in helix parameters, particularly in helix twist and in base-pair roll, slide, and propeller twist (Fig. 2.32). The extent of variation from base to base is much less than for the corresponding DNA hexanucleotide and seems to be dominated by the needs of the structure to achieve very nearly optimal base-stacking. This picture supports experimental studies which indicate that base-stacking and hydrogen-bonding are equally important as determinants of RNA helix stability.

Antisense RNA is defined as a short RNA transcript that lacks coding capacity, but has a high degree of complementarity to another RNA which enables the two to hybridize. The consequence is that such antisense, or complementary, RNA can act as a repressor of the normal function or expression of the targeted RNA. Such

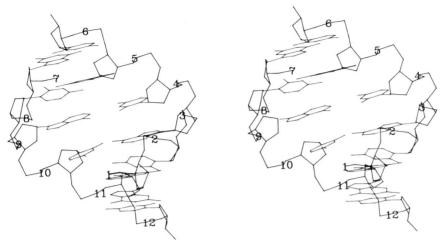

Fig. 2.32 3-D Structure of 5′r(GCAUGC)$_2$ (adapted from Happ, S. C. *et al.* (1988). *Biochemistry*, **27**, 1735–43. Copyright (1988) American Chemical Society).

species have been detected in prokaryotic cells with suggested functions concerning RNA-primed replication of plasmid DNA, transcription of bacterial genes, and messenger translation in bacteria and bacteriophages. Quite clearly, such regulation of gene expression depends on the integrity of RNA duplexes.

2.4.3 RNA:DNA duplexes

Helices which have one strand of RNA and one of DNA are very important species in biology.

(a) They are formed when reverse transcriptase makes a DNA complement to the viral RNA.

(b) They occur when RNA polymerase transcribes DNA into complementary messenger RNA.

(c) They are a feature in DNA replication of the short primer sequences in Okazaki fragments (Section 5.4).

Such hybrids can also be formed *in vitro* by annealing together two strands with complementary sequences, such as poly(rA).poly(dT) and poly(rI).poly(dC). These two heteroduplexes adopt the A-conformation common to RNA and DNA, the former giving an 11-fold helix typical of A-RNA and the latter a 12-fold helix characteristic of A′-RNA.

A self-complementary decamer, r(GCG)d(TATACGC), also generates a hybrid duplex with Watson–Crick base-pairs. It has a helix rotation of 33° with a step-rise of 2.6 Å and C$^{3'}$-*endo* sugar pucker typical of A-DNA and A-RNA (see Table 2.1).

Such hybrids appear to be more stable to thermal denaturation than their DNA:DNA counterparts.

The greater stability of RNA:DNA heteroduplexes over DNA:DNA homoduplexes is the basis of the construction of **antisense DNA oligomers**. These are intended to enter the cell where they can pair with, and so inactivate, complementary mRNA sequences. Additional desirable features such as membrane permeability and resistance to enzyme-degradation have focused attention on oligodeoxynucleotides either having phosphorothioates, phosphorodithioates, or methylphosphonates replacing the anionic phosphate diester linkages (Section 3.4.6). In most cases, the resulting heteroduplexes have unfortunately proved to have weaker association constants than the natural DNA.RNA duplexes. However, the potential benefits that could accrue from silencing a particular RNA target are substantial and this subject is likely to expand further.

2.4.4 RNA bulges, hairpins, and loops

Physical studies on the stability of RNA double-helices have provided reliable data for the stacking energies of all combinations of neighbouring base-pairs (Table 2.5). They have also made it possible to estimate the energetics of bulges, hairpins, and interior loops. While a hairpin of five or six non-bonded bases costs about +25 kJ mol^{-1}, a bulge costs about +9 kJ mol^{-1}, and a single looped-out base about +13 kJ mol^{-1}. It is thus possible to compute alternative secondary structures for single-stranded RNA species, as illustrated for a segment of R17 viral RNA (Fig. 2.33). RNA structures incorporating these features undoubtedly play a major role in the activity of ribozymes (Section 6.5).

Table 2.5. Thermodynamic parameters for the formation of an RNA base-pair at the end of a duplex

Base pair	5′-neighbour	$\Delta H°$ kJ mol^{-1}	$\Delta S°$ kJ mol^{-1} deg^{-1}	$\Delta G°$ kJ mol^{-1} [a]
U=A	U.A	−34.4	−0.10	− 5.05
U=A	A.U	−27.3	−0.069	− 6.7
C=G	A.U	} −24.8	−0.053	− 8.8
A=U	C.G			
G=C	C.G	−54.6	−0.141	−12.6
C=G	G.C	−61.7	−0.147	−18.1
C=G	C.G	−57.5	−0.125	−20.2
U=G	G.U			− 1.3
G=U	X.Y			0

[a] At 25°C

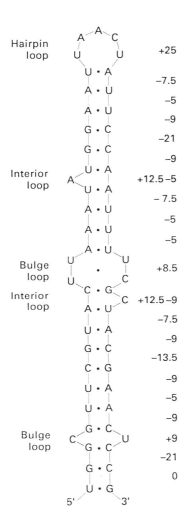

Fig. 2.33 A possible secondary structure for a 55-nucleotide fragment from R17 virus illustrating hairpin loop, bulge, and interior loop structures. This structure gives an overall $\Delta G°$ of –90 kJ mol^{-1}.

2.4.5 Triple-stranded nucleic acids

The first triple-stranded nucleic acid was described in 1957 when poly(rU).poly(rA) was found to form a stable 2:1 complex in the presence of magnesium chloride. A similar complex is formed from 2poly(rC) with oligo(rG) as long as the cytosine bases in one strand are protonated to complete the hydrogen-bond pattern, CH$^+$.G.C (Fig. 2.34). More complex triple-stranded structures are formed from the association of poly(rU–rC) or poly(dT–dC) with the DNA duplex poly(dT–dC). poly(dG–dA). Once again, the triple-helix is only stable below pH 6 and in the presence of MgCl$_2$.

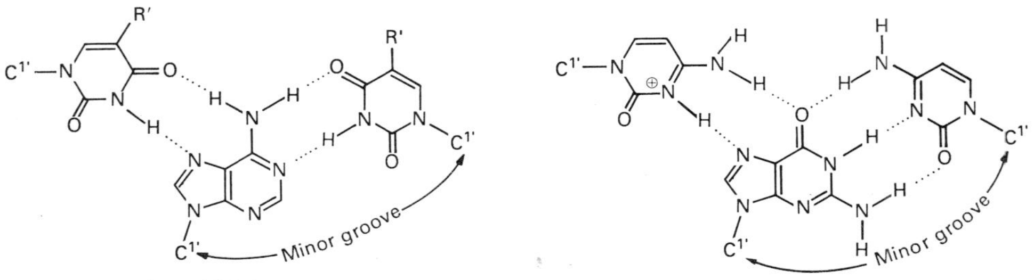

Base triplets
TAT (R' = Me)
UAU (R' = H)

CH⁺GC Base triplet

Fig. 2.34 Isomorphous base-triplets of TAT and CH⁺GC. The extra pyrimidine base is bound by Hoogsteen hydrogen-bonds in the major groove of the duplex.

X-ray diffraction patterns at low resolution suggest that the two polypyrimidine strands are antiparallel, which means that the 'Hoogsteen strands' must run parallel. An overall A'-like conformation enables the third strand to occupy the large, deep major groove of the Watson–Crick-paired strands and with relatively small base-tilting.

The importance of added cations to overcome the repulsion between the anionic chains of the Watson–Crick duplex and the polypyrimidine strand is an essential feature of triple-helix formation. As well as Mg^{2+}, $Co^{3+}(NH_3)_6$ and spermine are also effective. Knorre, Dervan, and Orgel have each described the use of oligopyrimidines equipped with a DNA-cleaving moiety to produce sequence-specific cleavage of single-stranded DNA. A triple-helix is formed with two identical polypyrimidine strands which must run antiparallel. This means that the DNA single strand has to be palindromic (i.e. has to read the same backwards and forwards).

In the case of dsDNA, non-palindromic oligopyrimidines can be used to generate a triple-stranded helix for sequence-specific cleavage, as shown by Dervan for d(T*TTTTCTCTCCTCT), where the first thymidine is linked to an EDTA arm to effect oxidative strand-scission of plasmid **pDMAG10** DNA at a designated locus (Section 7.3).

There is growing evidence that intramolecular DNA triplexes can be formed in supercoiled plasmids which have specific mirror repeat oligopurine–oligopyrimidine sequences, for example $(GAA)_9.(CTT)_9$ and $(AG)_{12}.(TC)_{12}$. What seems to happen is that a portion of the pyrimidine-rich strand unpairs from the complementary strand so that it can fold back into the major groove of the preceding section of duplex, where it forms a triplex with Hoogsteen base-pairs. The biological consequences of such structures have yet to be determined.

Summary

RNA primary structure is built on the regular 3′→5′ phosphodiester linkage, as for DNA. Oligoadenylates with a 2′→5′ linkage are generated in mammalian cells as a result of viral infection.

While RNA species vary in size from 65 nucleotides upwards, their helical structures are restricted to A-form duplexes with 11–12 residues per turn. Sequence-dependent modulation of structure is much less marked than for B-DNA. A-type heteroduplexes with one RNA and one DNA strand can be formed; they have higher melting points than their double-stranded DNA equivalents.

Reliable thermodynamic data for free energies of base-stacking, of hairpin loops, of bulges, and of interior loops permit the computation of possible stable secondary structures for single-stranded RNAs.

Triple-stranded nucleic acids use Hoogsteen base-pairing to add a polypyrimidine third strand to a polypurine.polypyrimidine duplex. This requires protonation of one cytosine in each base-triplet and the two pyrimidine strands must run antiparallel.

2.5 Dynamics of nucleic acid structures

Any over-emphasis on the stable structures of nucleic acids runs the risk of playing down the dynamic activity of nucleic acids which is intrinsic to their function. Pairing and unpairing, breathing, and winding are integral features of the behaviour of these species.

Established studies on structural transitions of nucleic acids have for a long time used classical physical methods which include light absorption, NMR spectroscopy, ultracentrifugation, viscometry, and X-ray diffraction. More recently, these techniques have been augmented by a range of powerful computational methods. In each case, the choice of experiment is linked to the time-scale and amplitude of the molecular motion under investigation.

2.5.1 Helix-coil transitions of duplexes

Double-helices have a lower molecular absorptivity for UV light than would be predicted from the sum of their constituent bases. This **hypochromicity** is usually measured at 259 nm, while C:G base-pairs can also be monitored at 280 nm. It results from coupling of the transition dipoles between neighbouring stacked bases and is larger in amplitude for A:U and A:T pairs than for C:G pairs. As a result, the UV absorption of a DNA duplex **increases** typically by 20–30 per cent when it is denatured. This transition from a helix to an unstacked, strand-separated coil has a strong entropic component and so is temperature dependent. The mid-point of this thermal transition is known as the **melting temperature** (T_m).

Such dissociation of nucleic acid helices in solution to give single-stranded DNA is a function of base composition, sequence, and chain length as well as of temperature, salt concentration, and pH of the solvent. In particular, early observations of the relationship between T_m and base composition for different DNAs showed that A:T pairs are less stable than C:G pairs, a fact which is now expressed in a linear correlation between T_m and the gross composition of a DNA polymer by the equation:

$$T_m = \mathbf{X} + 0.41(\%C + G) \qquad /^\circ C$$

The constant $\mathbf{X}$ is dependent on salt concentration and pH and has a value of 69.3°C for 0.3 M sodium ions at pH 7 (Fig. 2.35).

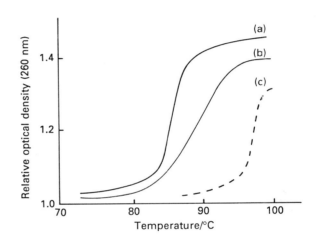

Fig. 2.35 Thermal denaturation of DNAs as a function of base composition (per cent G:C) for three species of bacteria: (a) *Pneumococcus* (38 per cent C:G); (b) *E. coli* (52 per cent G:C); (c) *M. phlei* (66 per cent G:C) (adapted from Marmur, J. and Doty, P. (1959), *Nature*, **183**, 1427–9. Copyright (1959) Macmillan Magazines Ltd.).

A second consequence is that the steepness of the transition also depends on base-sequence. Thus, melting curves for homopolymers have much sharper transitions than those for random-sequence polymers. This is because A:T rich regions melt first to give unpaired regions which then extend gradually with rising temperature until, finally, even the pure C–G regions have melted (Fig. 2.36). In some cases, the shape of the melting curve can be analysed to identify several components of defined composition melting in series.

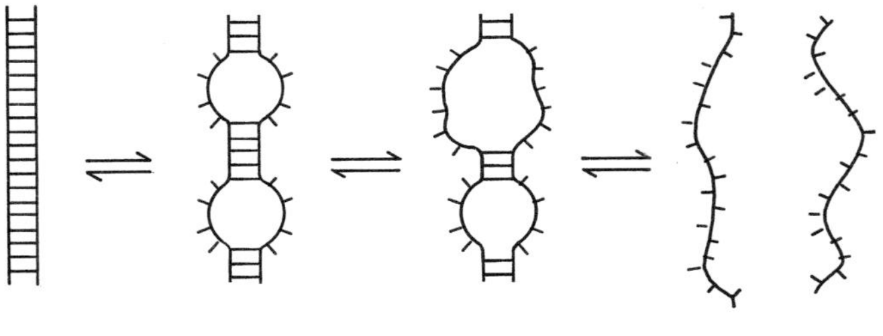

Fig. 2.36 Scheme illustrating the melting of A:T-rich regions (colour) followed by mixed regions, then by C:G-rich regions (black) with rise in temperature (left→right).

Because of end-effects, short homo-oligomers melt at lower temperatures and with broader transitions than longer homo-polymers. For example for poly$(rA)_n$-poly$(rU)_n$, the octamer melts at 9°C, the undecamer at 20°C, and long oligomers at 49°C in the same sodium cacodylate buffer at pH 6.9. Consequently, in the design of synthetic, self-complementary duplexes for crystallization and X-ray structure determination, C:G pairs are often placed at the ends of hexamers and octamers to stop them 'fraying'. Lastly, the marked dependence of T_m on salt concentration is seen for DNA from *Diplococcus pneumoniae* whose T_m rises from 70°C at 0.01 M KCl to 87°C for 0.1 M KCl and to 98°C at 1.0 M KCl.

Data from many melting profiles have been analysed to give a 'stability matrix' for nearest neighbour stacking (Table 2.6). This can be used to predict T_m for a B-DNA polymer of known sequence with a general accuracy of 2–3°C.

Table 2.6. Thermal stability matrix for nearest-neighbour stacking in base-paired dinucleotide fragments with B–DNA geometry

5'-neighbour	3'-neighbour			
	A	C	G	T
A	54.50	97.73	58.42	57.02
C	54.71	85.97	72.55	58.42
G	86.44	136.12	85.97	97.73
T	36.73	86.44	54.71	54.50

Numbers give T_m values in °C at 19.5 mM Na$^+$

The converse of melting is the renaturation of two, separated, complementary strands to form a correctly paired duplex. In practice, the melting curve for denaturation of DNA is only reversible for relatively short oligomers, where the rate-determining process is the formation of a nucleation site of about three base-pairs followed by rapid zipping-up of the strands and where there is no competition from other impeding processes.

When solutions of unpaired, complementary large nucleic acids are incubated at 10–20°C below their T_m, renaturation takes place over a period of time. For short DNAs of up to several hundred base-pairs, nucleation is rate-limiting at low concentrations and each duplex zips to completion almost instantly (>1000 bp s^{-1}). The nucleation process is bimolecular, so renaturation is concentration dependent with a rate constant around 106 M^{-1} s^{-1}. It is also dependent on the complexity of the single strands. Thus, for the simplest cases of homopolymers and of short heterogeneous oligonucleotides, nucleation sites will usually be fully extended by rapid zipping-up. This gives us an 'all-or-none' model for duplex formation. By contrast, for bacterial DNA each nucleation sequence is present only in very low concentration and the process of finding its correct complement will be slow. Lastly, in the case of eukaryotic DNA, the existence of repeated sequences means that locally-viable nucleation sites will form and can be propagated to give relatively stable structures. These will not usually have the two strands in their correct overall register. Because such pairings become more stable as the temperature falls, complete renaturation may take an infinitely long time.

Longer nucleic acid strands are able to generate intrastrand hairpin loops, which optimally have about six bases in the loop and paired sections of variable length. They are formed by rapid, unimolecular processes which can be 100 times faster than the corresponding bimolecular pairing process. Although such hairpins are thermodynamically less stable than a correctly-paired duplex, their existence retards the rate of renaturation so that propagation of the duplex is now the rate-limiting process (Fig. 2.37). One notable manifestation of this phenomenon is seen when a solution of melted DNA is quickly quenched to +4°C to give stable denatured DNA.

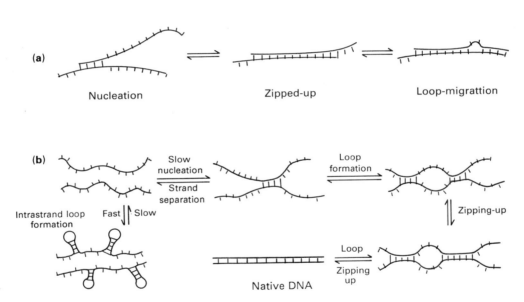

Fig. 2.37 Renaturation processes (**a**) for short oligonucleotide and longer homopolymers and (**b**) for natural DNA strands.

With longer DNA species, Britten and Kohne have shown that the rate of recombination, which is monitored by UV hypochromicity, can be used to estimate the size of DNA in a homogeneous sample. The time t for renaturation at a given temperature for DNA of single-strand concentration C and total concentration C_0 is related to the rate constant k for the process by an equation which in its simplest form is:

$$C/C_0 = (1 + kC_0t)^{-1}$$

In practice, where C/C_0 is 0.5 the value of C_0t is closely related to the complexity of the DNA under investigation.

This annealing of two complementary strands has found many applications. For DNA oligomers, it provided a key component of Khorana's chemical synthesis of a gene (Section 3.5). It is now an integral feature of the insertion of chemically synthesized DNA into vectors. For RNA:DNA duplexes, it has provided a tool of fundamental importance for gene identification (Section 10.2.3) and is being explored in the applications of antisense DNA.

2.5.2 DNA breathing

Complete separation of two nucleic acid strands in the melting process is a relatively slow, long-range process that is not easily reversible. By contrast, the hydrogen-bonds between base-pairs can be disrupted at temperatures well below the melting temperature to give local, short-range separation of the strands. This readily reversible process is known as 'breathing'.

The evidence for such dynamic motion comes from chemical reactions which take place at atoms that are completely blocked by normal base-pairing. Those used include tritium exchange studies on hydrogen-bonded protons in base-pairs, the reactivity of formaldehyde with base N–H groups, and NMR studies of imino-proton exchange with solvent water. This last technique can be used on a time scale from minutes down to ten milliseconds which shows that in linear DNA the base-pairs open singly and transiently with a lifetime around 10 milliseconds at 15°C.

Because NMR can distinguish between imino- and amino-proton exchange it can also be used to identify breathing in specific sequences. Some of the most detailed work of this sort has come from studies on tRNA molecules which shows that, with increasing temperature, base-triplets (Fig. 2.34) are destabilized first followed by the ribothymidine helix and then the dihydrouridine helix. Finally the acceptor helix 'melts' after the anticodon helix (Section 6.3.2).

Another possible motion which might be important for the creation of intercalation sites is known as 'soliton excitation'. The concept here is of a stretching vibration of the DNA chain which travels like a wave along the helix axis until, given sufficient energy, it leads to local unstacking of adjacent bases with associated deformation of sugar pucker and other bond conformations.

Such pre-melting behaviour may well relate to the process of drug intercalation, to the association of single-strand specific DNA binding proteins (Section 9.2.2), and to the reaction of small electrophilic reagents with imino and amino groups such as cytosine-N^3 (Section 7.5).

2.5.3 Energetics of the B–Z transition

The isomerization equilibrium between the right-handed B-form and the left-handed Z-form of DNA is determined by three factors:

(a) chemical structure of the polynucleotide (sequence, modified bases);

(b) environmental conditions (solvent, pH, temperature, etc.); and

(c) degree of topological stress (supercoiling, cruciform formation).

Many quantitative data have been obtained from spectroscopic, hydrodynamic, and calorimetric studies and linked to theoretical calculations. While these have not yet defined the kinetics or complex mechanisms of the B–Z transition, it is evident that the small transition enthalpies involved lie within the range of the thermal energies available from the environment. So, for example, the intrinsic free energy difference between the Z- and B-forms is close to 2 kJ mol^{-1} for polyd(G–C) base-pairs, only 1 kJ mol^{-1} for polyd(G–m^{5}C) pairs, and greater than 5 kJ mol^{-1} for polyd(A–T) pairs. It thus appears that local structural fluctuations may be key elements in the mediation of biological regulatory functions through the B–Z transition.

2.5.4 Rapid DNA motions

Rotations of single bonds, either alone or in combination, are responsible for a range of very rapid DNA motions with time-scales down to fractions of a nanosecond. For example, the twisting of base-pairs around the helix axis has a lifetime around 10^{-8}s while crankshaft rotation of the β, α, ζ, and ε C–O–P–O–C bonds (see Fig. 2.11) leads to an oscillation in the position of the phosphorus atom on a millisecond time-scale. Various calculations on the interconversion of 3'-*endo* and 2'-*endo* sugar pucker have given low activation energy barriers for their interconversion, in the range 3 to 20 kJ mol^{-1}, showing that these conformers are in rapid, although weighted, equilibrium at 37°C. Lastly, rapid fluctuations in propeller twist can result from oscillations of the glycosidic bond.

Summary

The dynamic motions of nucleic acids range from relatively high energy processes involving breaking multiple hydrogen-bonds to local conformational mobilities of very low activation energy.

Helix-coil transitions result from complete unpairing of the two strands of a duplex with rising temperature. The half-way stage is known as the melting temperature T_m and depends on the C:G content of the duplex, its length and sequence, and the ionic strength of the solution. A C:G pair is worth twice as much energetically as an A:T pair.

The coil–helix transition is fully reversible for short oligonucleotides and homopolymers, but can be an infinitely slow process for very long complementary DNA strands. This intermolecular process has to compete with rapid intramolecular formation of hairpin loops where there is a suitable base sequence.

Local and reversible unpairing of bases is known as breathing and can be monitored by chemical and spectroscopic methods at temperatures below T_m.

The best studied short-range motions are the interconversions of *syn-* and *anti-* glycosidic linkages and of the 3'-*endo* and 2'-*endo* sugar puckers and the crankshaft rotations of the C–O–P–O–C phosphate esters. These and other motions must be involved in the B–Z transition of left-handed into right-handed DNA but, though the energetics of this process have been measured, its detailed mechanism is still uncertain.

2.6 Higher order DNA structures

The way in which eukaryotic DNA is packaged in the cell nucleus is one of the wonders of macromolecular structure. In general, higher organisms have more DNA than lower ones (Table 2.7) and this calls for correspondingly greater

Table 2.7. Cellular DNA content of various species

Organism	Number of base-pairs	DNA length mm	Number of chromosomes
Escherichia coli	4×10^6	1.4	1
Yeast (*Saccharomyces cerevisiae*)	1.4×10^6	4.6	16
Fruit fly (*Drosophila melanogaster*)	1.7×10^7	56.0	4
Man (*Homo sapiens*)	3.9×10^9	990.0	23

Values are provided for haploid genomes

condensation of the double-helix. Human cells contain a total of 7.8×10^9 base-pairs which corresponds to an extended length of about two metres. It is packed into forty-six cylindrical chromosomes of total length 200 μm, which gives a net packaging ratio of about 10^4 for such metaphase human chromosomes. The overall process has been broken down into two stages: the formation of nucleosomes and the condensation of nucleosomes into chromatin.

2.6.1 Nucleosome structure

The first stage in the condensation of DNA is the nucleosome, whose core has been crystallized by Aaron Klug and John Finch and analysed by X-ray diffraction. The DNA duplex is wrapped around a block of eight histone proteins to give 1.75 turns of a left-handed superhelix (Chapter 9, Figs 9.4 and 9.5). This process achieves a packing ratio of seven. The number of base-pairs involved in nucleosome structures varies from species to species, being 165 bp for yeast, 183 bp for HeLa cells, 196 bp for rat liver, and 241 bp for sea urchin sperm. Such nucleosomes are joined by linker DNA whose length ranges from zero bp in neurons to 80 bp in sea urchin sperm but usually averages 30–40 bp. The details of packaging the histone proteins are discussed later (Section 9.2.1).

2.6.2 Chromatin structure

Chromatin is too large and heterogeneous to yield its secrets to X-ray analysis so that electron microscopy is the chosen experimental probe. At intermediate salt concentration (~1 mM NaCl), the nucleosomes are revealed as 'beads on a string'. Spherical nucleosomes can be seen with a diameter of 7–10 nm joined by variable length filaments, often about 14 nm long. If the salt concentration is increased to 0.1 M NaCl, the spacing filaments get shorter and a zig-zag arrangement of nucleosomes is seen in a fibre 10–11 nm wide (Chapter 9, Fig. 9.6). At even higher salt concentration and in the presence of magnesium, these condense into a 30 nm diameter fibre, called a solenoid, which is thought to be either a right-handed or a left-handed helix made up of close-packed nucleosomes with a packing ratio of around 40 (see Fig. 9.6).

For the further stages in DNA condensation, one of the models proposed suggests that loops of these 30 nm fibres, each containing about 50 solenoid turns and possibly wound in a supercoil, are attached to a central protein core from which they radiate outwards. Organization of these loops around a cylindrical scaffold could give rise to the observed miniband structure of chromosomes, which is some 0.84 μm diameter and 30 nm thickness. A continuous helix of loops would then constitute the chromosome. These ideas are illustrated in a possible scheme (Fig. 2.38).

It is clear from all of the relevant biological experiments that the single DNA duplex has to be continuously accessible despite all this condensed structure in

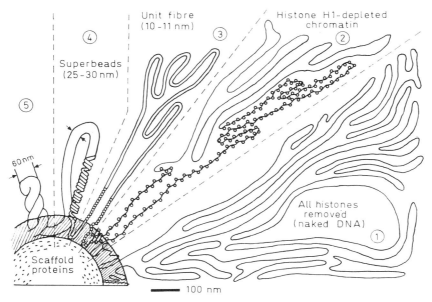

Fig. 2.38 Schematic drawing to illustrate the gradual organization of DNA into highly condensed chromatin. (1) DNA fixed to the protein scaffold; (2) DNA complexed with all histones except H1; (3) aggregation into 100 Å fibre; (4) formation of 'superbeads'; and (5) contraction into 600 Å knob (from Rindt, K.-P. and Nover, L. (1980). *Biol. Zentralblat.* **99**, 641–73).

order for replication to take place. Some of the most exciting electron micrographs of DNA have been obtained from samples where the histones have been digested away leaving only the DNA as a tangled network of interwound superhelices radiating from a central nuclear region where the scaffold proteins remain intact (Fig. 2.39).

Even then, the most condensed packing of nucleic acid is found in the sperm cell. Here a series of arginine-rich proteins called protamines bind to DNA, probably with their α-helices in the major groove of the DNA where they neutralize the phosphate charge and so enable very tight packing of DNA duplexes.

Bacterial DNA is also condensed into a highly organized state (Section 9.2.1). First, there is condensation involving histone-like proteins. This is followed by a further condensation into supercoiled domains. The process differs in several respects from assembly of chromatin in eukaryotes:

(1) there is no apparent regular repeating structure equivalent to the eukaryotic nucleosome, although short DNA segments of 60–120 base-pairs are organized by means of their interaction with abundant DNA-binding proteins;

(2) there is no prokaryotic equivalent to the solenoid structure;

(3) bacterial DNA seems to be torsionally strained *in vivo* and organized into independently supercoiled domains of about 100 kb.

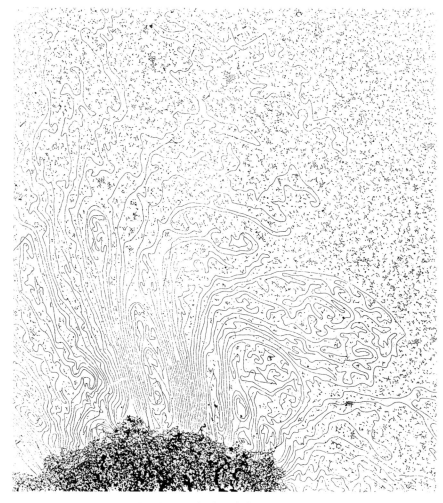

Fig. 2.39   Electron micrograph of a histone-depleted chromosome showing that the DNA is attached to the scaffold in loops (from Paulson, J. K. and Laemmli, U. K. (1977). *Cell*, **12**, 817–28. Copyright (1977) Cell Press).

Summary

DNA has to be packaged very efficiently to fit into the cell nucleus. For human chromosomes, the reduction in length is close to 10^4. In the nucleosome, some 180 base-pairs of DNA wind round a core of a histone octamer in 1.75 turns of a left-handed superhelix. These are joined by linker DNA, whose length is about 30–40 bp, and condense into fibres of 10–11 nm diameter.

Models of higher order structure have such fibres wound into a solenoid of 30 nm diameter and 11 nm pitch, giving a packing ratio of 40. How further stages of condensation give the chromosome is a subject for speculative modelling.

Further reading

2.1 and 2.2

Arscott, P. G., Lee, G., Bloomfield, V. A., and Evans, D. F. (1989). Scanning tunnelling microscopy of Z-DNA. *Nature*, **339**, 484–6.

Dickerson, R. E., Drew, H. R., Connor, B. N., Wing, R. M., Fratini, A. V., and Kapka, M. L. (1982). The anatomy of A-, B-, and Z-DNA. *Science*, **216**, 475–85.

Jurmak, F. A. and McPherson, A. (ed.) (1984). *Biological macromolecules and assemblies*, Vol 1. John Wiley, New York.

Saenger, W. (1984). *Principles of nucleic acid structure*. Springer Verlag, New York.

2.3

Blommers, M. J. J., Walters, A. L. I., Haasnoot, C. A. G., Aelen, J. M. A., van der Marel, G. A., van Boom, J. H., and Hilbers, C. W. (1989). Effects of base sequence on the loop folding in DNA hairpins. *Biochemistry*, **28**, 7491–8.

Calladine, C. R., Drew, H. R., and McCall, M. J. (1988). The intrinsic curvature of DNA in solution. *J. Mol. Biol.*, **210**, 127–37.

Drew, H. R. and Travers, A. A. (1984). DNA structural variations in the *E. coli tyrT* promoter. *Cell*, **37**, 491–502.

Kennard, O. (1988). In *DNA and its drug complexes, structure and expression* (ed. M. H. Sarma and R. H. Sarma), Vol. 2. Adenine Press, New York.

Lilley, D. M., Sullivan, K. M., Murchie, A. I. H., and Furlong, J. C. (1988). Cruciform extrusion in supercoiled DNA—mechanisms and contextual influence. In *Unusual DNA structures* (ed. R. D. Wells and S. C. Harvey), Springer Verlag, Heidelberg, pp. 55–72.

Liv, L. F. (1989). DNA topoisomerase poisons as antitumour drugs. *Ann. Rev. Biochem.*, **58**, 351–75.

Wang. J. C. (1985). DNA topoisomerases. *Ann. Rev. Biochem.*, **54**, 665–97.

Wells, R. D. (1988). Unusual DNA structures. *J. Biol. Chem.*, **263**, 1095–8.

Yoon, C., Privé, G. G., Goodsell, D. S., and Dickersen, R. E. (1988). Structure of an alternating B-DNA helix and its relationship to A-tract DNA. *Proc. Nat. Acad. Sci. USA*, **85**, 6332–6.

2.4

Cohen, J. S. (ed.) (1989). *Oligodeoxynucleotides, antisense inhibitors of gene expression*. Macmillan, London.

Clore, G. M., Oschkinat, H., McLaughlin, L. W., Benseler, F., Happ, C. S., Happ, E., and Gronenborn, A. M. (1988). Refinement of structure of the DNA dodecamer 5'd(CGCGPATTCGCG)$_2$ containing a stable purine–thymine base pair: combined use of nuclear magnetic resonance and restrained molecular dynamics. *Biochemistry*, **27**, 4185–97.

Happ, S. C., Happ, E., Nilges, M., Gronenborn, A. M., and Clore, G. M. (1988). Refinement of the solution structure of the ribonucleotide 5'r(GCAUGC)$_2$. *Biochemistry*, **27**, 1735–43.

Houba-He'rin, N. and Inouye, M. (1987). Antisense RNA. In *Nucleic acids and molecular biology* (ed. F. Eckstein and D. M. J. Lilley), Vol. 1. Springer Verlag, Berlin, pp. 210–21.

Moser, H. E. and Dervan, P. B. (1987). Sequence-specific cleavage of double-helical DNA by triple-helix formation. *Science*, **238**, 645–50.

Neuhaus, D. and Williamson, M. P. (1989). The nuclear Overhauser effect in structural and conformation analysis, Chapters 5 and 12. VCH, Weinheim.

Schimmel, P. R., Söll, D., and Abelson, J. N. (1979). *Transfer RNA: structure, properties, and recognition*. Cold Spring Harbor Laboratory, USA.

van Knippenberg, P. H. and Hilbers, C. W. (ed.) (1986). *Structure and dynamics of RNA*. NATO ASI Series. Plenum, New York.

2.5

Breslauer, K. J., Frank, R., Blöcker, H., and Marky, L. A. (1986). Predicting DNA duplex stability from base sequence. *Proc. Natl. Acad. Sci. USA*, **83**, 3746–50.

James, T. L. (1984). Relaxation behaviour of nucleic acids. In *Phosphorus-31 NMR* (ed. D. G. Gorenstein). Academic Press, New York, pp. 349–400.

McCammon, J. A. and Harvey, S. C. (1987). *Dynamics of proteins and nucleic acids*. Cambridge University Press, Cambridge.

Soumpasis, D. M. and Jovin, Th. M. (1987). Energetics of the B-Z transition. In *Nucleic acids and molecular biology* (ed. F. Eckstein and D. M. J. Lilley), Vol. 1. Springer Verlag, Heidelberg, pp. 85–111.

2.6

Cold Spring Harbor Symposia (1978). Chromatin. *Cold Spring Harbor Symp. Quant. Biol.*, Vol. 42.

Pederson, D. S., Thoma, F., and Simpson, R. T. (1986). Core particles, fibre and transcriptionally active chromatin structure. *Ann. Rev. Cell Biol.*, **2**, 117–47.

Schmid, M. B. (1988). Structure and function of the bacterial chromosome. *Trends Biol. Sci.*, **13**, 131–5.

Selker, E. U. (1990), DNA methylation and chromatin structure: a view from below. *Trends Biol. Sci.*, **15**, 103–7.

Travers, A. A. (1989). DNA conformation and protein binding. *Ann. Rev. Biochem.*, **58**, 427–52.

Travers, A. A. and Klug, A. (1987). The bending of DNA in nucleosomes and its wider implications. *Phil. Trans. Roy. Soc. Lond. B*, **317**, 537–61.

Chemical synthesis

3

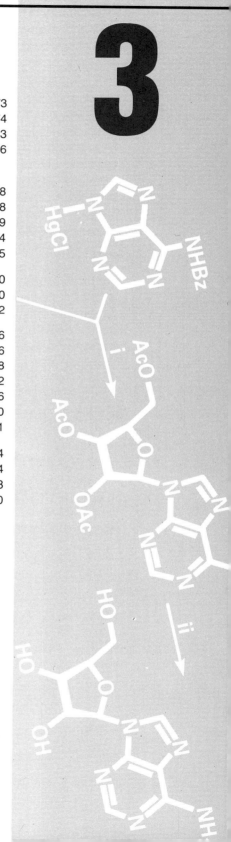

3.1 Synthesis of nucleosides

The first nucleoside syntheses were planned to prove the structures of adenosine and the other ribo- and deoxyribo-nucleosides. Modern syntheses have been aimed at producing nucleoside analogues, frequently for use as inhibitors of nucleic acid metabolism (Section 4.7). In spite of advances in stereospecific synthesis, the major nucleosides remain more economical to produce by degrading nucleic acids than by total synthesis.

Many modified nucleosides occur naturally. All species of tRNA contain unusual minor bases (Section 6.4.1) and many bacteria and fungi provide rich sources of nucleosides modified in the base, in the sugar, or in both base and sugar residues. Since some of these have been found to show a wide and useful range of biological activity, hundreds if not thousands more nucleoside analogues have been synthesized in pharmaceutical laboratories across the world. In recent times, industrial targets for this work have been anti-viral and anti-cancer agents (Section 4.7). For instance, the arabinose analogues of adenosine and cytidine, *ara*A and *ara*C (Fig. 3.1) are useful as anti-viral and anti-leukaemia drugs while 5-iodouridine is valuable for treating *Herpes simplex* infections of the eye.

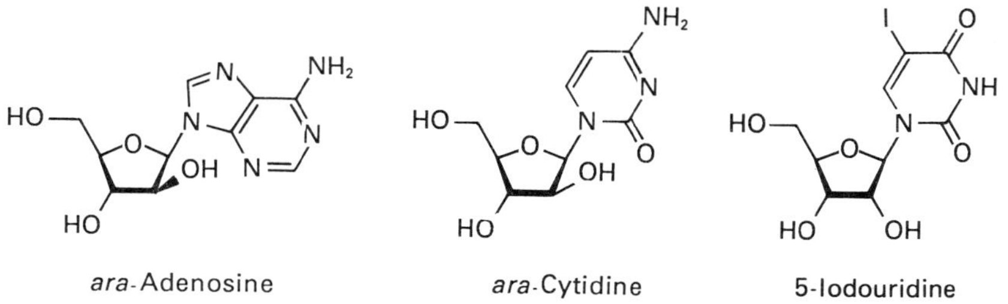

ara-Adenosine   ara-Cytidine   5-Iodouridine

Fig. 3.1

D-Ribose and other pentoses are relatively inexpensive starting materials which are especially useful in stereochemically controlled syntheses of modified sugars. Two principal strategies for the synthesis of modified nucleosides have been developed. These are illustrated by retrosynthetic analysis (Fig. 3.2). First, **disconnection A** identifies formation of the glycosidic bond by joining the sugar on to a preformed base. In practice, this uses the easy displacement of a leaving group from C–1 of an aldose derivative by a nucleophilic nitrogen (or carbon) atom of the heterocyclic base. Secondly, the **double disconnection B** identifies the process of building a heterocyclic base on to a pre-formed nitrogen or carbon substituent at C–1 of the sugar moiety. We shall now explore each of these routes in turn.

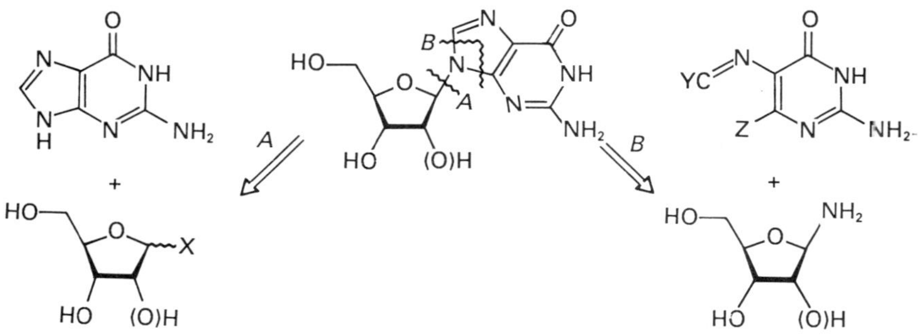

Fig. 3.2 Disconnection analysis of nucleoside synthesis.

3.1.1 Formation of the glycosidic bond

Two general methods have evolved and been refined over the years for the synthesis of the *N*-glycosidic bond in nucleosides. In turn, each has been applied to the preparation of *C*-glycosides in which the sugar residue is joined from C–1 to a **carbon** atom in the base.

Heavy metal salts of bases

Fischer and Helferich and Koenigs and Knorr introduced the use of a heavy metal salt [initially silver(I)] of a purine to catalyse the nucleophilic displacement of a halogen substituent from C–1 of a protected sugar. In a later modification, Davoll and Lowy used mercury(II) salts to improve the yields of products. Typically, chloromercuri-6-benzamidopurine reacts with 2,3,5-tri-*O*-acetyl-D-ribofuranosyl chloride or bromide to give a protected nucleoside from which adenosine is obtained by removal of the protecting groups (Fig. 3.3). These syntheses almost invariably gave the desired regioselectivity, bonding to N–9 of the purine base, and

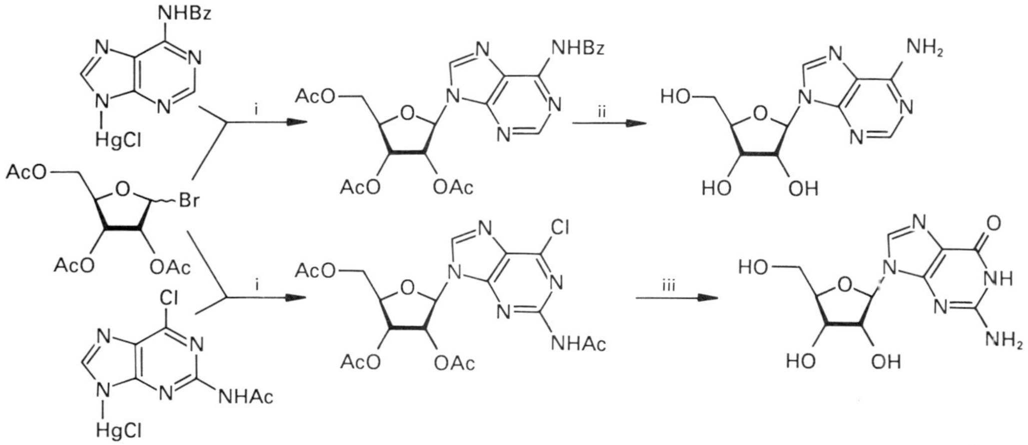

Fig. 3.3 Chloromercuri route for synthesis of purine nucleosides.
Reagents: (i) xylene, 120°C; (ii) NH_3, MeOH; (iii) NaOH aq.

more often than not they gave the desired stereoselectivity, providing predominantly the β-anomer at C–1 of the sugar. The chloromercuri salts of a range of purines can be used, providing that nucleophilic substituents are protected. Thus, amino groups have to be protected by acylation as shown in a synthesis of guanine nucleosides using 2-acetamido-6-chloropurine followed by appropriate hydrolysis (Fig. 3.3).

   The chloromercuri derivatives of suitable pyrimidines can be used in much the same way, as illustrated by a synthesis of cytidine from 4-ethoxypyrimidin-2-one. (Fig. 3.4). This reaction probably gives an *O*-glycoside first (as kinetic product) which then undergoes a mercuric halide-catalysed rearrangement to an *N*-glycoside (the thermodynamic product).

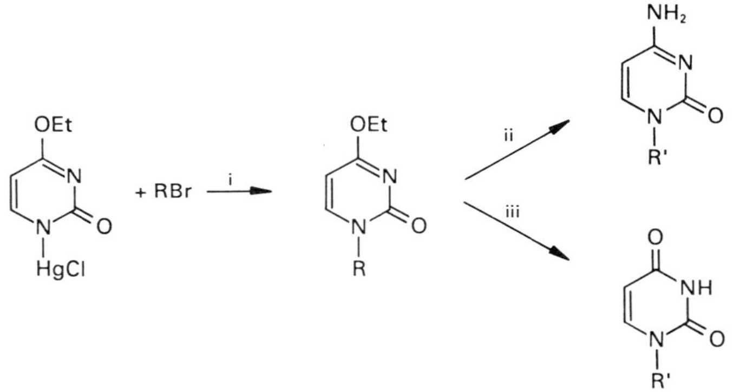

Fig. 3.4 Chloromercuri route for synthesis of pyrimidine nucleosides.
Reagents: (i) xylene, 120°C; (ii) NH₃, MeOH; (iii) NaOH aq. R = protected ribofuranosyl; R' = 1-β-D-ribofuranosyl.

Fusion synthesis

Two disadvantages of the above method are the poor solubility of the mercuri-derivatives and the instability of the halogeno-sugar derivative. One early improvement was the combination of 1-acetoxy sugars with Lewis acids such as $TiCl_4$ or $SnCl_4$ as a means of generating the reactive halogeno-sugar *in situ*. That led to the fusion process, in which a melt of the 1-acetoxy sugar and a suitable base *in vacuo*, often with a trace of an acid catalyst, can give acceptable yields of nucleosides. Thus 1,2,3,5-tetra-*O*-acetyl-D-ribofuranose fused with 2,6,8-trichloropurine or 3-bromo-5-nitro-1,2,4-triazole gives useful yields of the corresponding acylated nucleosides (Fig. 3.5). This method is at its best for weakly basic heterocycles having low melting points.

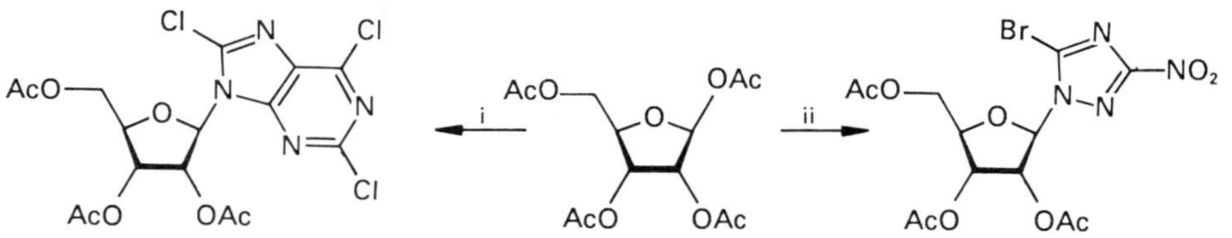

Fig. 3.5 The fusion method of nucleoside synthesis.
Reagents: (i) 2,6,8-trichloropurine, melt at 150°C; (ii) 3-bromo-5-nitro-1,2,4-triazole, melt at 150°C.

The quaternization procedure

Hilbert and Johnson noticed that substituted pyrimidines are sufficiently nucleo-philic to react directly with halogeno-sugars without any need for electrophilic catalysis. The method, which bears their name, involves the alkylation of a 2-alkoxypyrimidine with a halogeno-sugar. The initial product is a quaternary salt which at higher temperatures eliminates an alkyl halide to give an intermediate condensation product. Further chemical modification of substituents on the pyrimidine ring can lead to a range of natural and artificial bases (Fig. 3.6). Such condensations frequently give mixtures of α- and β-anomers although the use of HgBr$_2$ increases the proportion of the β-anomer.

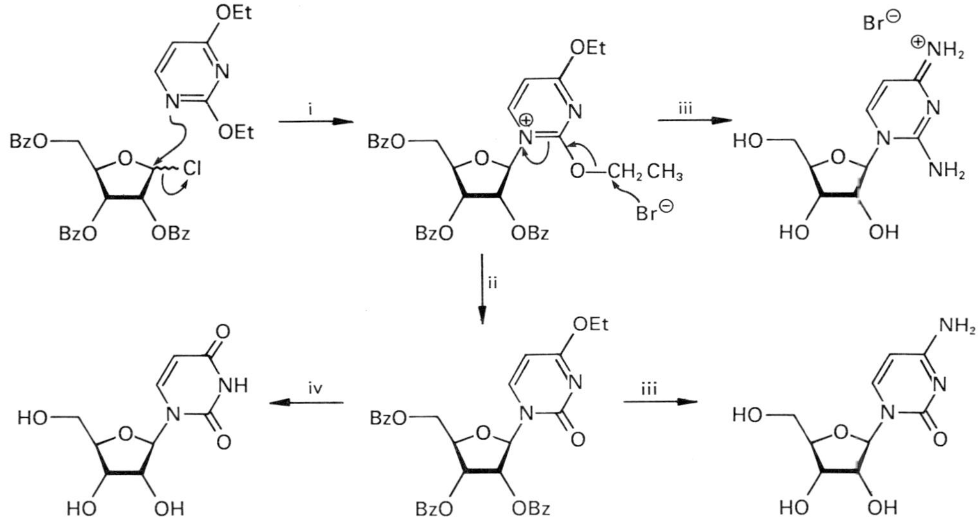

Fig. 3.6 The quaternization method of nucleoside synthesis.
Reagents: (i) CH$_3$CN, 10°C, (ii) CH$_3$CN, reflux; (iii) NH$_3$, MeOH, (iv) NaOH aq.

Silyl base procedure

A major improvement in this method came from the utilization of silylated bases, developed independently by Nishimura, by Birkofer, and by Wittenberg. Such bases have three advantages: (i) they are easily prepared, (ii) they react smoothly with sugars in homogeneous solution, and (iii) they give intermediate products that can easily be converted into modified bases. The early use of mercuric oxide as a catalyst gave way to Lewis acid catalysts (e.g. SnCl$_4$ or Hg(OAc)$_2$) and they, in turn, have been superseded by the use of silyl esters of strong acids. notably tri-methylsilyl triflate, trimethylsilyl nonaflate, or trimethysilyl perchlorate. Typically, the reaction is carried out in acetonitrile or 1,2-dichloromethane as polar solvent at around −20° to +50°C, depending on the reactivity of the components. Some examples are listed (Fig. 3.7). Very often the silylated base is generated *in situ* by the use of bis(trimethylsilyl)acetamide.

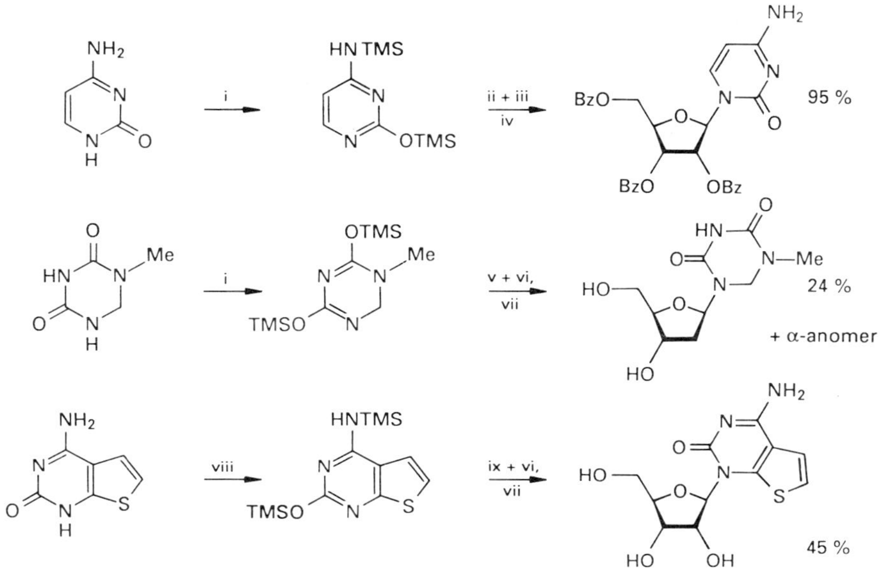

Fig. 3.7 Examples of the silyl base method of nucleoside synthesis.
Reagents: (i) (Me$_3$Si)$_2$NAc; (ii) 1-*O*-acetyl-2,3,5-tri-*O*-benzoyl-D-ribose; (iii) CF$_3$SO$_3$SiMe$_3$; (iv) H$_2$O; (v) 3,5-di-*O*-toluoyl-2-deoxyribofuranosyl chloride; (vi) SnCl$_4$; (vii) NH$_3$, MeOH; (viii) (Me$_3$Si)$_2$NH; (ix) 1,2,3,5-tetra-*O*-acetyl-D-ribose. TMS = Me$_3$Si.

This silyl-Hilbert–Johnson method works very well for a large number of nucleoside analogues which have modified bases that are difficult to prepare by other methods. It suffers, as do Koenigs–Knorr procedures, from a lack of precise control of regio- and stereoselectivity. That is because normally it has the characteristics of an S$_N$1 reaction and depends on the capture of a carbocation at C–1 of the sugar moiety by the most electronegative nitrogen on the base. The problem is well illustrated by the condensation of a 2-azidoarabinose derivative with the trimethylsilyl derivative of 6-chloropurine (Fig. 3.8). Four products have been identified (by UV and NMR spectra) as 9-α (13.7 per cent), 9-β (22.3 per cent), 7-α (10.3 per cent), and 7-β (1.6 per cent) along with other minor species. This lack of regioselectivity is even more marked in the condensation of 2,3,5-tri-*O*-benzoyl-1-*O*-acetyl-D-ribofuranose with a 1,2,4-triazole base when the sugar can become linked to any of the three azole nitrogen atoms.

Transglycosylation
It is often relatively easy to convert one of the natural nucleosides, typically 2′-deoxythymidine, into a nucleoside with a modified sugar residue, for instance the

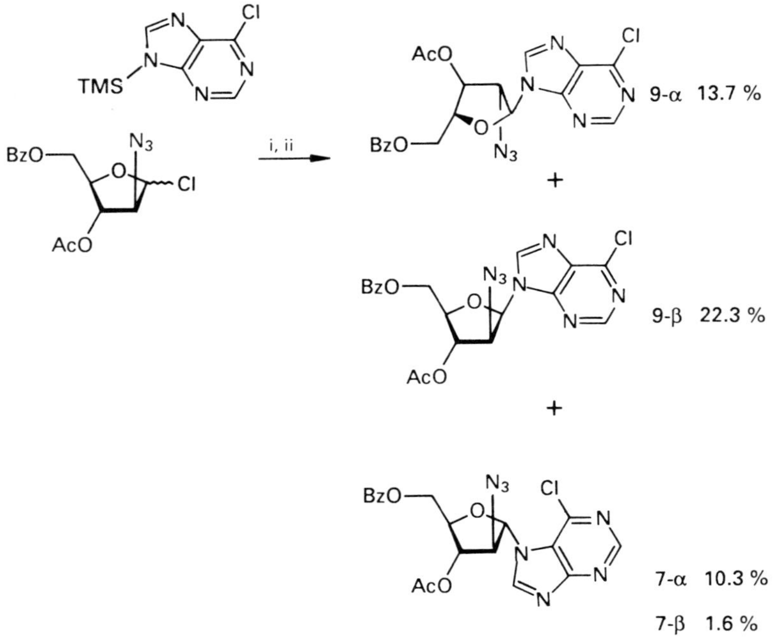

Fig. 3.8 Variable regio- and stereospecificity in silyl base condensation synthesis of nucleosides. **Reagents:** (i) Hg(OAc)$_2$; (ii) ClCH$_2$CH$_2$Cl, 50°C.

drug 3′-azido-2′,3′-dideoxythymidine (AZT, Section 4.7.2). At the same time, it can be difficult to achieve the same transformation of 2′-deoxyadenosine into AZdA. In such cases the sugar moiety can be transferred from one base to another by a process known as **transglycosylation**. The reaction is particularly effective for transferring sugars, including quite complex species, from pyrimidines (which are π-deficient heterocycles) to the more basic purines (π-excessive heterocycles) (Fig. 3.9).

This reaction has all the hallmarks of an S_N1 ionization process, as shown both by the intramolecular transfer of a sugar residue from N–7 to N–9 of 6-chloro-1-deaza-purine and by the anomerization of β- into α-nucleosides. Transglycosylation is now a favoured method for the preparation of α-anomers of pyrimidine nucleosides from their natural isomers. It also provides a useful synthesis of α-anomers of pur-ine nucleosides (Fig. 3.10). The mixture of α- and β-species which is usually formed can be separated by chromatography. However, only experience is able to predict the thermodynamically favoured regioselectivity of these processes.

Control of anomeric stereochemistry

The fact that sugars which have a 2-acyloxy substituent invariably give *N*-glycosides with the 1,2-*trans*-configuration led Baker to suggest that neighbouring group parti-cipation is responsible. Essentially, the ionization of the leaving group at C–1 of the

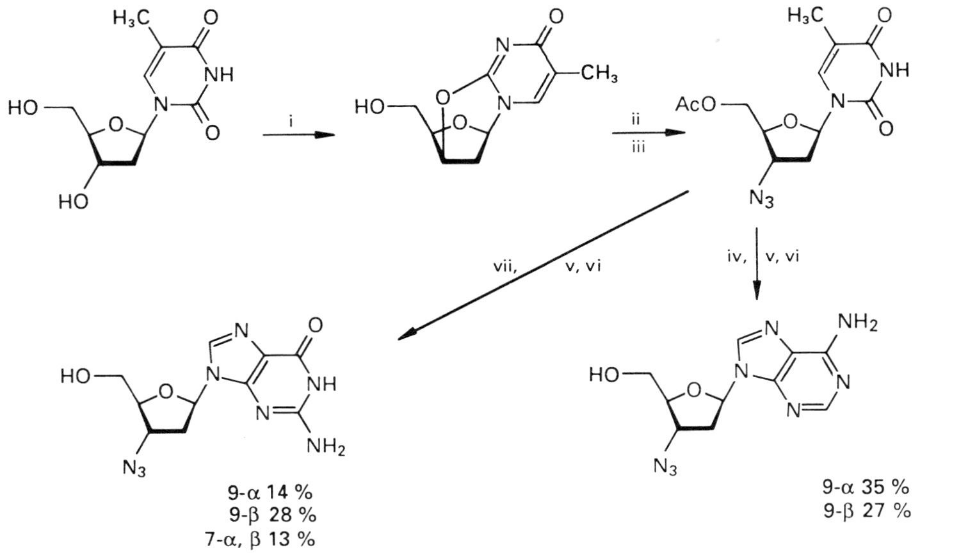

Fig. 3.9 Transglycosylation synthesis of nucleosides showing variable regio- and stereospecificity.
Reagents: (i) CHClFCF$_2$NEt$_2$; (ii) LiN$_3$; (iii) Ac$_2$O; (iv) trimethylsilyl-6-octanoyladenine;
(v) CF$_3$SO$_3$SiMe$_3$; (vi) NH$_3$, MeOH; (vii) trimethylsilyl-2-palmitoylguanine.

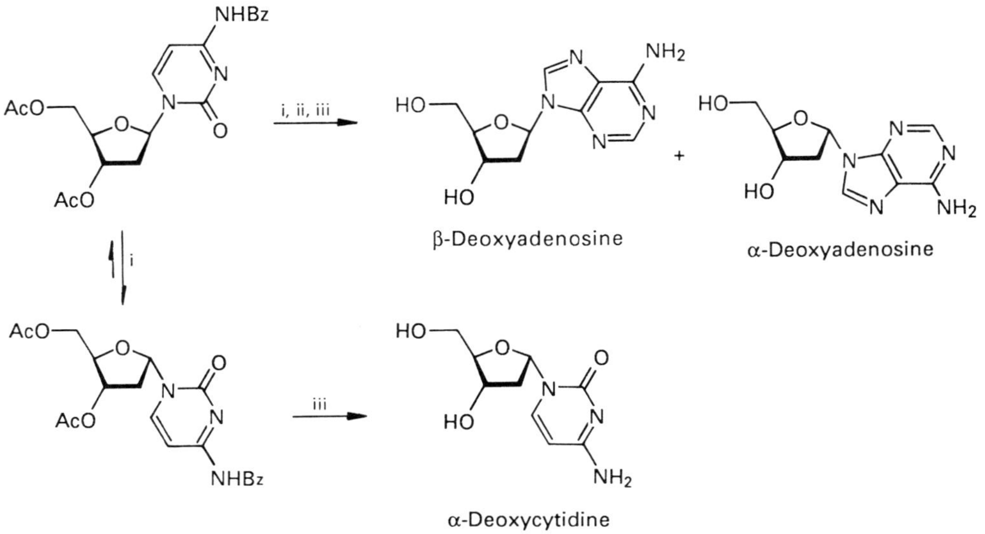

Fig. 3.10 Transglycosylation synthesis of α-deoxyribonucleosides.
Reagents: (i) (Me$_3$Si)$_2$NAc, CF$_3$SO$_3$SiMe$_3$; (ii) 6-benzoyladenine; (iii) NH$_3$, MeOH.

sugar (from either face) generates a carbocation which is 'captured' by the carbonyl oxygen of the adjacent acyl group. This bicyclic intermediate can only be attacked by the nucleophilic base from the **opposite side** of the furanose ring to the C–2 substituent: hence **Baker's 1,2-*trans* rule** (Fig. 3.11).

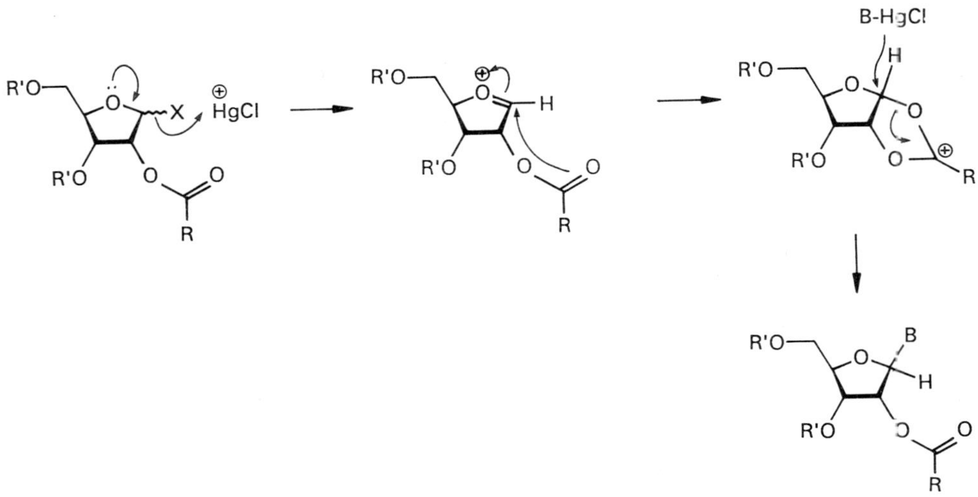

Fig. 3.11   Mechanistic basis of the 1,2-*trans* Rule. R = alkyl or aryl; R' = acyl, benzyl, trimethylsilyl, etc.; B = heterocyclic base (as chloromercuri salt or trimethylsilyl derivative).

It follows that arabinose and lyxose sugars with a 2-acyloxy substituent give α-anomers while ribose and xylose sugars give β-anomers. In cases where the hydroxyl group at C–2 is protected by a benzyl ether or by an isopropylidene or carbonate group cyclized on to the adjacent 3-hydroxy group, then neighbouring group participation is not possible. As a result, mixtures of anomers are formed. Similarly, for 2-deoxy-sugars and 2-deoxy-2-fluoro- or 2-deoxy-2-azido-sugars, there is no anomeric control.

There seem to be only a few exceptions to Baker's 1,2-*trans* rule, and these can be explained rationally. For instance, when the acyl group at position-2 is *p*-nitrobenzoyl then an α:β ratio of 1:3 is observed. This has been attributed to the electron-withdrawing effect of the nitro group which impairs participation by the carbonyl oxygen. Also, condensations of tri-*O*-acetyl-5-deoxy-5-(diethoxy-phosphonomethyl)ribofuranose with purines by the silylation procedure are known to give mixtures of anomers. Montgomery has attributed this fact to competing participation of the phosphoryl oxygen for the carbocation on the β-face of the sugar which leads to the α-anomer (Fig. 3.12).

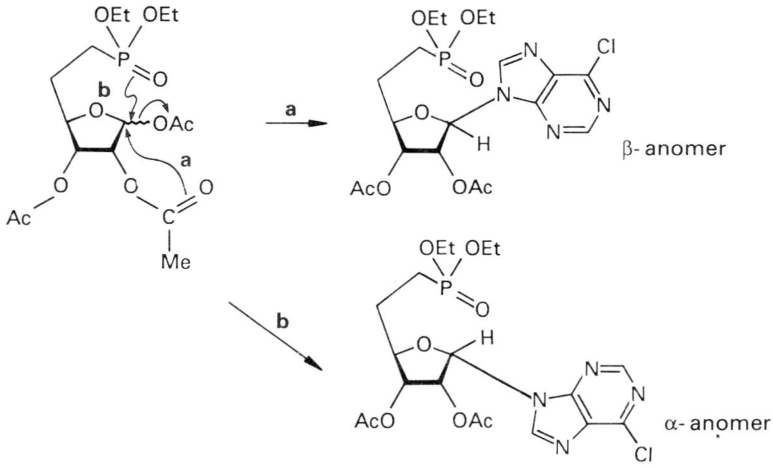

Fig. 3.12 Competing participation by the carbonyl oxygen (**a**) giving a β-anomer and by the phosphoryl oxygen (**b**) giving the α-anomer.

A stereoselective synthesis for α-ribonucleosides has been developed which starts from 1-hydroxy sugars. The preferential reaction of the β-anomer (in equilibrium with the α-isomer) with 2-fluoro-1-methyl-pyridinium tosylate fixes the configuration of the anomeric carbon. Subsequent reaction with a silylated base proceeds by an S$_N$2 displacement with predominant inversion at C–1 to give the α-anomer product (Fig. 3.13)

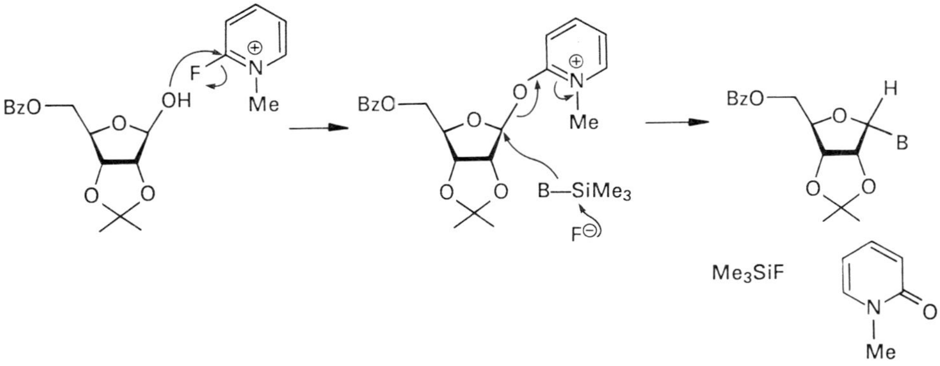

Fig. 3.13 Stereoselective synthesis of α-ribonucleosides.

C-nucleosides

A few *C*-nucleosides have been made by carbanion displacement reactions at C–1 of a suitably protected sugar, although the high basicity of the carbanion can often lead to an unwanted 1,2-elimination process. A classic example is Brown's synthesis of pseudouridine, a common component of tRNA species (Section 6.4.1), by the reaction of 2,4-bis-(*t*-butoxy)-5-lithiopyrimidine with 2,3,5-tri-*O*-benzoyl-ribofuranosyl chloride. As expected from the discussion above, when the sugar component of the reaction is changed to 3,5-di-*O*-benzoyl-2-deoxyribofuranose, the product is a mixture of 25 per cent α- and 22 per cent β-pseudodeoxyuridines (Fig. 3.14). More often, displacement reactions of carbanions at C–1 of protected sugars are designed to provide intermediates for construction of the base on to the sugar moiety. A useful procedure of this sort is the reaction of 1-fluoropentoses with enol silyl ethers (Fig. 3.14).

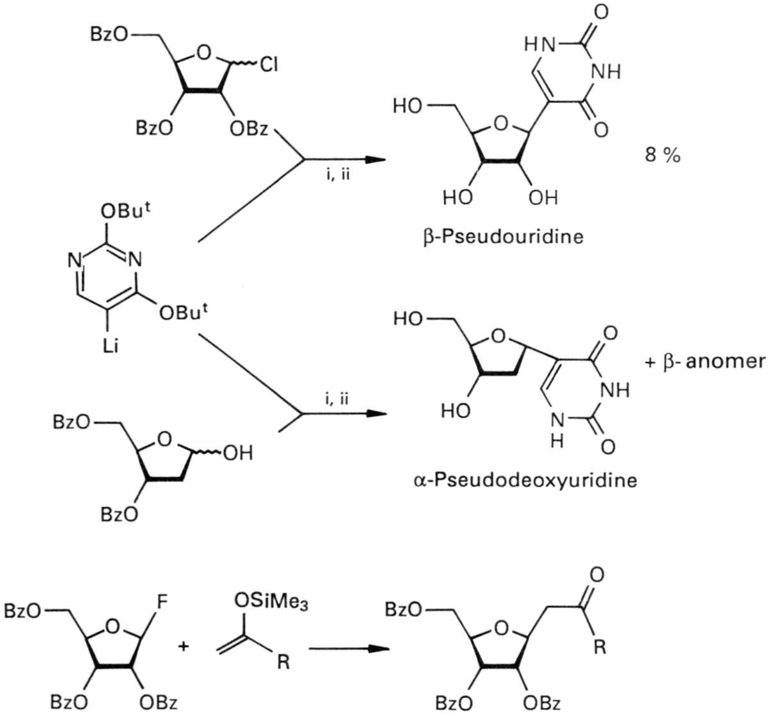

Fig. 3.14 Carbanion and enolate condensations at C–1 of pentose derivatives.
Reagents: (i) THF, –10°C; (ii) mild acidic hydrolysis.

For 2-deoxynucleosides, the direct S$_N$2 displacement of halogen from 1-α-chloro-3,5-di-*O*-*p*-toluoyl-D-ribofuranose by the sodium salt of a purine or related heterocycle gives good yields of the β-2′-deoxynucleosides (though the reaction is not always regiospecific) (Fig. 3.15).

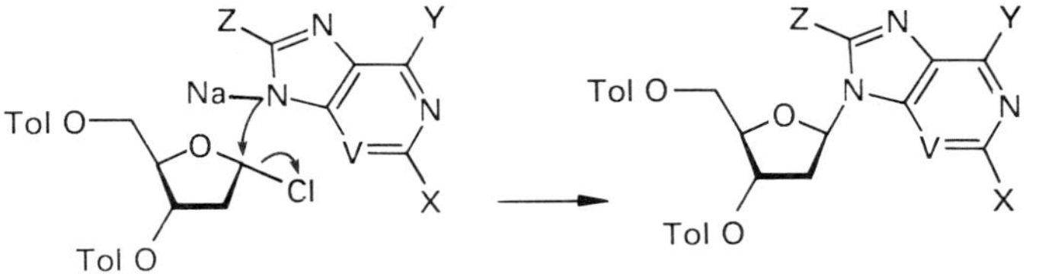

Fig. 3.15 Synthesis of β-2-deoxynucleosides by S_N2 displacement of chloride from 1-α-chloro-3, 5-di-O-toluoyl-D-2-deoxyribofuranose:

V	X	Y	Z	Yield β (%)
N	H	Cl	H	71
N	H	Cl	Cl	63 (+19% N–1β)
N	Me	Cl	Cl	59
N	MeS	Cl	Cl	87
N	MeS	Cl	Me	80
CH	Cl	Cl	H	66

3.1.2 Building the base on to a C–1 substituent of the sugar

This approach to nucleoside synthesis has three important features. Historically it was used in Todd's group for a regiospecific synthesis of adenosine (Fig. 3.16). Later, it became the preferred route for the synthesis of C-nucleosides and some unusual N-nucleosides. Most recently, it has emerged as the most flexible pathway for the synthesis of nucleosides with highly modified sugars linked to normal or to modified bases.

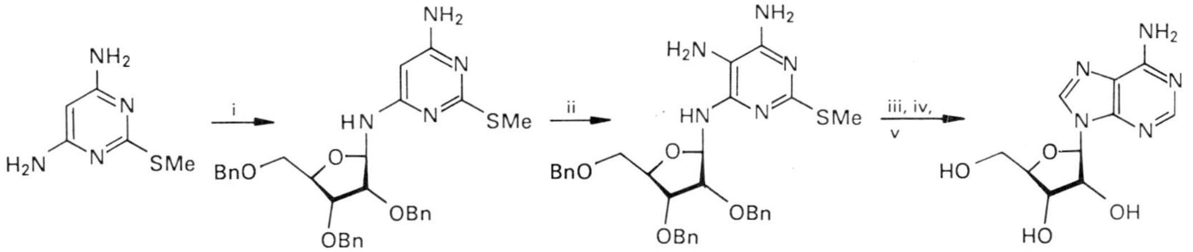

Fig. 3.16 Todd's synthesis of 9-ribofuranosyladenine.
Reagents: (i) 2,3,5-tri-O-benzyl-D-ribose; (ii) diazonium coupling or nitrosation followed by reduction; (iii) thiourea; (iv) Raney nickel desulphurization; (v) H_2/PdC debenzylation.

Nucleosides with modified bases

A good example of the use of this route is the synthesis of the fluorescent base Wyosine, which is found in the anticodon loop of some species of tRNA (Section 6.4.1). In this case, the isocyanate function is the foundation for construction of the tricyclic imidazopurine base. The same isocyanate precursor has been used in a synthesis of 5-azacytidine (Fig. 3.17). This nucleoside is elaborated by a *Streptomyces* species and has been used in the treatment of certain leukaemias.

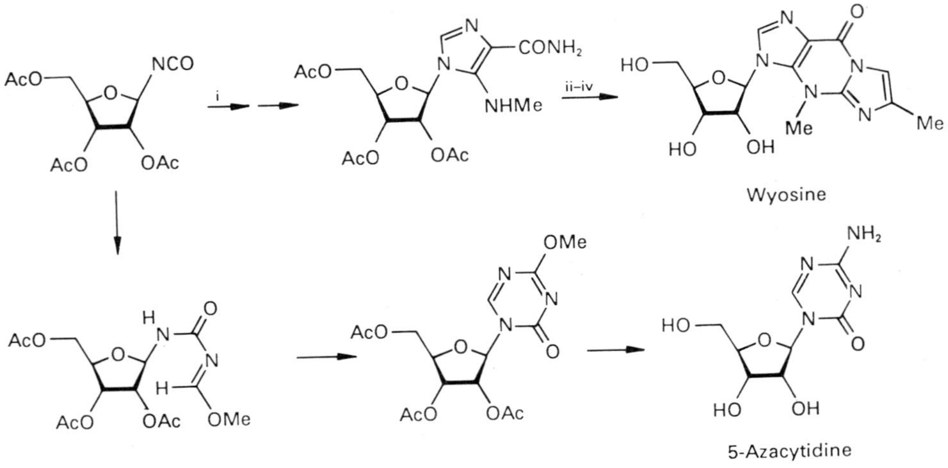

Fig. 3.17 Building the Wye base and 5-azacytosine on to a C–1 isocyanate.
Reagents: (i) 3-carbon fragment; (ii) CNBr; (iii) NaOEt, EtOH; (iv) BrCH$_2$COCH$_3$.

In a similar way, a cyanomethyl group at C–1 of D-ribose supports synthesis of 9-deazainosine, the antibiotic oxazinomycin, and of pseudouridine (Fig. 3.18). Oxazinomycin has both child growth-promoting properties and some antitumour activity.

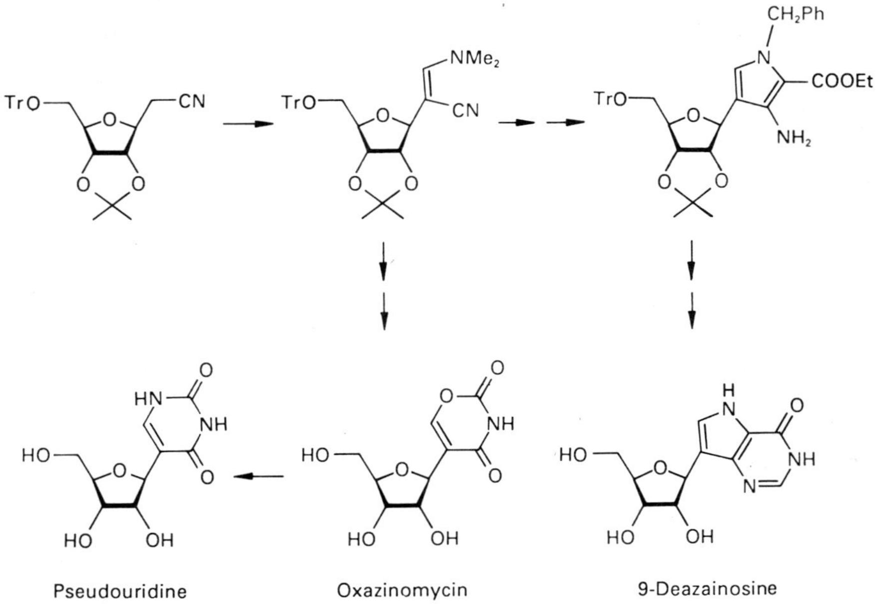

Fig. 3.18 Synthesis of some C-nucleosides.

C-nucleosides

With the growing availability of chemical reactions endowed with a high degree of stereochemical selectivity, the synthesis of *C*-nucleosides by this route has moved away from sugars as starting materials. Showdomycin is a product of *Streptomyces showdowensis* and has useful cytotoxic and enzyme inhibitory properties. A route starting from a tricyclic precursor can branch to give either showdowmycin or pseudouridine in a stereospecific fashion (Fig. 3.19).

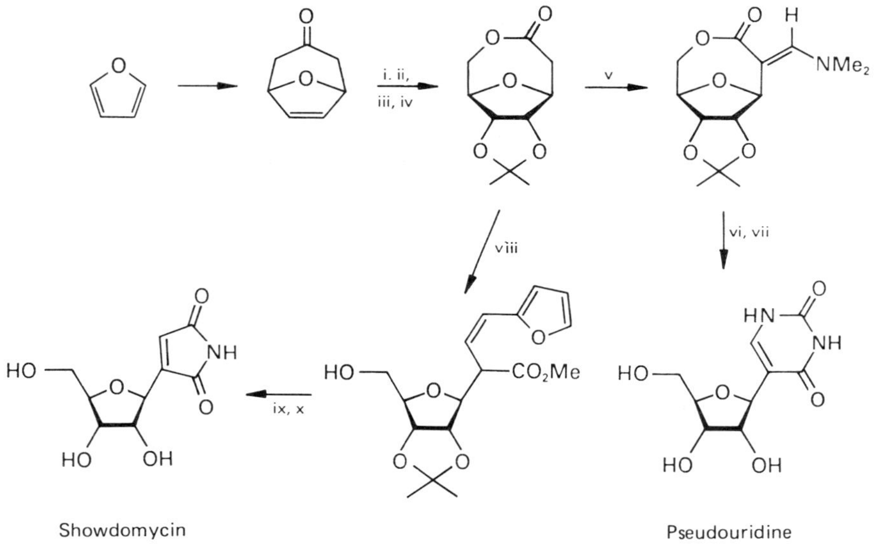

Fig. 3.19 Synthesis of showdomycin and pseudouridine from furan.
Reagents: (i) OsO$_4$, H$_2$O$_2$; (ii) acetone, H$^+$; (iii) CF$_3$CO$_3$H; (iv) resolution; (v) dimethylformamide; (vi) urea; (vii) H$_3$O$^+$; (viii) furfural, MeONa; (ix) ozone; (x) Ph$_3$P=CHCONH$_2$.

Carbocyclic nucleosides

The formal replacement of the ring oxygen in the sugar by a methylene group gives a carbocyclic nucleoside analogue. These species are being explored intensively in the search for anti-viral activity and increased resistance to enzyme degradation. Many syntheses use the key 'carbocyclic ribofuranosylamine', which is made from cyclopentadiene in five steps and then built into pyrimidine or purine carbocyclic nucleosides by standard methods. The adaptation of this route for the introduction of a fluorine atom into the 6-position (which may mimic an oxygen lone pair of electrons in binding to a receptor) presents a nice example of the development of such syntheses to highly modified sugars (Fig. 3.20).

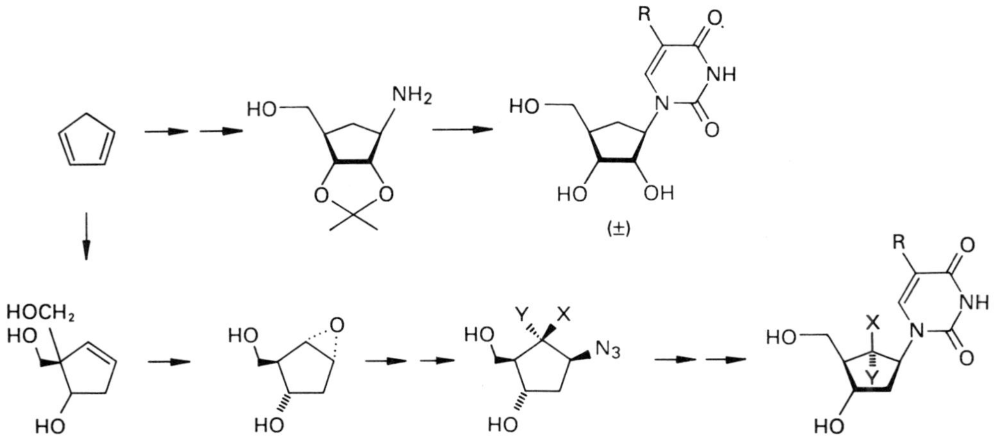

Fig. 3.20 Synthesis of carbocyclic analogues of deoxy- and ribouridine (R = H) and thymidine (R = Me).

3.1.3 Synthesis of acyclonucleosides

The success of 'Acyclovir' for the treatment of genital herpes infections has stimulated much work in this area. In these acyclonucleosides (or seco-nucleosides) the base is usually adenine, guanine, or a related purine base which can be converted into adenine or guanine as a result of metabolic deamination or hydroxylation (i.e. prodrugs, Section 4.7.2). In principle, four sections of the sugar ring can be 'cut away' and promising biological results have been found in three of these areas. Formally, one can excise (i) C–2′, (ii) C–3′, (iii) C–2′ + C–3′, and (iv) O–4′ + C–4′ + C–5′ (Fig. 3.21). The syntheses of all of these types of acyclonucleoside are invariably based on N–9 alkylation of the desired purine or its precursor, with subsequent manipulation of the necessary protecting groups (Fig. 3.22).

Fig. 3.21 Relationship of various acyclonucleosides to natural prototypes.

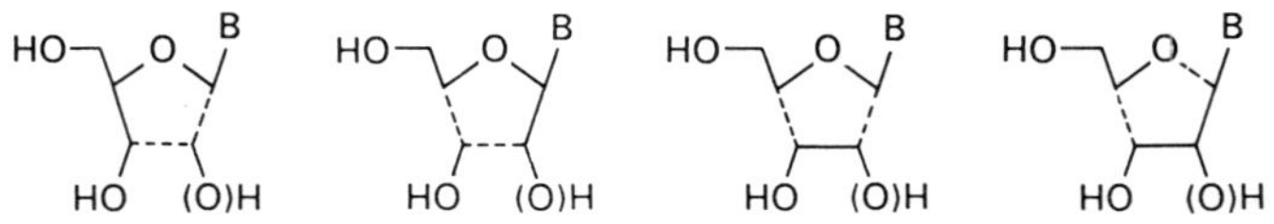

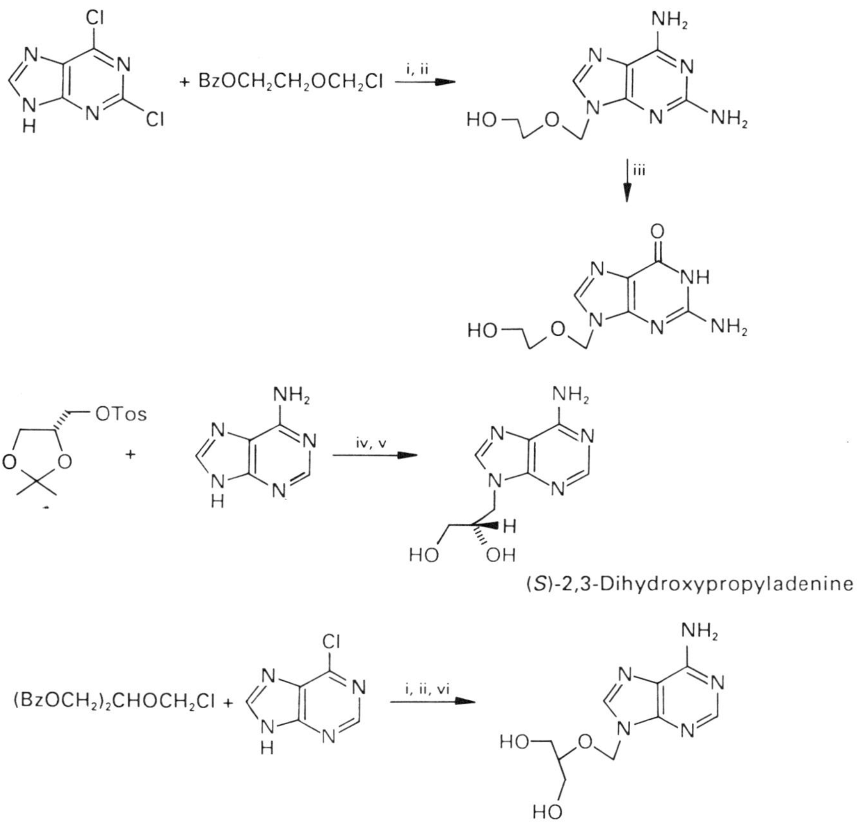

Fig. 3.22 Synthesis of some representative acyclonucleosides.
Reagents: (i) Et₃N, dimethylformamide; (ii) NH₃, MeOH; (iii) HNO₂; (iv) NaOEt; (v) H₃O⁺; (vi) H₂/PdC.

Summary

The early need to prove the structure of the major nucleosides led to syntheses based on formation of the glyosidic bond. These have been refined to give good control of regio- and stereoselectivity. The discovery of *C*-nucleosides resulted in syntheses based on construction of the base on to C–1 of a modified pentose moiety. They have also provided good routes for the preparation of base-modified nucleosides and carbocyclic nucleosides. The search for new anti-viral agents has led to syntheses of acyclonucleoside analogues which lack one or more atoms from the pentofuranose ring.

3.2 Chemistry of esters and anhydrides of phosphorus oxyacids

3.2.1 Phosphate esters

The predominant forms of phosphorus in biology are as orthophosphoric acid, H_3PO_4, its esters, anhydrides, and some amides. Its oxidation state is P(V) and it has a co-ordination number 4 (CN.4). Orthophosphates are tetrahedral. The 'single' P–O bonds use sp^3 hybrid orbitals at phosphorus and are ~ 1.6 Å long. In triesters the P=O 'double' bond is shorter, ~ 1.46 Å, and involves a phosphorus d-orbital in a p_π–d_π hybrid. Because phosphorus has five $3d$ orbitals, it can participate in such bonding simultaneously to more than one oxygen ligand (the corresponding bonding to neutral nitrogen ligands in phosphoramidates is rather weak so the nitrogen remains moderately basic).

Orthophosphoric acid   P—O d_π-p_π bonding

Phosphate triesters

Triesters have all three hydrogens of phosphoric acid replaced by alkyl or aryl groups. They are non-ionic, soluble in many organic solvents, and sufficiently stable to be purified by chromatography. The P=O bond is effectively transparent in the UV region and has an IR absorption at 1280 nm. When all three ester groups are different, as in butyl ethyl phenyl phosphate (Fig. 3.23), the phosphorus atom is a centre of asymmetry and optical isomers are possible.

Phosphate diesters

Diesters have two hydrogens replaced by alkyl or aryl groups. The remaining OH ligand is strongly acidic, pK_a ~ 1.5. Consequently, phosphate diesters exist as monoanions at pH > 2, and are usually water-soluble. The negative charge is shared equally between the two unsubstituted oxygens (Fig. 3.23). When the two

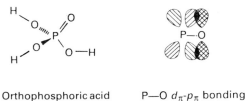

Fig. 3.23   Chiral phosphate triester   Prochiral oxygens in a phosphate diester

ester groups are different, the phosphorus atom is a prochiral centre and the two unsubstitued oxygens are non-equivalent (i.e. diastereotopic).

Phosphate monoesters

Monoesters have a single alkyl or aryl group and two ionizable OH groups. These have $pK_{a_1} \sim 1.6$ and $pK_{a_2} \sim 6.6$, so there is an equilibrium in neutral solution (effectively from pH 5 to pH 8) between the monoanion and the dianion. The equivalent oxygens share the negative charge in both mono- and dianions and there is partial double-bonding to each. In the monoanion, the hydrogen atom trans-locates rapidly between the three oxygens making them all equivalent in solution. These three oxygens are pro-pro-chiral. Thus the use of the three isotopes of oxygen, ^{16}O, ^{17}O ($=\emptyset$), and ^{18}O ($=\bullet$) has been widely used in stereochemical synthesis and in analysis of substitution reactions of phosphates (Fig. 3.24).

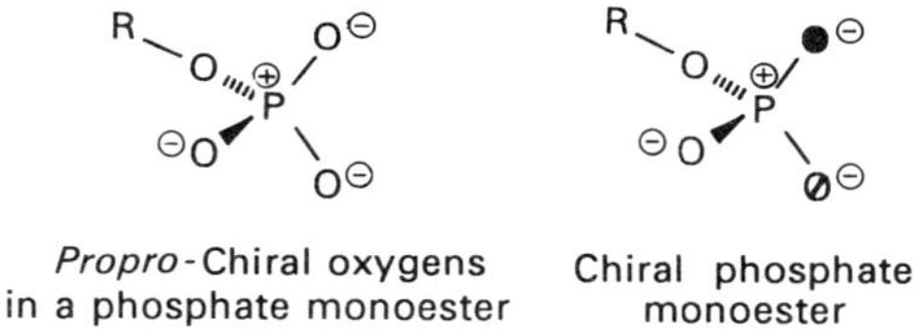

Propro-Chiral oxygens   Chiral phosphate
in a phosphate monoester   monoester

Fig. 3.24

3.2.2 Hydrolysis of phosphate esters

The great stability to hydrolysis under physiological conditions of phosphate diesters and, to a lesser extent, of monoesters is an essential feature of the chemistry of nucleosides and nucleic acids and intrinsic to life itself. Studies of mechanisms for their hydroysis have had to be carried out at elevated temperatures (up to 100°C) and often at extremes of pH and the data then extrapolated to 37°C and pH 7. Reactivity can also be enhanced by using aryl esters, especially the chromophoric *p*-nitrophenyl esters.

Both C–O and P–O cleavage reactions have been identified. In general, associative, $S_N2(P)$, mechanisms are more common than dissociative, $S_N1(P)$, ones and they are usually linked to inversion of configuration at phosphorus. The associative process is best described by a five-co-ordinate species (CN.5) as intermediate or transition state in which ligand positional interchange, often called **pseudorotation**, is usually slower than breakdown of the trigonal bipyramidal species to form products (Fig. 3.25). In dissociative reactions, a planar, three-co-ordinate species, often described as a monomeric metaphosphate, has the opportunity of capturing the incoming nucleophile on either face and so racemization at phosphorus results (Fig. 3.25).

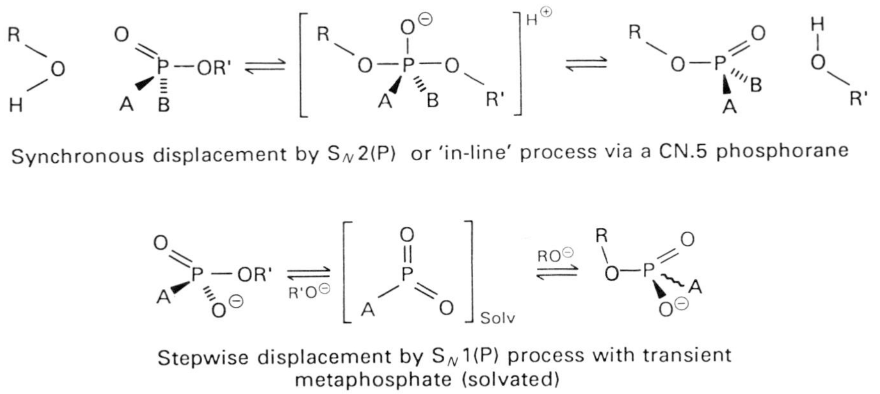

Synchronous displacement by $S_N2(P)$ or 'in-line' process via a CN.5 phosphorane

Stepwise displacement by $S_N1(P)$ process with transient
metaphosphate (solvated)

Fig. 3.25 Mechanisms of displacement reactions for phosphates.

Hydrolysis of alkyl triesters

Trimethyl phosphate is slowly hydrolysed in alkaline solution in an $S_N2(P)$ process ($k_{OH^-} = 3 \times 10^{-2}$ l mol^{-1} s^{-1} at 100°C). With $H_2^{18}O$ as solvent, no isotope exchange is observed into the P=O group of unreacted ester and no Me^{18}OH is formed. This shows that there is exclusive P–O cleavage. Other 'hard' nucleophiles such as F$^-$ react similarly and, indeed, fluoride catalysis of the transesterification of triesters is a useful process (Fig. 3.26). The intramolecular migration of phosphorus in a triester to a vicinal hydroxyl group is especially easy and must be avoided in the synthesis of oligoribonucleotide and inositol phosphate precursors.

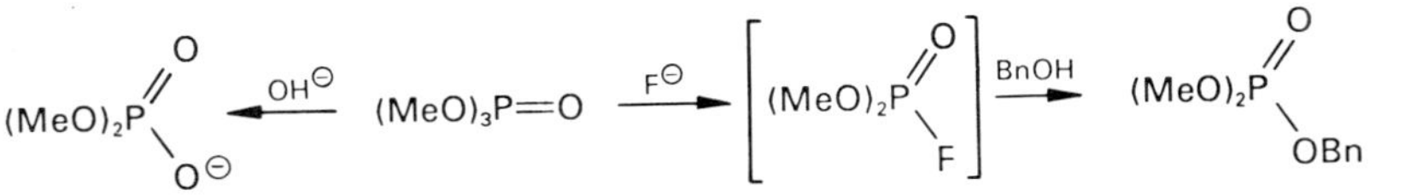

Fig. 3.26 P–O cleavage reactions with hard nucleophiles for triesters.

Trimethyl phosphate is hydrolysed extremely slowly in neutral and in acidic conditions ($k_w = 1.6 \times 10^{-7}$ s^{-1} at 44.7°C) with C–O cleavage. Soft nucleophiles, such as RS$^-$, Br$^-$, or I$^-$, also dealkylate phosphate triesters with C–O cleavage. Such reactions are typical S_N2 processes and show a clear preference for dealkylation of Me > Et > R$_2$CH. This characteristic is particularly well exploited in the thiophenolate deprotection of methyl phosphate triesters in oligonucleotide synthesis (Section 3.4.4) (Fig. 3.27).

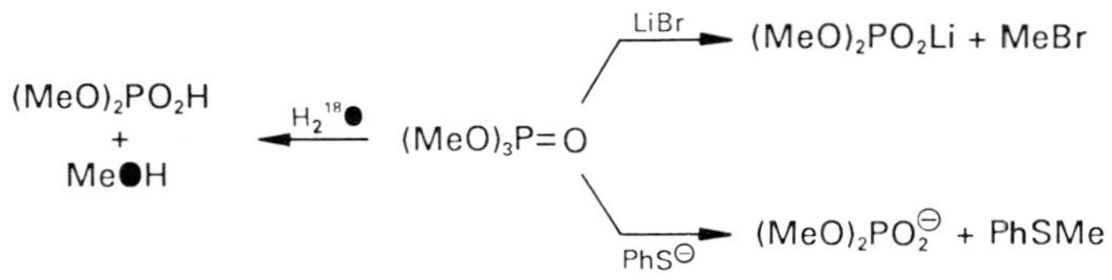

Fig. 3.27 C–O cleavage reactions with soft nucleophiles for triesters.

$(MeO)_2PO_2H$
+
Me•H

$(MeO)_3P=O$

LiBr → $(MeO)_2PO_2Li$ + MeBr

PhS$^-$ → $(MeO)_2PO_2^-$ + PhSMe

Alkyl phosphate triesters are also sensitive to β-elimination processes. While such reactions appear to have no role in nucleic acid metabolism, they have been usefully adapted for the selective deprotection of phosphate triesters in oligo-nucleotide synthesis. The 2-cyanoethyl group and its congeners have an acidic β-hydrogen and so are susceptible to cleavage under mildly basic conditions (Section 3.4.4). In a similar fashion, the 2,2,2-trichloroethyl ester group can be eliminated by reduction either using a zinc–copper couple or at a reducing anode using an appropriate electrode potential (Fig. 3.28).

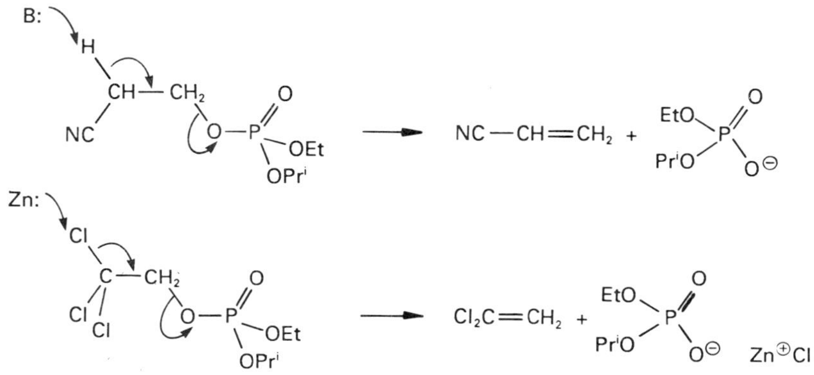

Fig. 3.28 Selective cleavage of alkyl phosphate triesters by β-elimination.

Hydrolysis of aryl triesters

Because aryl phosphates are much more reactive than alkyl phosphate triesters, it is possible to achieve selective, nucleophilic displacement of the phenolic residue in a dialkyl aryl phosphate on account of its better leaving group ability ($pK_a > 5$). One of the best nucleophiles for this purpose is the oximate anion (Fig. 3.29) (Section 3.4.4).

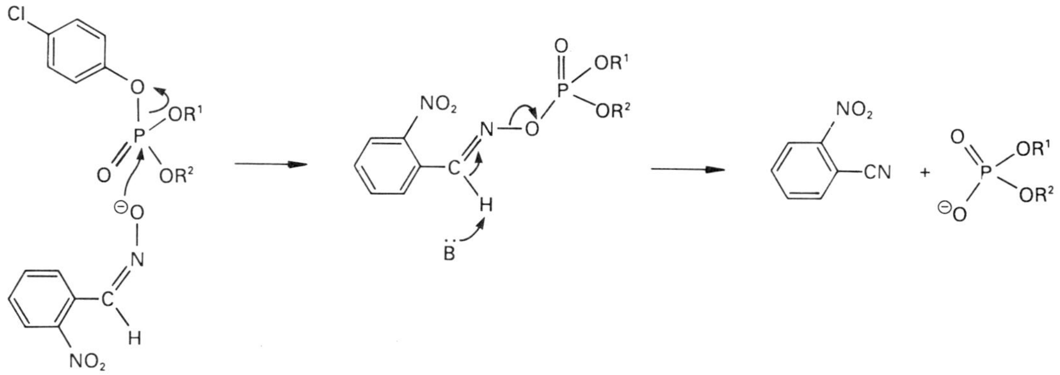

Fig. 3.29 Selective nucleophilic displacement in an aryl triester.

Hydrolysis of phosphate diesters

At pH > 2, phosphate diesters exist as their monoanions, which are stable in boiling water (Table 3.1). Even in strong alkaline conditions, diesters hydrolyse more slowly than triesters; (anion–anion repulsion) with predominant (>90 per cent) C–O cleavage. In acidic conditions their hydrolysis is rather similar to that of trialkyl phosphates. The diaryl esters are rather more reactive under alkaline conditions, as is to be expected for reaction involving a better leaving group.

Table 3.1. Rate constants $[\log (10^6 \times k)]$ for the hydrolysis of phosphate esters (100°C) and (patterns of bond cleavage)

	Conjugate acid (P=OH$^+$) k_{H^+}/s^{-1}	Neutral k_w /s^{-1}	Monoanion k /s^{-1}	Alkaline k_{OH^-} /s^{-1}
$(RO)_3PO$	1 (C–O)	1 (C–O)	—	4.5 (P–O)
$(PhO)_3PO$	1.7 (P–O)	−1.2 (P–O)	—	5.6 (P–O)
$(MeO)_2PO_2H$	0.5 (C–O)	0.4	<<−3 (C–O)	−1 (C–O > P–O)
$(PhO)_2PO_2H$	0.2 (P–O)	—	−2.5 (P–O)	1.6 (P–O)
$(RO)_2PO_2H$	1 (C–O)	−0.3	0.9 (P–O)	<−2
$PhOPO_3H_2$	1 (P–O)	1.5	2.5 (P–O)	−0.4 (P–O)
R–O–P(=O)(O$^-$)–O–P(=O)(O$^-$)–O–P(=O)(O$^-$)–O$^-$	5.6 (P–O)^a	—	—	2.7 (P–O)^b

(a) 30°C
(b) 25°C

This marked stability of the phosphate diester linkage to hydrolysis is a vital feature of the biological role of DNA. It is dramatically changed for esters of 1,2-diols. Here, the vicinal hydroxyl group enormously enhances the rate of hydrolysis of di- and tri-esters and cyclic phosphates of 1,2-diols hydrolyse some 10^7 times faster than do their acyclic or 6- and 7-membered cyclic relatives. This corresponds to a decrease in $\Delta G^{\ddagger}$ of 36 kJ mol^{-1}. About 60 per cent of this acceleration is attributed to relief of strain in the 5-membered cyclic ester, which has a 98° O–P–O angle and an enhanced enthalpy of hydrolysis of −20 kJ mol^{-1}.

The essential observations are that there is acceleration of (i) ring closure, (ii) ring opening, and (iii) exocyclic P–O bond cleavage (in both phosphate and phosphorane species) (Fig. 3.30).

How can ring-strain accelerate both endocyclic and exocyclic substitution at phosphorus? This problem has been investigated by Westheimer and his associates over some 25 years. They have found that the hydrolysis of ethylene phosphate shows incorporation of isotope into P–O bonds from $H_2^{18}O$ in both acidic and alkaline conditions, and these reactions must involve an addition–elimination (i.e. an associative) process. It is generally agreed that a transient CN.5 phosphorane intermediate is both stabilized and made kinetically more accessible relative to its

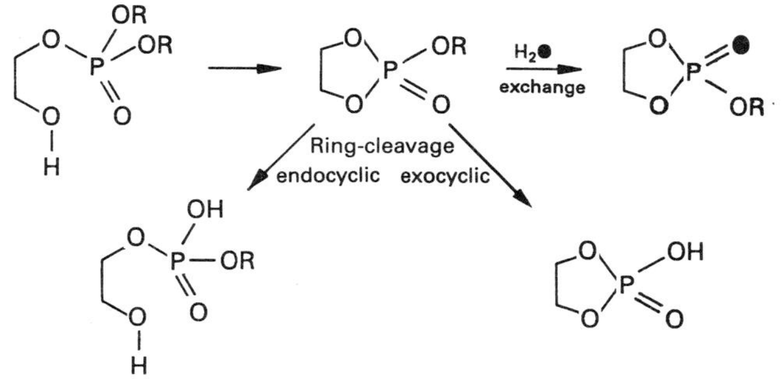

Fig. 3.30 Accelerated P–O cleavages associated with 5-membered ring phosphate esters in acidic and in alkaline solution.

acyclic counterpart because of the geometry of the 5-membered ring. It is then reasonable to invoke topoisomerism (in this case the pseudorotation of a trigonal bipyramidal species) to explain most of the phenomena associated with this remarkably enhanced reactivity (Fig. 3.31).

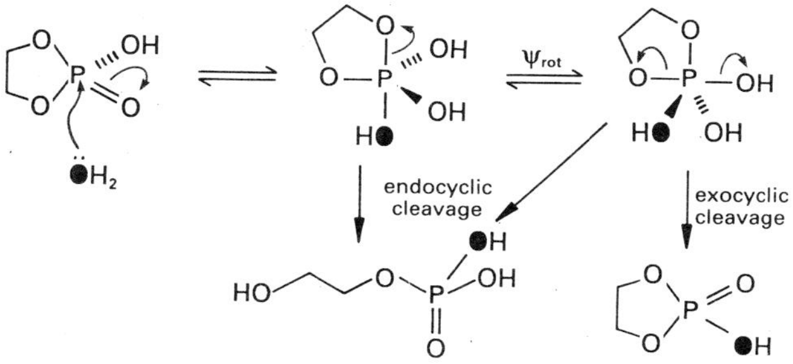

Fig. 3.31 Role of trigonal bipyramidal pseudorotation (ψ_{rot}) in ^{18}O isotope (●) incorporation into ethylene phosphate.

Regardless of the subtleties of the mechanism, this phenomenon is clearly involved in the hydrolysis of RNA by alkali and by ribonucleases. In both cases, the 5-membered 2′,3′-cyclic phosphates of nucleosides are formed by the displacement of the 5′-*O*-nucleoside residue. The enzymatic reaction is completed by the regio-selective ring-opening of the cyclic phosphate to give a 3′-nucleoside phosphate while alkaline hydrolysis leads to a mixture of 2′- and 3′-phosphates (Fig. 3.32).

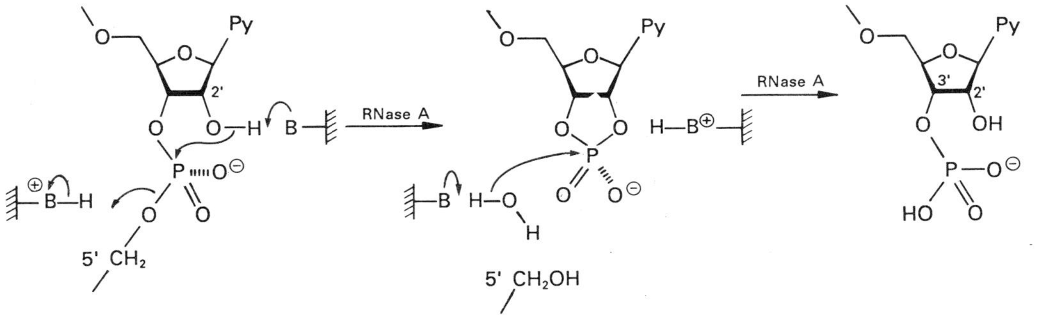

Fig. 3.32 Ribonuclease (and alkaline) hydrolysis of RNA via 2′,3′-cyclic phosphate.

Both reactions exhibit retention of configuration at the phosphorus centre. This is interpreted as a double inversion of stereochemistry as a result of two, successive 'in-line' displacement processes.

Full details of the mechanism of action of RNase A and RNase T1 have emerged from a combination of X-ray crystallography, site-directed mutagenesis, and mechanistic analysis (Section 9.2.3).

It must be emphasized that this remarkable reactivity is exclusive to 5-membered cyclic phosphate esters and esters of 1,2-diols. That contrasts totally with the stability of esters of 1,3-diols and 6-ring cyclic phosphates. The most notable example is 3′,5′-cAMP, whose key role as the second messenger in cell signalling is dependent on its kinetic stability to non-enzymatic hydrolysis.

Phosphate monoesters

The hydrolysis of monoalkyl phosphates at low pH proceeds via the conjugate acid, $ROP^+(OH)_3$, and is similar in mechanism to that of triesters (Table 3.1). These esters are very resistant to alkaline hydrolysis, as a result of anionic repulsion, but they show an unusually large reactivity for the neutral hydrolysis of the monoanion. This proceeds by P–O cleavage and has all the characteristics of a dissociative process via a CN.3 metaphosphate intermediate. A better description has invoked the idea of an 'exploded transition state' in which there is a very loose association of incoming and outgoing ligands (Fig. 3.33). Similar phenomena have been analysed for spontaneous hydrolyses of acetyl phosphate, creatine phosphate, and ATP (loss of the γ-phosphate), all of which have good leaving groups on a terminal phosphate.

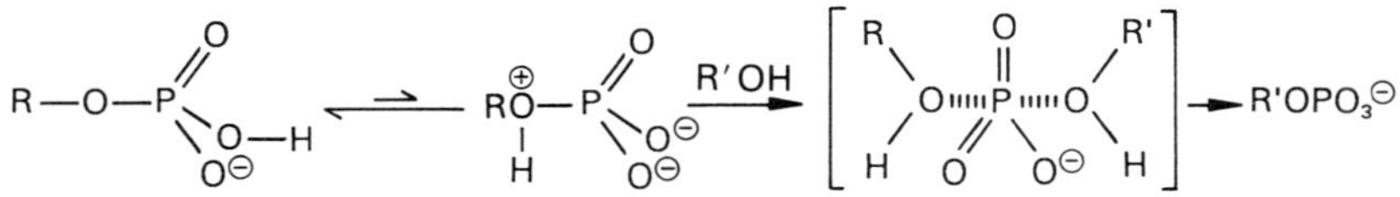

Fig. 3.33 Dissociative hydrolysis of phosphate monoester monoanions.

3.2.3 Condensed phosphates

Phosphoric acid can form chains of alternating oxygen–phosphorus linkages, which

$$-O-\overset{|}{\underset{|}{P}}-(O-\overset{|}{\underset{|}{P}}-)_nO-$$

are relatively stable in neutral aqueous solution. The major condensed phosphates of biological importance are pyrophosphoric acid, $(HO)_2P(O)OP(O)(OH)_2$, its esters, and esters of tripolyphosphoric acid. The stability of such species can be related to that of the corresponding phosphates after due allowance for (i) the changed stability of the anionic charge and (ii) improved leaving group characteristics.

Thus, tetraethyl pyrophosphate is an ethylating agent towards hard nucleophiles and also a phosphorylating agent; tetraphenyl pyrophosphate is exclusively a phosphorylating agent. P^1,P^2-Dialkyl pyrophosphates have considerable stability towards hydrolysis at ambient pHs where they exist exclusively in the form of the dianion. This feature is very important for the stability of the pyrophosphate link as a structural feature in many co-enzymes, including NADH, FAD, and CoA (Section 3.3). Similarly, P^1,P^3-diesters of tripolyphosphoric acid are stable components of the 'cap' structure of eukaryotic mRNA (Section 6.4.3) and P^1,P^4-diadenosyl tetraphosphate, Ap$_4$A, is a stable minor nucleotide species found in low concentration in all mammalian tissues (Fig. 3.34).

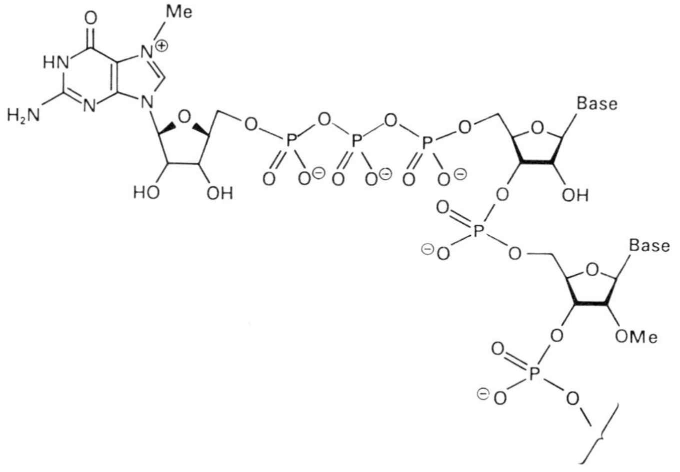

Fig. 3.34 P^1,P^3-Dinucleosidyl triphosphate in the 'cap' structure at the 5'-end of eukaryotic mRNA.

3.2.4 Synthesis of phosphate esters

The most common approaches to dinucleoside phosphate ester synthesis use phosphorylation reactions in which a 3'-nucleotide component is converted into a reactive phosphorylating species by a condensing agent. One of the major problems is that the more reactive condensing agents not only activate the nucleotide, but may also react with nucleoside bases or with the product, leading to unacceptable loss in yield. The ideal condensing agent should have a high rate of activation of the phosphate species and a low rate of reaction with the hydroxylic component or with the N-protected bases.

Interrelationships of esters of phosphorus oxyacids

The formal relationship between phosphorus halides (X = Cl) and the mono-, di-, and tri-alkyl esters of phosphorus oxyacids is shown schematically (Fig. 3.35). Actual reaction conditions have to be controlled carefully to avoid the formation of by-products, especially of alkyl halides. Many of the interconversions shown are best accomplished using nitrogen ligands at phosphorus (X = NPr$^i_2$ or an azole).

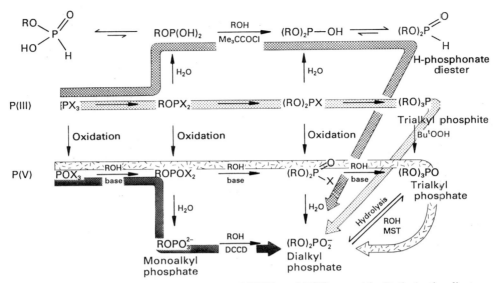

Fig. 3.35 Formal relationship between esters of P(III) and P(V) oxyacids. Path A, the diester route for nucleotide synthesis; Path B, the P(V) triester route; Path C, the P(III) route and Path D, the H-phosphonate route.

Early syntheses worked mainly with mild condensing agents such as dicyclohexyl carbodiimide (DCCD). More powerful reagents, such as the arenesulphonyl chlorides, were introduced next and made more selective by building in steric factors. Since 1980, the demand for faster reactions for oligonucleotide synthesis has switched attention to P(III) chemistry. In a very recent variation, the use of H-phosphonates as a four-co-ordinate P(III) species has built upon the pioneering studies of Todd in the 1950s.

Syntheses via phosphate diesters

In the diester route to oligonucleotides (Fig. 3.35, Path A), the key step is the condensation of a phosphate monoester with an alcohol using DCCD. The reactions

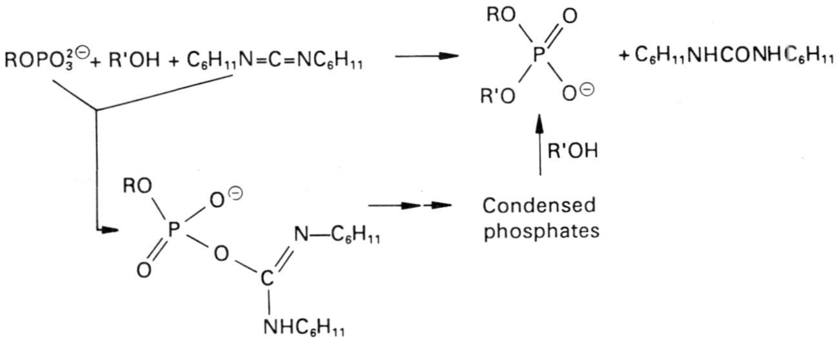

Fig. 3.36 Synthesis of phosphate diesters using dicyclohexylcarbodiimide.

are slow, but at room temperature there is no formation of triesters. The mechanism is complex: an initial imidoyl phosphate adduct of DCCD and the 3'-nucleotide is probably converted into condensed phosphates before final formation of the phosphodiester (Fig. 3.36).

Syntheses via phosphate triesters

DCCD was superseded by mesitylenesulphonyl chloride as a faster and more efficient condensing agent. In particular, its greater reactivity enables it to make triesters from dialkyl phosphates and an alcohol, and so it formed the basis of the first triester syntheses of oligonucleotides (Fig. 3.35, Path B). The key step here is the condensation of a suitable nucleotide diester, as $(RO)_2POX$, with a 5'-nucleoside to give a triester, $(RO)_3PO$. To avoid problems arising from the nucleophilicity of chloride anion, the condensing agents now used are mesitylenesulphonyl tetrazolide or nitrotriazolide (Section 3.4.3). A mixed phosphoryl–sulphonyl anhydride is produced initially in which the methyl groups of the mesitylene ring provide steric hindrance to reaction at sulphur and ensure condensation at phosphorus (Fig. 3.37). Subsequent complex reactions, which may also involve condensed phosphate intermediates, lead to the triester product. The final conversion of the triester into the desired diester uses one of the specific cleavages described above (Section 3.2.1).

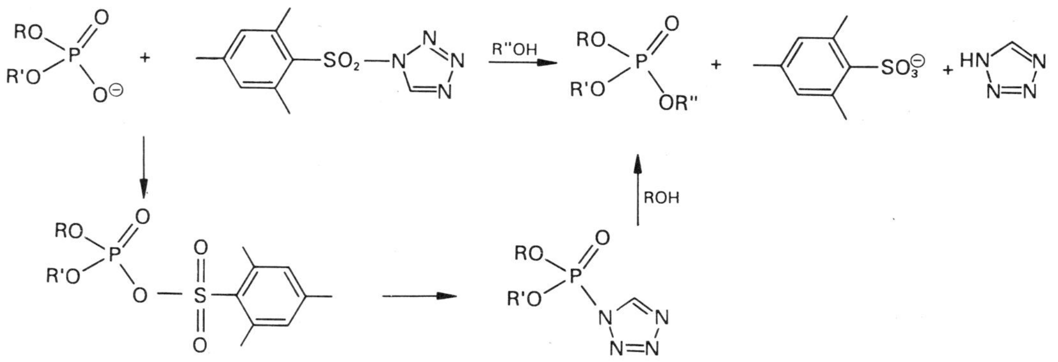

Fig. 3.37 Synthesis of phosphate triesters using mesitylenesulphonyl tetrazolide.

Syntheses via phosphite triesters

The P(III) triester route to oligonucleotides (Fig. 3.34, Path C) harnesses the intrinsically greater reactivity shown by PCl_3 compared to $POCl_3$ to achieve faster coupling steps. A major breakthrough was achieved by Matteucci and Caruthers, who established the value of alkyl phosphoramidites ($X = NPr^i_2$) as stable 3'-derivatives of nucleosides which nonetheless react rapidly and efficiently with nucleoside 5'-hydroxyl groups in the presence of azole catalysts (Section 3.4.3). The resulting product is an unstable phosphite triester which must be oxidized immediately to give the stable phosphate triester in a process that can be cycled up to 100 times on a solid-phase support (Fig. 3.38).

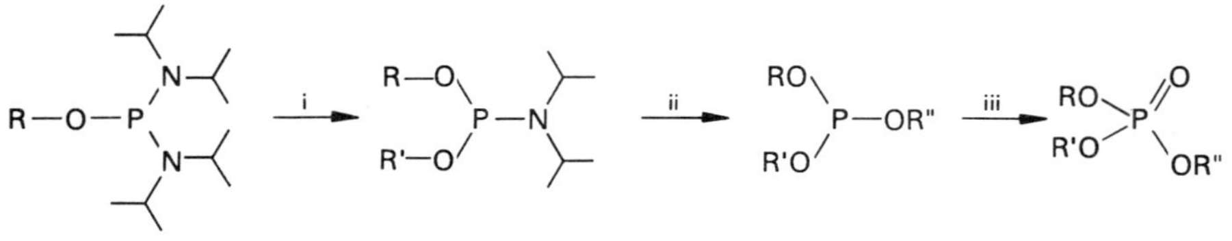

Fig. 3.38 Synthesis of phosphate triesters via phosphite triesters.
Reagents: (i) R′OH, tetrazole; (ii) R″OH, $Pr^i_2NH^{\oplus}_2$ tetrazolide$^{\ominus}$; (iii) oxidize, I_2, H_2O, THF, Et_3N.

Syntheses via H-phosphonate diesters

The H-phosphonate monoesters of protected 3′-nucleosides are readily prepared using PCl_3 and excess imidazole followed by mild hydrolysis. The intermediates do not require further protection at phosphorus and are rapidly and efficiently activated by a range of condensing agents, such as pivaloyl chloride. This can be carried out before or after the addition of the second nucleoside if excess condensing agent is avoided. A dinucleoside H-phosphonate is rapidly formed in high yield (Fig. 3.39). The procedure can be repeated many times before a single oxidation step finally converts the H-phosphonate diesters into phosphate diesters. Studies on the mechanism of the condensation indicate that a mixed phosphonate anhydride is a likely intermediate.

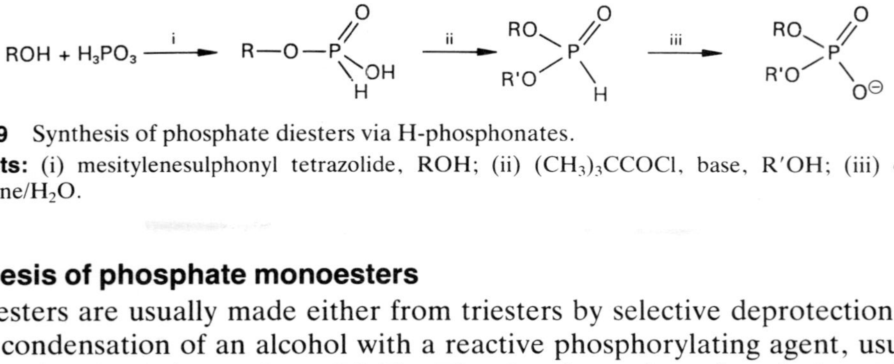

Fig. 3.39 Synthesis of phosphate diesters via H-phosphonates.
Reagents: (i) mesitylenesulphonyl tetrazolide, ROH; (ii) $(CH_3)_3CCOCl$, base, R′OH; (iii) oxidize, I_2/pyridine/H_2O.

Synthesis of phosphate monoesters

Monoesters are usually made either from triesters by selective deprotection or by direct condensation of an alcohol with a reactive phosphorylating agent, usually a polyfunctional species such as $POCl_3$. The first-formed product is immediately hydrolysed to give the desired monoester.

Triester procedures almost always call for selective protection of the nucleoside hydroxyl groups to leave only the reaction centre free. Reagents such as dibenzyl phosphorochloridate (with deprotection via catalytic hydrogenolysis), bis (2,2,2-trichloroethyl) phosphorochloridate (deprotection by hydrolysis to the diester then oxidation removal of catechol) give good yields of phosphate monoester (Fig. 3.40a).

Phosphoramidates are also useful for this purpose, as illustrated by *bis*(anilino) phosphorochloridate (with deprotection by nitrosation with amyl nitrite).

Several sterically selective phosphorylating species have been designed which condense preferentially with the primary 5'-hydroxyl function in unprotected ribo- and deoxyribo-nucleosides. These are illustrated by *bis*(2-*t*-butylphenyl) phosphorochloridate (deprotection by acidic hydrolysis) and *bis*(2,2,2-trichloro-1,1-dimethylethyl) phosphorochloridate, which is used in the presence of 4-dimethylaminopyridine and can be cleaved by reduction with cobalt(I) phthalocyanin (Fig. 3.40b).

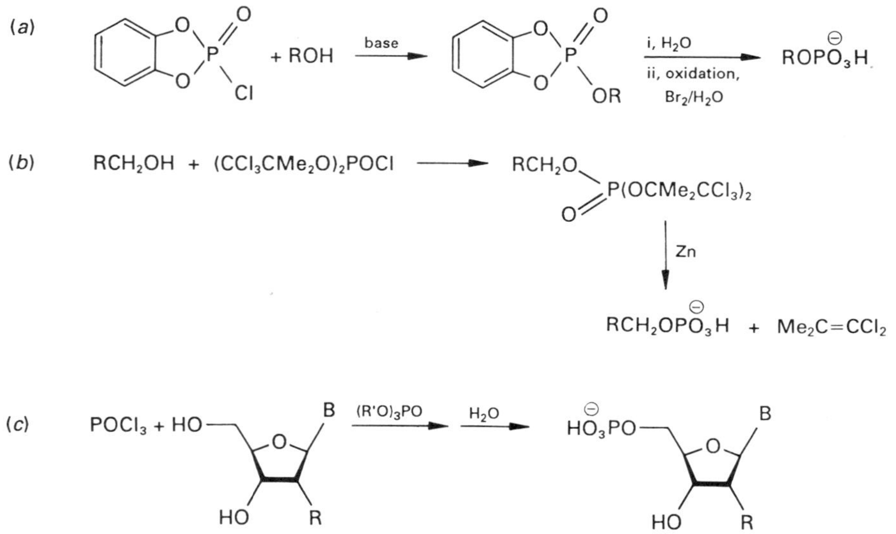

Fig. 3.40 Synthesis of phosphomonoesters. In (*c*) R = H, OH; R' = Me, Et; B = A,C,G,T,U, or modified base.

For a wide range of sugars and bases, direct phosphorylation at the 5'-position of unprotected nucleosides and 2'-deoxynucleosides has proved especially easy using $POCl_3$ in a trialkyl phosphate solvent system (the Yoshikawa method). This procedure has been refined by the use of aqueous pyridine in acetonitrile solution (the Sowa–Ouchi method) which provides yields greater than 80 per cent with over 90 per cent regioselectivity (Fig. 3.40c). In the same way, the use of $PSCl_3$ leads to phosphorothioates while isotopic oxygen can be readily introduced by the generation of $P^{18}OCl_3$ *in situ*.

Summary

The monoesters and diesters of orthophosphoric acid are relatively unreactive towards hydrolysis and so are stable structural components of nucleotides and nucleic acids. Much of this stability results from the repulsion of nucleophiles by the anionic nature of these esters at neutral pH. A neighbouring vicinal hydroxyl group can have a dramatic effect in lowering this stability, as shown in the relative lability of RNA to hydrolysis by alkali and by ribonucleases. By contrast, non-ionic phosphate triesters and fully-esterified condensed phosphates are much more easily hydrolysed.

Most ester syntheses require activation of the phosphate by a condensing reagent which both suppresses its ionization and generates a good leaving group. Some preparations make use of the alkylation of phosphate anions. Phosphate triesters have become the preferred intermediates for many syntheses. That is in part because they can be purified by standard techniques of organic chemistry and in part because methods have been devised for highly selective deprotection to give the desired diesters or monoesters.

3.3 Nucleoside esters of polyphosphates

3.3.1 Structures of nucleoside polyphosphates and co-enzymes

Monoalkyl esters are the most ubiquitous examples of P^1-nucleoside esters of polyphosphates. They include the ribo- and deoxyribo-nucleoside esters of pyrophosphoric acid (NDPs and dNDPs) and of tripolyphosphoric acid (NTPs and dNTPs) (Fig. 3.41). These esters are metabolically labile and participate in a huge range of C–O and P–O cleavage processes. Thiamine pyrophosphate is a co-enzyme which is a metabolically stable monoester of pyrophosphate.

Among the minor nucleoside polyphosphates, the 'magic spot' nucleotides MS1 (ppGpp) and MS2 (pppGpp) are species formed by stringent strains of *Escherichia coli* during amino acid starvation.

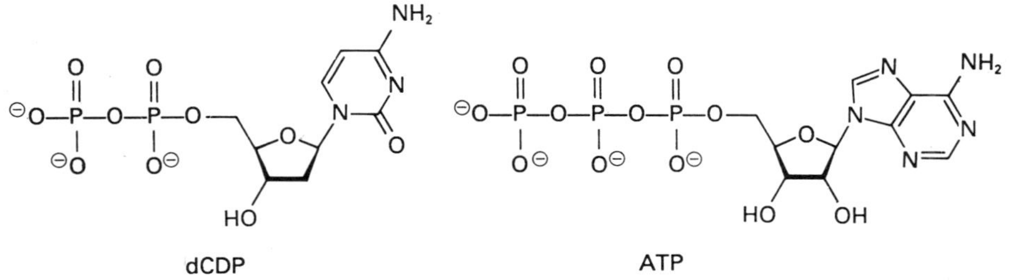

Fig. 3.41 Structures of nucleoside 5′-di- and 5′-triphosphates.

Dialkyl esters are biologically significant for di-, tri-, tetra-, and penta-polyphosphoric acids. In every case, the esters are located on the two terminal phosphate residues leaving (as described below) an ionic phosphate at every position which ensures stability to spontaneous hydrolysis. Several of the co-enzymes, such as Co-enzyme A, flavine adenine dinucleotide (FAD), and nicotinamide adenine dinucleotide (NAD$^+$) are stable P^1,P^2-diesters of pyrophosphoric acid (Fig. 3.42a). Their biosynthesis involves the condensation of ATP with a monoalkyl phosphate and the pyrophosphate appears to act generally as a structural unit providing coulombic binding to appropriate enzyme residues. By contrast, the active forms of many hexoses are found as pyrophosphate esters of uridine 5′-diphosphate. These include UDP-glucose, UDP-galactose, and UDP-N-acetylglucosamine (Fig. 3.42b). While they are also formed biosynthetically from UTP and the hexose-1-α-phosphate, the pyrophosphate ester is **metabolically labile** and used for catabolic processes involving C–1 of the sugar residue.

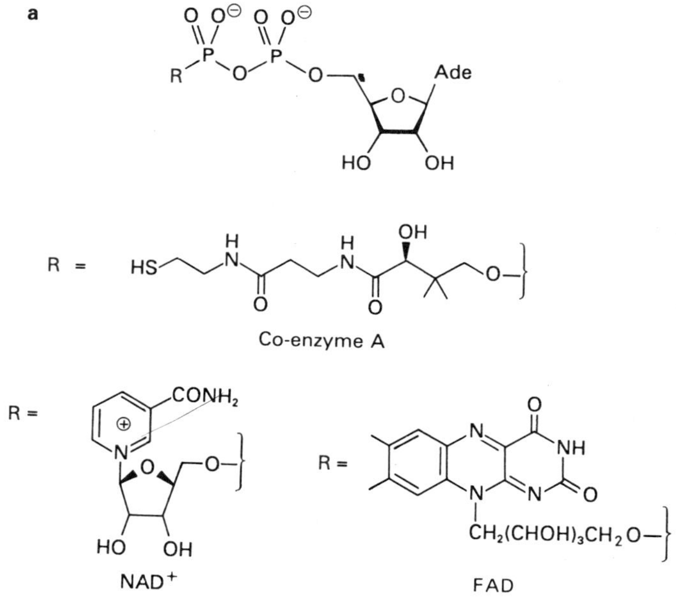

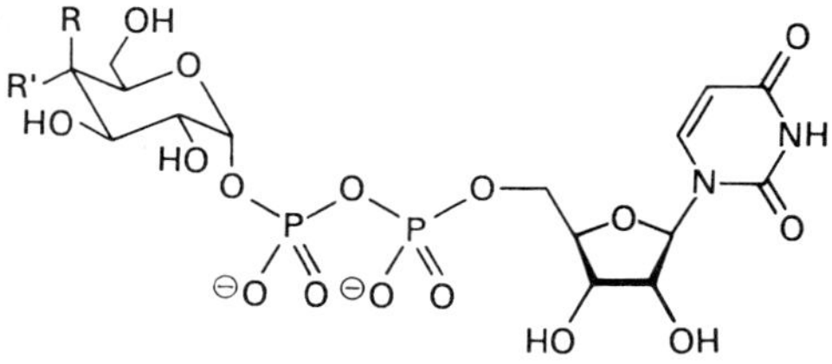

Fig. 3.42 (a) Structures of adenosine co-enzymes. (b) Structures of UDP-hexoses.
R = H, R′ = OH   UDP-glucose;
R = OH, R′ = H   UDP-galactose.

The P^1,P^4-dinucleosidyl tetraphosphates, Ap$_4$A and Ap$_4$G, are found in all cells, especially under conditions of metabolic stress (Fig. 3.43). They are produced as a result of the phosphorolysis of aminoacyl adenylates, particularly tryptophanyl and lysyl adenylates, with ATP or GTP. While these minor nucleotides were discovered by Zamecnik in 1966, their purpose remains uncertain, though it is believed they may have a role in the initiation of DNA biosynthesis. Somewhat related structures are found in the 'caps' at the 5'-ends of eukaryotic mRNAs (Section 6.4.3), which have a 7-methylguanosin-5'-yl residue linked to the 5'-triphosphate (Fig. 3.34). Both of these species, and their analogues, have been targets for synthesis as a means of discovering their biological function.

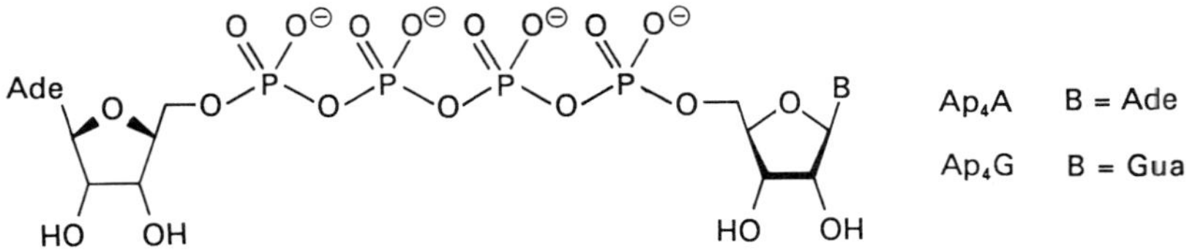

Ap$_4$A   B = Ade

Ap$_4$G   B = Gua

Fig. 3.43 P^1,P^4-dinucleoside tetraphosphate structures.

3.3.2 Synthesis of monoalkyl polyphosphate esters

All of the naturally occurring nucleoside polyphosphates have at least one negative charge on each phosphate residue. This is because uncharged residues in a string of phosphates are readily hydrolysed. As a result, most syntheses have avoided the formation of fully esterified intermediates, though an early synthesis of UTP was achieved (in low yield) by the catalytic hydrogenolysis of its tetrabenzyl ester. Generally, syntheses of monoalkyl esters fall into two classes: they involve C–O bond or P–O–P bond formation.

Poulter has made good use of the alkylating properties of nucleoside 5'-O-tosyl-ates towards pyrophosphate or tripolyphosphate anions and their methylene ana-logues. This has made possible direct syntheses of nucleoside 5'-diphosphates and -triphosphates (Fig. 3.44).

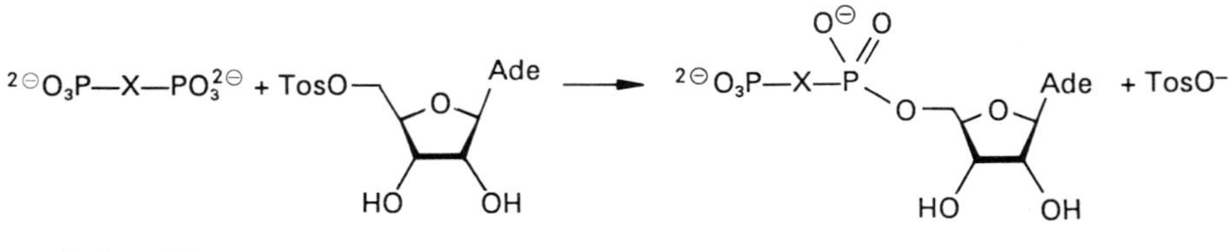

X = O, CH$_2$, CF$_2$.

Fig. 3.44   Synthesis of ADP and its analogues by C–O bond formation.

A more general procedure depends on the activation of a nucleoside 5'-phosphate or 5'-thiophosphate which is then able to condense with phosphate, pyrophosphate, or a methylenebisphosphonate. Among the condensing agents which have been used widely are DCCD, diphenyl phosphorochloridate, and carbonyl bisimidazole. This procedure is well suited to introduce isotopic oxygen into nucleotides in a non-stereochemically controlled fashion (Fig. 3.45) for subsequent use in positional isotope exchange (PIX) studies.

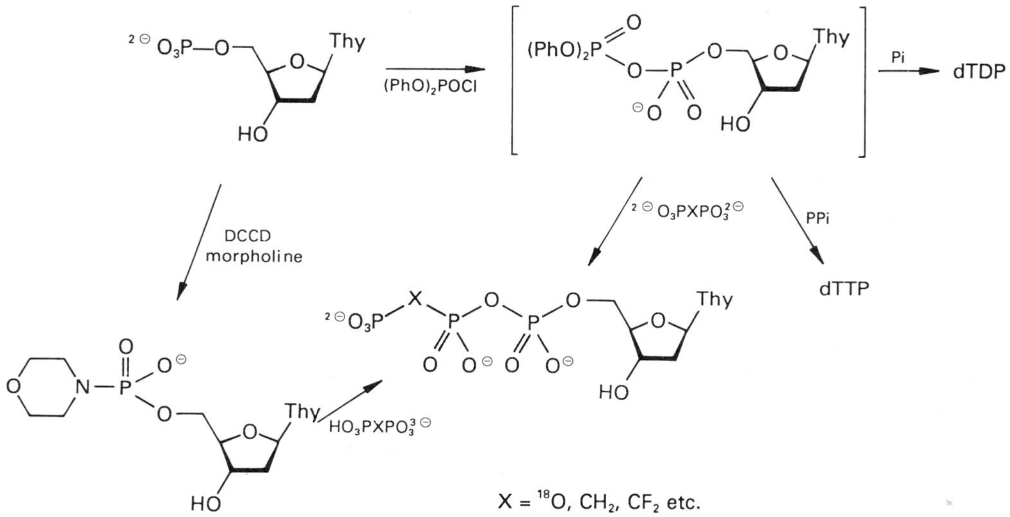

Fig. 3.45 Synthesis of nucleoside diphosphates and triphosphates and analogues by P–O bond formation.

A great deal of attention has been focused on the stereochemically-defined introduction of sulphur and/or oxygen isotopes into nucleotides. These have become prime tools for the investigation of the stereochemistry of enzyme-catalysed phosphoryl transfer processes. For example, the (Sp) isomer of adenosine 5'-O-thiotriphosphate, ATPαS, is readily made from AMPS by the combined action of adenylate kinase and pyruvate kinase (both enzymes can be immobilized on a polymer support for large scale syntheses) and using phosphoenol pyruvate and a little ATP to start the cycle. This synthesis illustrates the stereospecificity of adenylate kinase. The ^{31}P NMR of this product has been used to identify the (Rp) and (Sp) diastereoisomers of dATPαS, which have been synthesized (Fig. 3.46) and separated by ion-exchange chromatography. Such species have been employed *inter alia* to show that DNA polymerase I, T4 RNA ligase, and adenylate cyclase all operate on adenine nucleotides with inversion of configuration at Pα.

For such purposes, ATP has been made with incorporation of either oxygen-17 or oxygen-18 in just about every possible position in the three phosphate residues.

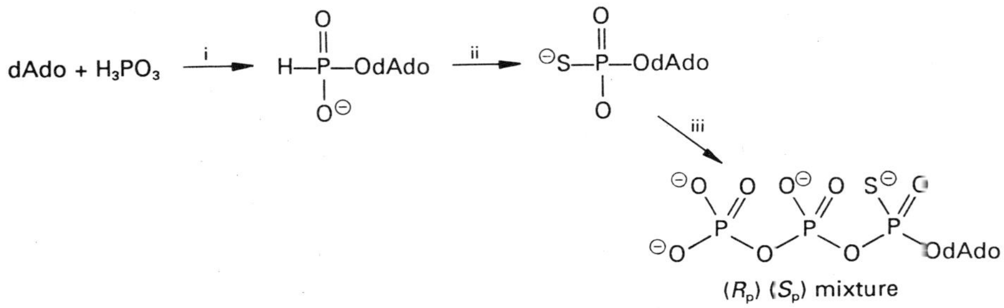

Fig. 3.46 Synthesis of mixed stereoisomers of dATPαS.
Reagents: (i) N,N-bis-(4-tolyl)-carbodiimide; (ii) S_8, pyridine, TMSCl; (iii) $(PhO)_2POCl$ pyridine then pyrophosphate.

The more useful species for nucleic acid chemistry are the α-phosphate substituted nucleotides. These can be made either by *ab initio* synthesis or by the stereochemically-controlled replacement of sulphur from an α-thiphosphate residue by isotopic oxygen. This transformation is best carried out by controlled bromine oxidation in ^{17}O- or ^{18}O-enriched water (Fig. 3.47). While this reaction proceeds with inversion of configuration, similar oxidations with N-bromosuccinimide or cyanogen bromide have been found to be less stereoselective. In some cases, careful choice of substrates has been necessary to avoid the apparent migration of the oxygen isotope to a second phosphoryl centre. An alternative procedure, though less widely applied, has used $[^{18}O]$styrene oxide, when the substitution of sulphur by oxygen proceeds with exclusive retention of stereochemistry at phosphorus.

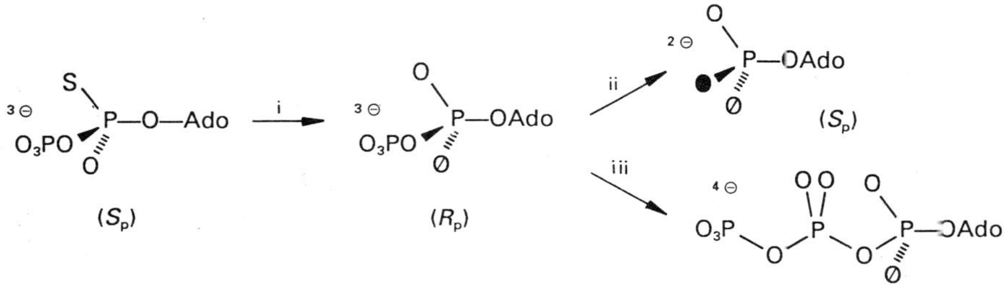

Fig. 3.47 Synthesis of (R_p)-$[\alpha$-$^{17}O]$ ATP and (S_p) AMP.
Reagents: (i) Br_2, $H_2{}^{17}O$; (ii) snake venom phosphodiesterase, $H_2{}^{18}O$ (retention); (iii) pyruvate kinase, Mg^{2+}, K^+, phosphoenolpyruvate.

The P^1,P^2-diesters of pyrophosphoric acid are most often made by coupling together two phosphate monoesters using DCCD, by a morpholidate procedure, or by diphenyl phosphorochloridate. A classical example is Khorana's synthesis of Co-enzyme A. The same methods have worked well for syntheses of Ap_4A and its analogues, where the use of an excess of activated AMP and limiting pyrophosphate (or one of its analogues) gives acceptable yields of P^1,P^4-dinucleosidyl tetraphosphate or analogue (Fig. 3.48).

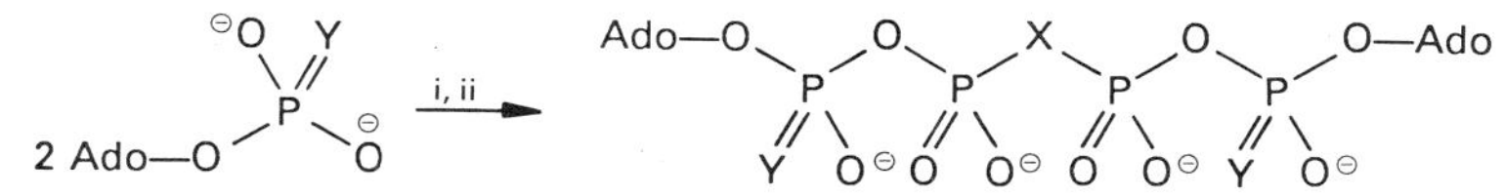

Fig. 3.48 Synthesis of Ap$_4$A and some analogues.
Reagents: (i) (PhO)$_2$POCl, pyridine; (ii) $^{2-}$O$_3$PXPO$_3{}^{2-}$ (X = O, CF$_2$, etc; Y = O or S).

For the P^1,P^3-dialkyl triphosphates of the mRNA 'cap' structures, it has proved necessary to devise more sophisticated coupling procedures. This is partly on account of the lability of the glycosidic bond in the 7-MeGuo residues and partly because of the unsymmetrical character of the diester (Fig. 3.49). In general, the major problem encountered in the syntheses of all these species has arisen during purification because there appears to be no good alternative to ion-exchange chromatography.

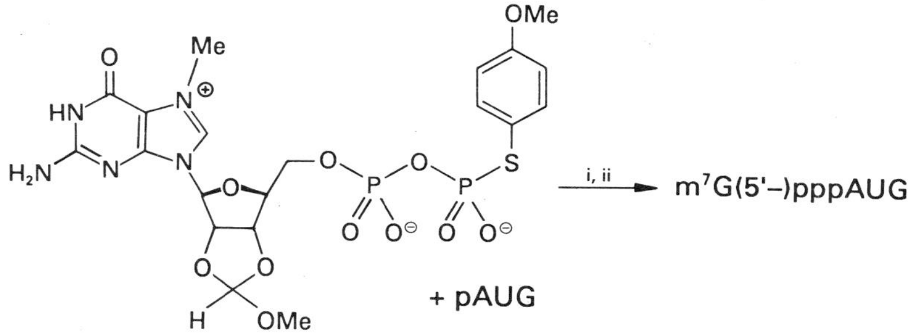

Fig. 3.49 Synthesis of the 'cap' structure of mRNA.
Reagents: (i) Ag$^+$, imidazole; (ii) H$_3$O$^+$

Summary

The nucleoside mono- and diesters of condensed phosphates serve on the one hand as stable structural components of co-enzymes and minor nucleotides and on the other as key catabolic species for the biosynthesis of nucleic acids and for energy-coupling purposes. Most syntheses of these esters can be accomplished by the selective activation of a phosphate monoester followed by coupling to a second phosphate or condensed phosphate. Much use has been made in nucleoside mono-, di- and triphosphates of stereochemical labelling of their prochiral oxygens by sulphur and/or isotopic oxygen. These chiral nucleotides have been employed for the analysis of stereochemistry of a wide range of enzyme-catalysed phosphate transfer processes which invariably proceed by single-inversion and occasionally by double-inversion processes.

3.4 Synthesis of oligodeoxyribonucleotides·

An oligonucleotide is a single-stranded chain consisting of a number of nucleoside units linked together by phosphodiester bridges. In general the phosphodiesters are formed between a 3'-hydroxyl group of one nucleoside and a 5'-hydroxyl group of another, just as is found in the case of naturally occurring nucleic acids. In the context of nucleic acids the prefix 'oligo' is generally taken to denote a few nucleoside residues, whereas the prefix 'poly' means many. Recently it has become common practice to refer to all chemically synthesized, single-stranded nucleic acid chains of defined length and sequence as oligonucleotides, even if they are well beyond 100 residues in length. The term 'polynucleotide' is used to refer to a synthetic single-stranded nucleic acid of less defined length and sequence obtained by a polymerization reaction (e.g. polycytidylic acid, polyC).

No review of chemical synthesis of oligonucleotides has appeared since 1984. This is partly a reflection of the general belief that oligonucleotide synthesis is now a routine and reliable procedure for which commercial machines (known affectionately as 'gene machines') are available. Currently this is true for oligodeoxyribonucleotides, where the individual units are 2'-deoxyribonucleosides. Here the overall strategy of assembly is now firmly established. In the case of oligoribonucleotides (Section 3.5) and modified oligonucleotides (Section 3.4.6) reliable methods are only just emerging.

3.4.1 Overall strategy

Nucleic acids are highly sensitive to a wide range of chemical reactions and only the mildest of reaction conditions can be used in assembly of an oligonucleotide chain. The heterocyclic bases are prone to alkylation, oxidation, and reduction. The phosphodiester backbone is vulnerable to hydrolysis. In the case of DNA, acidic hydrolysis occurs more readily than alkaline hydrolysis (cf. RNA, Section 3.5), due to the lability of the C^1–N glycosidic bond to acids, especially in the case of purine nucleotides (depurination) (Section 7.1). Such considerations limit the range of chemical reactions in oligodeoxyribonucleotide synthesis to (i) mild alkaline hydrolysis; (ii) very mild acidic hydrolysis; (iii) mild nucleophilic displacement reactions; (iv) base-catalysed elimination reactions; and (v) certain mild redox reactions (e.g. iodine or Ag(I) oxidations, reductive eliminations using zinc).

The key step in synthesis of oligodeoxyribonucleotides is the specific and sequential formation of internucleoside 3'-5' phosphodiester linkages. The main nucleophilic centres on a 2'-deoxyribonucleoside are the 5'-hydroxyl group, the 3'-hydroxyl group and, in the case of dA, dC, and dG, the exocyclic amino groups. In order to form a specific 3'-5' linkage between two 2'-deoxyribonucleoside units, the nucleophilic centres not involved in the linkage must be chemically protected. The first (5')-unit must have a **protecting group** on the 5'-hydroxyl and also on the heterocyclic base, whereas the second (3')-unit must be protected on the

3′-hydroxyl and on the heterocyclic base. In the example of joining a 5′-dA unit to a 3′-dG unit (Fig. 3.50), R^1 and R^2 protect the 5′-dA unit and R^3 and R^4 protect the 3′-dG unit. One of the two units must now be phosphorylated or phosphitylated on the only available hydroxyl group and then coupled to the other nucleoside unit in a **coupling reaction**. The resultant dinucleoside monophosphate is now fully protected. Commonly the phosphate carries a protecting group, R^5, introduced during the phosphorylation (or phosphitylation) step, such that the internucleoside phosphate becomes a **triester**. To extend the chain, one of the two terminal protecting groups, R^1 or R^3, must be removed selectively to generate a free hydroxyl function to which a new partially protected unit can now be joined.

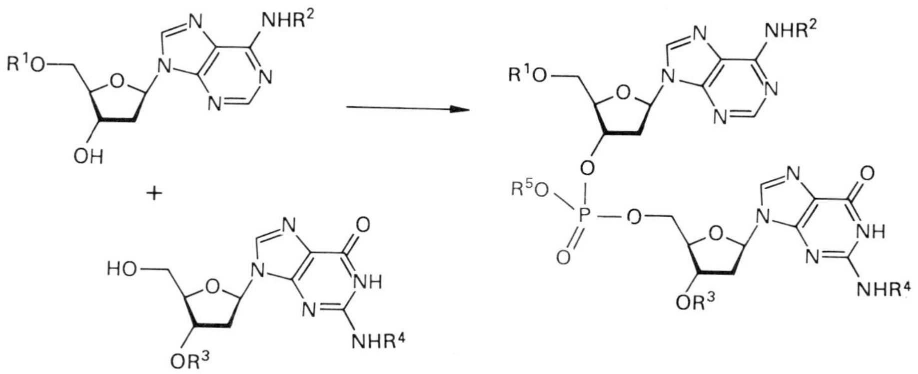

Fig. 3.50 Joining of a 5′-dA unit to a 3′-dG unit.

Where R^1 and R^3 are conventional protecting groups, oligonucleotide synthesis is referred to as **solution-phase**. This has now been largely superseded by the **solid-phase** method, where either R^1 or R^3 is an insoluble polymeric or inorganic support (Section 3.4.4). Whereas extension of the chain in solution-phase synthesis is possible in either the 3′-to-5′ or 5′-to-3′ direction, in solid-phase synthesis the oligonucleotide can only be extended in one direction. The conventional protecting group removed after each coupling step (R^1 or R^3, whichever is not the solid phase) is known as a **temporary** protecting group. R^2, R^4, and R^5 (and in a sense the solid support) are all **permanent** protecting groups. They must remain stable throughout the assembly and be removed at the end of the synthesis in order to generate the final deprotected oligonucleotide.

Summary
Oligodeoxyribonucleotides must be synthesized using only the mildest of reaction conditions. Specific coupling of two deoxyribonucleosides to form an internucleoside 3'-5' phosphodiester linkage occurs if all nucleophilic centres not involved in the linkage are properly protected. In solid-phase synthesis one end of the growing chain is attached to a solid support. A temporary protecting group is removed from the other end after each coupling step.

3.4.2 Protected 2'-deoxyribonucleoside units

The most convenient way to assemble an oligonucleotide is to utilize preformed deoxynucleoside phosphates as basic building units and to couple these sequentially to a terminal nucleoside attached to a solid support. Since the primary (5')- hydroxyl group is a more effective nucleophile than the secondary (3')-hydroxyl, the phosphate is best placed on the 3'-position. To achieve this selectively it is necessary to protect both the heterocyclic amino groups and the 5'-hydroxyl group.

Heterocyclic bases

Good permanent protecting groups for the exocyclic amino groups of adenine, cytosine, and guanine were first developed over 20 years ago by Khorana and co-workers for use in oligonucleotide synthesis. (N.B. No protection was found necessary for thymine since it does not carry an exocyclic amino group.) The acyl protecting groups chosen were designed to remain stable for long periods under mildly basic or acidic conditions and during chromatography, but to be removed by treatment with concentrated ammonia at the end of the synthesis. Despite substantial changes that have since arisen in the speed and in the type of chemistry used in assembly of oligonucleotides, these acyl protecting groups are still the most popular today. The benzoyl group is used to protect both adenine and cytosine and the isobutyryl group to protect guanine (Fig. 3.51).

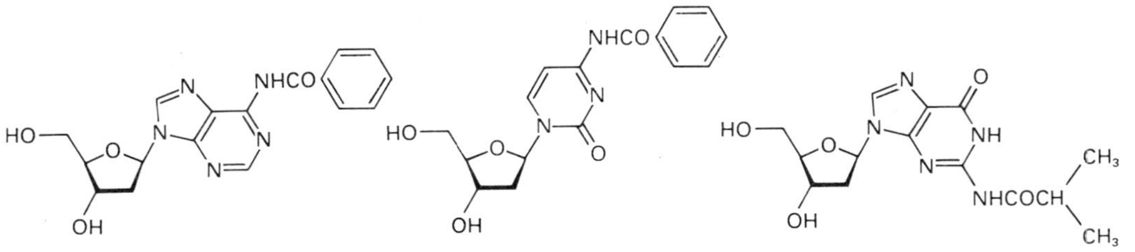

Fig. 3.51 Common protecting groups for the heterocyclic bases of dA, dC, and dG.

The protecting groups are introduced by acylation of the parent 2'-deoxynucleo-side. In the case of cytosine, selective acylation of the amino groups (as opposed to the sugar hydroxyl groups) is possible, but the amino groups of adenine and guanine are too weakly basic for selective reaction. Thus the two most common procedures for acylation are (a) **per-acylation** and (b) **transient protection**. Pre-acylation involves use of excess acylating agent to acylate both the hydroxyl and amino functions followed by selective deacylation of the hydroxyl groups. The selectivity is due to the greater stability of amides compared to esters at high pH. For example, Fig. 3.52 (Route **A**) shows per-acylation of 2'-deoxyadenosine. (N.B. The major product of acylation is the N^6,N^6-dibenzoyl derivative. During treatment with alkali one of the N-benzoyl groups is also removed.) In transient protection (Route **B**), the sugar hydroxyl groups are first silylated with trimethylsilyl chloride (TMSCl). Benzoylation now gives the N^6,N^6-dibenzoyl derivative, which is deacylated to N^6-benzoyl-2'-deoxyadenosine upon treatment with concentrated ammonia solution. Deoxycytidine is benzoylated similarly.

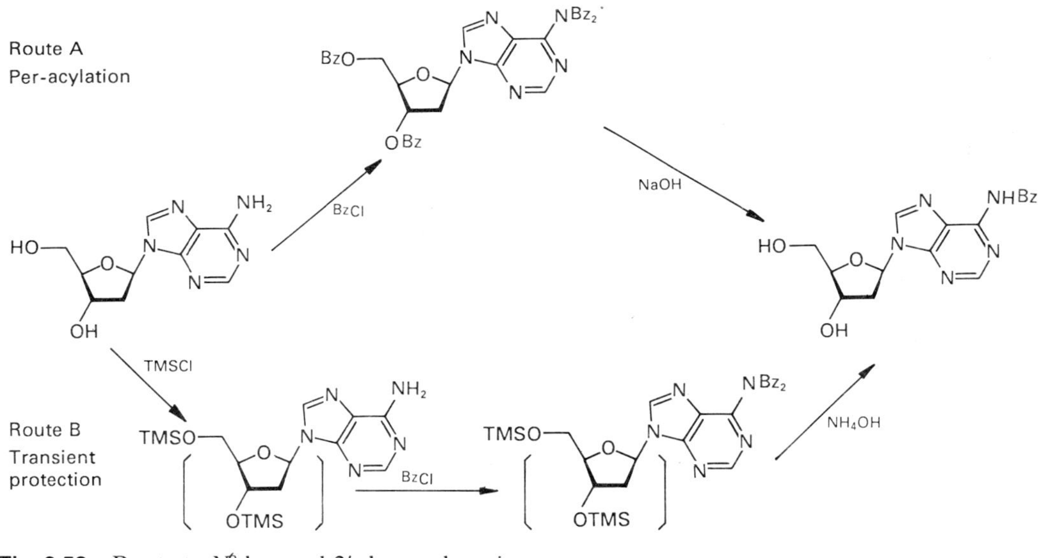

Fig. 3.52 Route to N^6-benzoyl-2'-deoxyadenosine.

In the case of deoxyguanosine, acylation of the amino group is effected with iso-butyric anhydride using either the per-acylation or transient protection route. A complication with deoxyguanosine is that the O^6-position is vulnerable to reaction with certain reagents used in oligonucleotide synthesis. This is particularly so in the case of coupling agents and phosphorylating agents used in the phosphotriester method of synthesis (see below). The O^6-position may be protected with a variety of alkyl or aryl protecting groups, but such protection is not necessary for the now widely used phosphoramidite method (Section 3.4.3).

5'-Hydroxyl group

By far the most useful protecting group for the 5'-position is the 4,4'-dimethoxy-triphenylmethyl group (dimethoxytrityl, DMTr). This is the most easily cleaved of a family of acid-labile protecting groups (Fig. 3.53), the labilities of which increase as the number of methoxy groups increases.

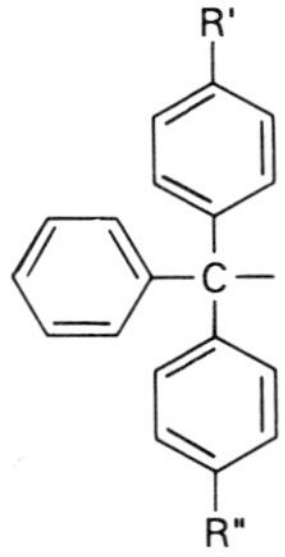

Fig. 3.53 The triphenylmethyl group and its derivatives. R' = R" = H, Triphenylmethyl (Trityl, Tr); R' = H, R" = OCH$_3$ 4-Methoxytriphenylmethyl (monomethoxytrityl, MMTr); R' = R" = OCH$_3$ 4,4'-Dimethoxytriphenylmethyl (dimethoxytrityl, DMTr).

The DMTr group is introduced on to the 5'-position of the *N*-acylated deoxy-nucleoside by reaction with DMTrCl in the presence of a mildly basic catalyst such as pyridine or 4-dimethylaminopyridine. For example, Fig. 3.54 shows 5'-protection of N^6-benzoyl-2'-deoxyadenosine. The reaction is regioselective for the primary (5')-hydroxyl compared to the secondary (3')-hydroxyl partly because of the bulk of the protecting group. The DMTr group is removed by treatment with acids such as dichloroacetic or trichloroacetic in a non-aqueous solvent, conditions just mild enough to prevent unwanted depurination. During deprotection the brightly orange coloured dimethoxytrityl cation is liberated, which can be used as a measure of the amount of deoxynucleoside attached (Section 3.4.4). The potential for depurination can be reduced considerably by use of an alternative protecting group, such as phthaloyl or di-*N*-butylaminomethylene, on the amino function of adenine.

Introduction of phosphate

In the original chemistry developed by Khorana (phosphodiester—Section 3.4.3) commercially available deoxynucleoside 5'-phosphates were the building blocks. In all other chemistry developed subsequently, 5'-*O*-dimethoxytrityl-(*N*-acylated)-2'-deoxynucleosides are phosphorylated or phosphitylated at the 3'-hydroxyl (Fig. 3.54). In these cases the products of synthesis after assembly of the oligonucleotide chain are phosphate triesters where the internucleoside phosphate carries a protecting group. In **phosphotriester** chemistry [P(V)] the best protecting groups are aryl (usually mono- or di-chlorophenyl derivatives). This is because an aryl phospho-diester is a much more reactive deoxynucleoside building block than an alkyl phosphodiester in a coupling reaction. For example, 5'-*O*-dimethoxytrityl-6-*N*-benzoyl-2'-deoxyadenosine gives the corresponding 3'-*O*-2-chlorophenylphosphodiester by

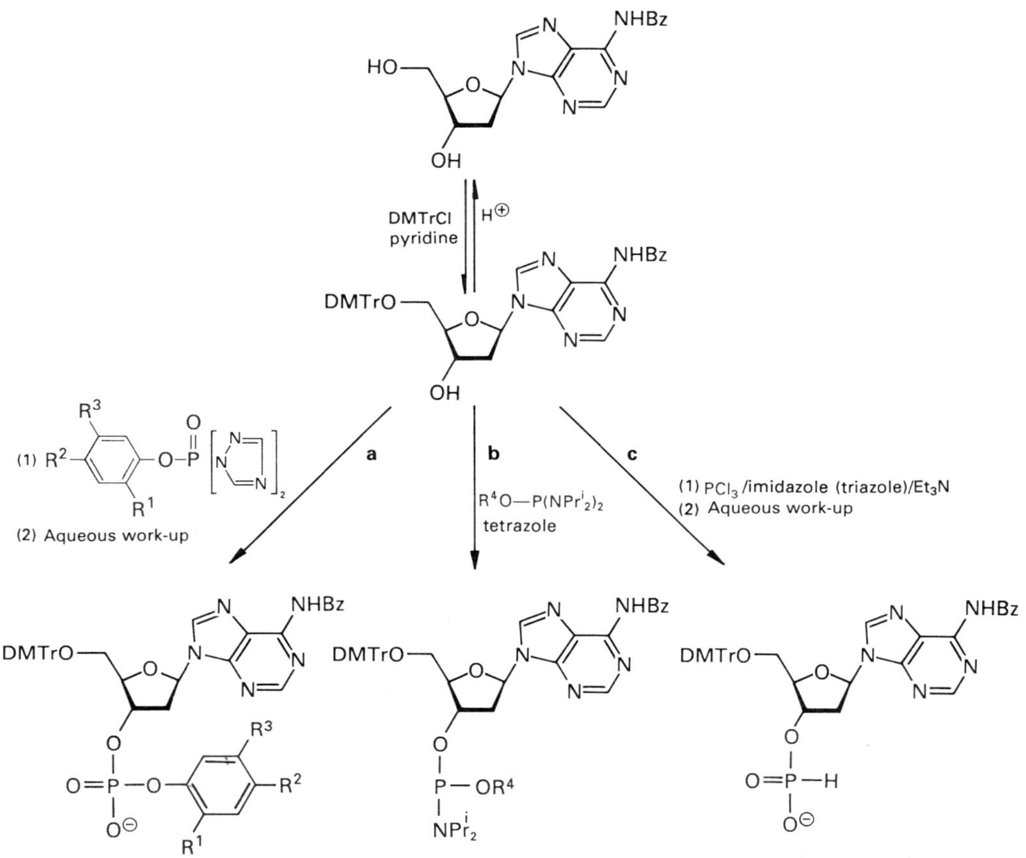

Fig. 3.54 Introduction of a 3'-phosphate by (**a**) phosphorylation, (**b**) phosphitylation, and (**c**) H-phosphonylation. R^1, R^3 = H, R^2 = Cl 4-chlorophenyl; R^2,R^3 = H, R^1 = Cl 2-chloro-phenyl; R^2 = H, R^1, R^3 = Cl, 2,5-dichlorophenyl; R^4 = methyl or 2-cyanoethyl.

reaction with 2-chlorophenyl phosphoro-bis(triazolide) (Fig. 3.54, Route **a**). Despite this being a bifunctional phosphorylating agent it acts like a monofunctional one in the absence of any stronger catalyst. In **phosphite-triester** chemistry [P(III)] both aryl and alkyl phosphites are highly reactive species. Here the methyl group or the 2-cyanoethyl group are the protecting groups of choice because they can be removed conveniently and selectively at the end of synthesis (Section 3.4.4). Once again a bifunctional reagent is used in a monofunctional capacity, but in order to obtain a sufficiently stable product, a phosphoramidite is prepared (Route **b**).

By contrast, **H-phosphonate** chemistry requires no phosphate protecting group, since an assembled oligonucleotide chain containing internucleoside H-phosphonate diester links is relatively inert to the conditions of coupling (Section 3.4.3). In a sense, a proton is the protecting group! A deoxynucleoside 3'-H-phosphonate is simply prepared by the reaction of the deoxynucleoside derivative with phosphorus trichloride and imidazole or triazole plus a basic catalyst such as *N*-methylmorpholine, followed by aqueous work-up (Route **c**).

Summary

Acyl groups are used to protect the exocyclic amino groups of adenine, cytosine, and guanine. An acid-labile dimethoxytrityl group is used to protect the 5′-hydroxyl. Phosphorylation or phosphitylation on the 3′-position now gives a suitable building block. In phosphotriester and phosphite-triester syntheses, a phosphoryl-oxygen also carries a protecting group. A proton acts as a virtual protecting group in H-phosphonate synthesis.

3.4.3 Ways of making an internucleotide bond

The development of an efficient way of making an internucleotide bond was for many years the most central issue in oligonucleotide synthesis. This problem has been effectively solved by the advent of phosphite triester (phosphoramidite) chemistry and, to some extent, H-phosphonate chemistry. However, an understanding of earlier phosphodiester and phosphotriester chemistry is important.

Phosphodiester

No discussion of oligonucleotide synthesis would be complete without mention of the pioneering gene-synthesis of Khorana and his colleagues in the 1960s and early 1970s. Here oligonucleotide synthesis involved coupling a 5′-protected deoxynucleoside derivative with a 3′-protected deoxynucleoside-5′-phosphomonoester (Fig. 3.55). The coupling agent (triisopropylbenzenesulphonyl chloride, TPS) activates the phosphomonoester, via a complex reaction mechanism, to give a powerful phosphorylating agent. Reaction of this with the 3′-hydroxyl group of the 5′-unit gives a dinucleoside phosphodiester. The main drawback is that the product phosphodiester is also vulnerable to phosphorylation by the activated deoxynucleoside phosphomonoester to give a trisubstituted pyrophosphate derivative. An aqueous work-up is necessary to regenerate the desired phosphodiester. Extension of the chain involves removal of the 3′-protecting group using alkali (for R = acetyl) or fluoride ion (for R = *t*-butyldiphenylsilyl, TBDPS) and coupling with another deoxynucleoside-5′-phosphate derivative. To prepare oligonucleotides beyond five units it is necessary to resort to coupling of preformed **blocks** containing two or more deoxynucleotide residues. Here the reactants contain several unprotected phosphodiesters which undergo considerable side-reactions that substantially reduce yields. These reactions require lengthy procedures for purification of products and substantial efforts in preparation of reactant blocks. Synthesis of an oligonucleotide of 10–15 residues (the effective limit of the method) takes upwards of three months.

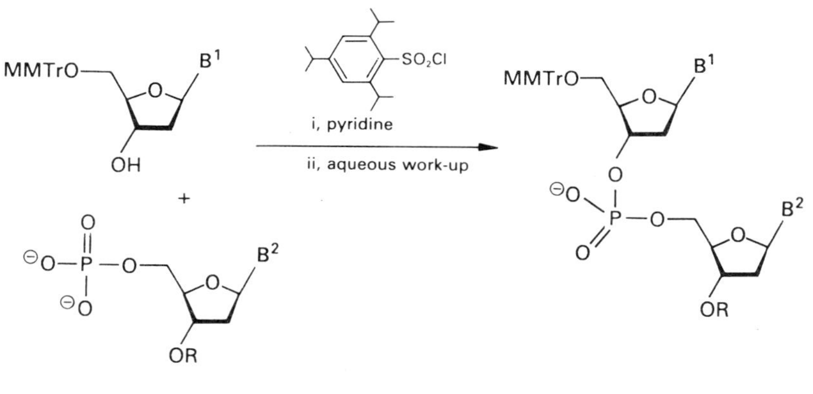

Fig. 3.55 Formation of an internucleotide bond by the phosphodiester method.
B = T, C^{bz}, A^{bz}, or G^{ib}.

In the late 1970s, phosphodiester chemistry was successfully applied to solid-phase synthesis. Here the 5'-MMTr group is replaced by linkage to a solid support (Section 3.4.4). Using only monomer units, the time of assembly is reduced to about two weeks and tedious purification of intermediates is avoided. However, the low yields intrinsic to phosphodiester chemistry remain.

Phosphotriester

Although this chemistry was first applied in solution-phase synthesis, it proved particularly successful when applied to solid-phase synthesis in the early 1980s. Here a 5'-*O*-dimethoxytrityl-(*N*-acylated)-deoxynucleoside-3'-*O*-(chlorophenyl phosphate) (Section 3.4.3) is coupled to a deoxynucleoside attached at its 3'-position to a solid support (Fig. 3.56). The coupling agent (mesitylenesulphonyl 3-nitro-1,2,4-triazolide, MSNT) is similar to that used in phosphodiester synthesis except that 3-nitrotriazolide replaces chloride. The coupling agent activates the deoxynucleoside 3'-phosphodiester and allows reaction with the hydroxyl group of the support-bound deoxynucleoside. The rate of reaction can be enhanced by addition of a nucleophilic catalyst such as *N*-methylimidazole. This participates in the reaction by forming a more activated phosphorylating intermediate (an *N*-methylimidazolium phosphodiester), since the *N*-methylimidazole provides a better leaving group. The product is a phosphotriester and accordingly is protected from further reaction with phosphorylating agents. The yield is therefore much better than in the case of a phosphodiester coupling, but phosphotriester chemistry could only be used satisfactorily after the development of selective reagents for cleavage of the aryl protecting group (Sections 3.2.2 and 3.4.4). To extend the chain, the DMTr group is removed by treatment with acid to liberate the hydroxyl group ready for further coupling. Note that the direction of extension is 3'-to-5', in contrast to the solid-phase phosphodiester method.

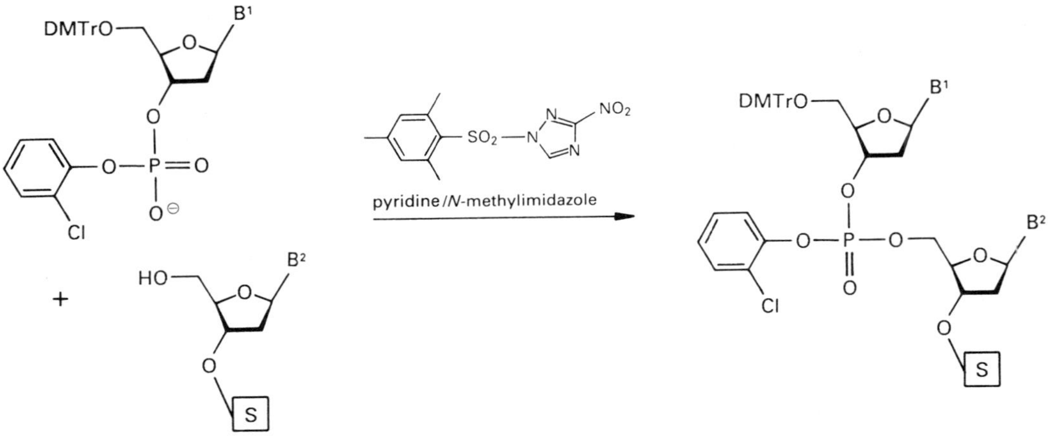

Fig. 3.56 Formation of an internucleotide bond by the solid-phase phosphotriester method.

Two side-reactions give rise to limitations. During coupling there is a competitive reaction (about 1 per cent) of sulphonylation of the 5'-hydroxyl group by the coupling agent. This limits the efficiency of phosphotriester coupling to 97–98 per cent and thus also the length of oligonucleotide attainable to about 40 residues. More seriously, deoxyguanosine residues are subject to both phosphorylation and nitrotriazole substitution at the O^6-position unless an extra protecting group is used. O^6-Phosphorylation is particularly serious since this is not easily reversible (in contrast to phosphitylation) and leads to chain branching and eventually chain degradation.

Phosphite triester

The development of phosphite triester (now often called phosphoramidite) chemistry by Caruthers and co-workers in the early 1980s transformed oligonucleotide synthesis from a manual or semi-manual procedure carried out by a few specialists into a commercialized process performed using a machine. The crux of this chemistry is a highly efficient coupling reaction between a 5'-hydroxyl group of a support-bound deoxynucleoside and an alkyl 5'-DMTr-(N-acylated)-deoxynucleoside 3'-O-(N,N-diisopropylamino)phosphite (the alkyl group being methyl or 2-cyanoethyl) (Fig. 3.57). In early development of this chemistry, a chlorophosphite was used in place of the N,N-diisopropylaminophosphite, but was found to be too unstable upon storage. By contrast, a phosphoramidite is considerably less reactive and requires protonation at nitrogen to make the phosphoramidite into a highly reactive phosphitylating agent. Tetrazole is just sufficiently acidic to do this without causing loss of the DMTr group. The product of coupling is a dinucleoside phosphite, which must be oxidized with iodine to the phosphotriester before proceeding with chain extension.

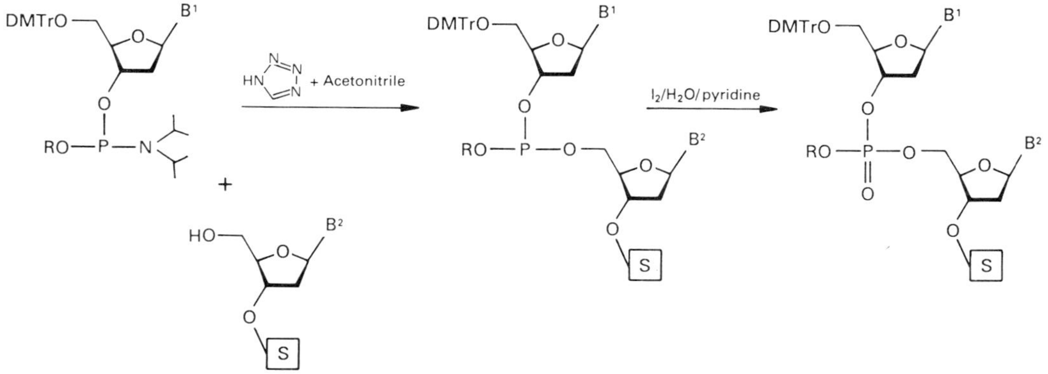

Fig. 3.57 Formation of an internucleotide bond by the solid-phase phosphoramidite method; R = methyl or 2-cyanoethyl.

The efficiency of coupling is extremely high (>98 per cent) and the only major side reaction is phosphitylation of the O-6 position of guanine. Fortunately treatment with acetic anhydride and, after coupling, N-methylimidazole (introduced originally to cap off any unreacted hydroxyl groups, see Section 3.4.4) completely reverses this side reaction.

H-Phosphonate

Although the origins of this chemistry lie with Todd and co-workers in the 1950s, it has emerged only very recently with high potential in oligonucleotide synthesis. A deoxynucleoside 3'-O-(H-phosphonate) is essentially a tetracoordinated P(III) species, preferring this structure to the tautomeric tricoordinated phosphite monoester. Activation is achieved with a hindered acyl chloride (e.g. pivaloyl chloride), which couples the H-phosphonate to a nucleoside hydroxyl group (Fig. 3.58). The resultant H-phosphonate diester is relatively inert to further phosphitylation, such that the chain may be extended without prior oxidation. Oxidation of all phosphorus centres is carried out simultaneously at the end of the synthesis. An advantage of this chemistry is that oxidation is subject to general base catalysis and this allows nucleophiles other than water to be substituted during oxidation to give a range of oligonucleotide analogues (Section 3.4.6).

Unfortunately, a serious side reaction occurs if an H-phosphonate is premixed with activating agent before coupling. The H-phosphonate rapidly dimerizes to form a symmetrical phosphite anhydride. Subsequent reaction of this with a hydroxyl group gives rise to a branched trinucleoside derivative. The complete elimination of this side reaction, even under optimal conditions, is probably impossible and may account for the marginally lower yields obtained in practice with this route.

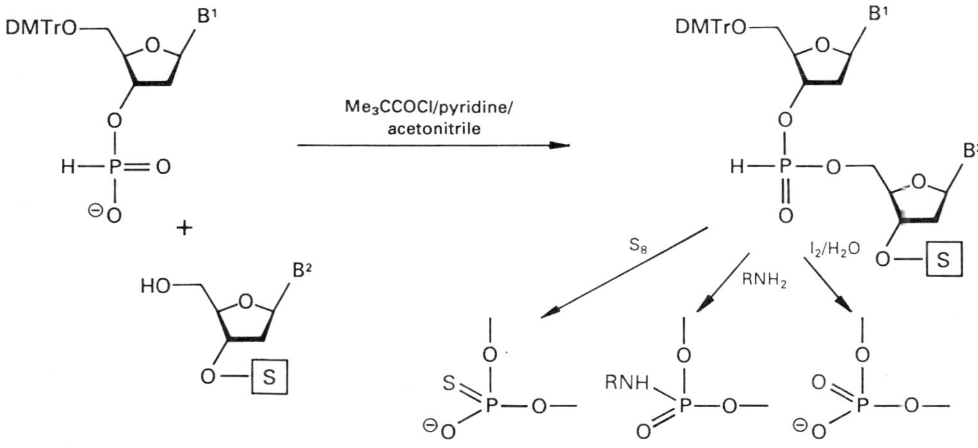

Fig. 3.58 Formation of an internucleotide bond by the solid-phase H-phosphonate method.

Summary

In phosphodiester synthesis, an activated deoxynucleoside 5'-phosphomonoester is reacted with a deoxynucleoside 3'-hydroxyl group. The lack of a protecting group on the resultant phosphodiester gives rise to serious side reactions. Much higher coupling yields are possible in phosphotriester and phosphite triester syntheses, since the triester products are protected from side reactions. Stable deoxynucleoside 3'-phosphoramidites are powerful phosphitylating agents in the presence of tetrazole and give high coupling yields. Whereas phosphite triesters must be oxidized to phosphotriesters after each coupling step, an H-phosphonate diester is sufficiently stable to allow oxidation of all linkages following the completion of chain assembly.

3.4.4 Solid-phase synthesis

The essence of solid-phase synthesis is the use of a heterogeneous coupling reaction between a deoxynucleotide derivative in solution and another residue bound to an insoluble support. This has the advantage that a large excess of the soluble deoxynucleotide can be used to force the reaction to high yield. The support-bound product dinucleotide can be removed from the excess of reactant mononucleotide simply by filtration and washing. Other reactions can also be carried out heterogeneously and reagents removed similarly. This process is far faster than a conventional separation technique in solution and easily lends itself to mechanization. There are four essential features of solid-phase synthesis.

Attachment of first deoxynucleoside to the support

Although many types of support have been used for solid-phase oligonucleotide synthesis, only one has proved to be useful under all conditions and with all chemistries. Controlled pore glass beads (CPG) are ideal in being rigid and non-swellable. They are manufactured with different particle sizes and porosities, and they are chemically inert to reactions involved in oligonucleotide synthesis. Currently, 500 and 1000 Å porosities are favoured. The silylation reactions involved in functionalization of glass (introduction of reactive sites) are beyond the scope of this chapter. It is sufficient here to note that a long spacer is used to extend the sites away from the surface and ensure accessibility to all reagents. One type of spacer is illustrated (Fig. 3.59). The loading of amino groups on the glass is best kept within a narrow band of 10–50 μmol g^{-1}, below which the reactions become unreproducible and above which they are subject to steric crowding between chains.

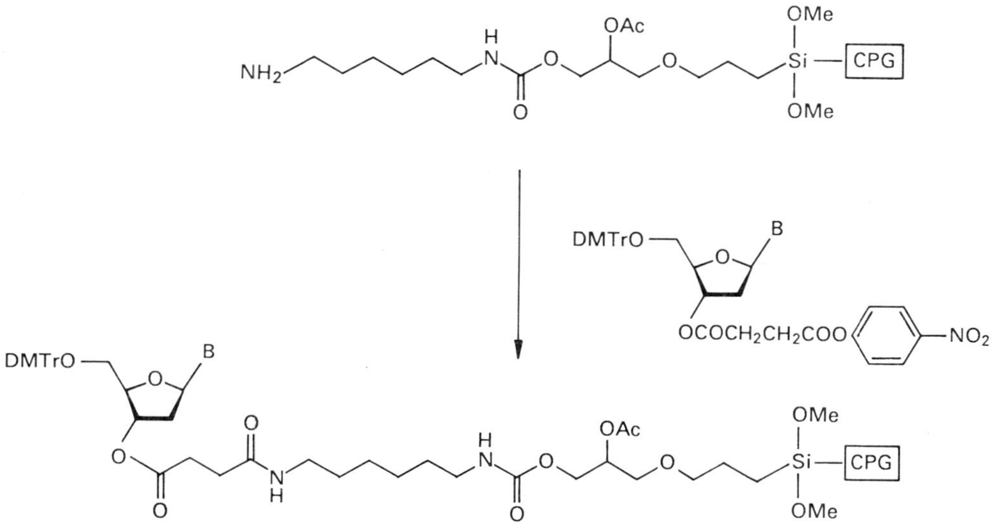

Fig. 3.59 Attachment of a nucleoside to a solid support of controlled pore glass (CPG) functionalized by a long chain alkylamine.

The 3'-terminal deoxynucleoside of the oligonucleotide to be synthesized is attached to the CPG support by conversion of its 5'-*O*-DMTr-(*N*-acylated)-derivative into the corresponding 3'-*O*-(4-nitrophenyl) succinate, which is subsequently reacted with amino groups on the support (Fig. 3.59).

Assembly of oligonucleotide chains

Assembly of the protected oligonucleotide chain is carried out by packing a small column of deoxynucleoside-loaded glass (as little as 10 mg can be used, 0.2 μmol) and flowing solvents and reagents through in predetermined order. This is most reproducibly accomplished using a commercial DNA synthesizer, but a manual flow system or even a small sintered glass funnel can be substituted. Machine specification varies considerably, but the basic steps involved in one cycle of nucleotide addition using the popular phosphoramidite chemistry are shown (Fig. 3.60).

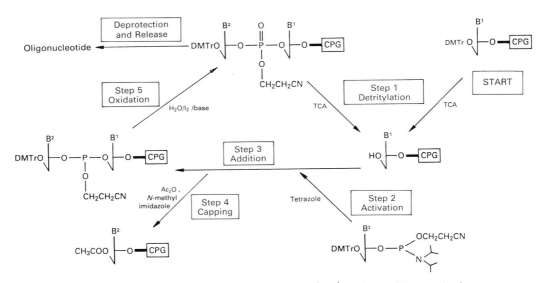

Fig. 3.60 Basic steps in a cycle of nucleotide addition by the phosphoramidite method.

Step 1. Detritylation (removal of dimethoxytrityl groups) is accomplished with dichloroacetic or trichloroacetic acid (TCA) in methylene chloride. The orange colour (dimethoxytrityl cation) liberated into solution is compared in intensity with the detritylation of the previous cycle to obtain the **coupling efficiency**.

Step 2. Activation of the phosphoramidite occurs when it is mixed with tetrazole in acetonitrile solution.

Step 3. Addition of activated phosphoramidite to the growing chain.

Step 4. Capping is a safety step introduced to block chains which are somehow not reacted during the coupling reaction and is designed to limit the number of failure sequences (those missing an internal residue). A fortuitous benefit of this step is that phosphitylation of the O–6 position of guanine is reversed.

Step 5. Oxidation of the intermediate phosphite to the phosphotriester is achieved with iodine and water. Pyridine or 2,6-lutidine is used to neutralize the hydrogen iodide produced.

The cycle is repeated the requisite number of times for the length of oligonucleotide required, with each deoxynucleotide phosphoramidite being added in the desired sequence and building from 3'-to-5'.

Deprotection and removal of oligonucleotide from the support

(i) The 5'-DMTr group is removed with the same detritylating agent as used in the assembly cycle.

(ii) The phosphate protecting groups are removed. In the phosphotriester method, an aryl protecting group is selectively displaced using *syn*-2-nitrobenzaldoximate ion or 2-pyridinecarbaldoximate ion. The product undergoes rapid elimination in the presence of water (see Fig. 3.29). In phosphoramidite synthesis, a methyl protecting group is removed with thiophenate ion (generated with thiophenol and triethylamine) which acts by nucleophilic attack on the methyl group, followed by hydrolysis. Alternatively, a 2-cyanoethyl group is removed by β-elimination using aqueous triethylamine or ammonia (Section 3.2.2).

(iii) All heterocyclic base protecting groups are removed with concentrated aqueous ammonia at room temperature for 24–48 h or at 50°C for 5 h.

(iv) The succinate linkage is cleaved with mild aqueous base.

In phosphoramidite chemistry using the 2-cyanoethyl group, steps (ii)–(iv) are carried out simultaneously using aqueous ammonia.

Purification of oligonucleotides

The importance of a good separation technique for synthetic oligonucleotides is often neglected. Since the impurities from a large number of reactions are stored up on the support (at least those pertaining to the growing chain), these must all be resolved, preferably in a single chromatographic step. Fortunately, powerful separation methods have been developed for purification of μg to mg quantities.

(i) **Polyacrylamide gel electrophoresis** separates oligonucleotides by virtue of their unit charge difference. This method is particularly useful for the purification of long oligonucleotides (>50 residues), but is limited to small scale (up to 1 mg). Single nucleotide resolution is possible to well beyond 100 residues.

(ii) **High performance liquid chromatography** is suitable for purification of perhaps up to 50 mg of oligonucleotide. Ion exchange chromatography resolves predominantly by charge difference and is useful both diagnostically and preparatively for oligonucleotides up to about 50 residues. Reversed phase chromatography separates according to hydrophobicity. Here the elution position of a fully deprotected oligonucleotide is hard to predict. More reliable identification of products is achieved by leaving the highly lipophilic 5′-DMTr group intact, such that the oligonucleotide is well resolved from shorter, non-DMTr containing impurities. The DMTr group is removed following chromatography.

Summary

The four essential steps in solid-phase synthesis are: (i) attachment of the first deoxynucleoside to the support, (ii) assembly of the oligonucleotide chain, (iii) deprotection and removal of the oligonucleotide from the support and (iv) purification. Assembly of the protected oligonucleotide chain can be carried out on a very small scale with the aid of a machine to add solvents and reagents reproducibly to the solid support, followed by filtration purification.

3.4.5 Duplex DNA and gene synthesis

The principles of construction of synthetic duplex DNA were developed 20 years ago by Khorana and his colleagues and have survived with minor modifications to the present day.

5′-Phosphorylation

In order to join the 3′-end of one oligonucleotide to the 5′-end of another, a phosphate group must be attached to one of them. The needs of an enzyme-catalysed ligation process (see below) require the phosphate to be added at a 5′-end. This can be achieved chemically using a monofunctional phosphorylating agent whilst the protected oligonucleotide is attached to the support. More commonly, and in order to introduce a radioactive tracer, phosphorylation is carried out enzymatically using T4 polynucleotide kinase (Section 9.4). The enzyme transfers the γ-phosphate of ATP to the 5′-end of an oligonucleotide (either ribo- or deoxyribo-) to give an oligonucleotide 5′-phosphate ('kinased' oligonucleotide).

Joining of oligonucleotides (ligation)

Figure 3.61 shows schematically the construction of a gene coding for a small bovine protein, caltrin (a protein believed to inhibit calcium transport into spermatozoa). Each synthetic oligonucleotide is denoted by the position of the arrows. These are so arranged such that annealing (heating to 90°C and slow cooling to ambient temperature) of all ten oligonucleotides simultaneously gives rise to a contiguous section of double-stranded DNA, the sequence of which corresponds to the desired protein sequence. In this example, the oligonucleotides are 24–38 residues long, but chains of 80 residues or more have been used in gene synthesis. Oligo-

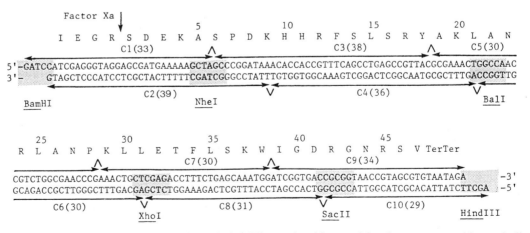

Fig. 3.61 A synthetic gene for bovine caltrin. Oligonucleotides used for the gene assembly are indicated by arrows and caret marks denote points of ligation. The amino acid sequence is shown above. Restriction enzyme recognition sites are shaded [from Heaphy *et al.* (1987). *Protein Engineering*, **1**, 425–31].

nucleotides C2–C9 are previously phosphorylated such that, for example, the 5'-phosphate group of C3 lies adjacent to the 3'-hydroxyl group C1. The duplex is only held together by virtue of the complementary base-pairing between strands. The enzyme T4 DNA ligase (Section 9.4) is now used to join the juxtaposed 5'-phosphate and 3'-hydroxyl groups (the caret marks denote the joins). The overlaps are such that each oligonucleotide acts as a splint for joining two others.

Note that oligonucleotides C1 and C10 are not phosphorylated. Each end corresponds to a sequence which would be generated by cleavage by a restriction enzyme (Section 9.3.3). Lack of a phosphate group prevents these self-complementary ends from joining to themselves during ligation. The ends are later joined to a vector DNA, previously cleaved by the same two restriction enzymes, to give a closed circular duplex ready for transformation and cloning in *E. coli* (Section 10.2).

Other features of the synthetic gene are internal restriction enzyme sites (shaded), which can be introduced artificially merely by judicious choice of codons specifying the required amino acid sequence. This particular gene is designed without a methionine initiation codon, since the protein is intended to be expressed as a fusion with another vector-encoded protein. This fusion can be cleaved to generate caltrin by treatment with the proteolytic enzyme *Factor X_a* (an enzyme important in the blood-clotting cascade and whose natural substrate is prothrombin) since the synthetic gene has been designed to include a section encoding the tetrapeptide recognition sequence for this enzyme.

3.4.6 Modified oligonucleotides

Oligonucleotide analogues

A number of oligonucleotide analogues have found application as inhibitors of translation of RNA into protein and as potential anti-viral agents. For example, a good inhibitor of translation ('**antisense oligonucleotide**', Section 2.4.3) should hybridize specifically to a target RNA sequence, be resistant to nuclease digestion, and block translation of that RNA (**hybridization arrest**). Of those analogues investigated, two types of phosphate-modified, nuclease-resistant oligonucleotide are showing particular promise.

(i) **Phosphorothioate** linkages can be made by substituting the oxidation step in either phosphite or H-phosphonate solid-phase synthesis by a sulphurization step using elemental sulphur (Fig. 3.62). In both these cases a mixture of (*S*p) and (*R*p) isomers is formed. Good chemical routes to chirally pure phosphorothioate oligonucleotides are not yet available except for short oligonucleotides. However, pure (*R*p) linkages in DNA can be prepared enzymatically using pure deoxynucleoside [*S*p]-α-thio triphosphates and a polymerase enzyme, where the reaction goes with inversion of configuration. Pure (*S*p)-containing phosphorothioate DNA has not yet been achieved. Since phosphorothioates are isopolar and isosteric with natural phosphates and since only one isomer is in general utilized as a substrate, nucleotide

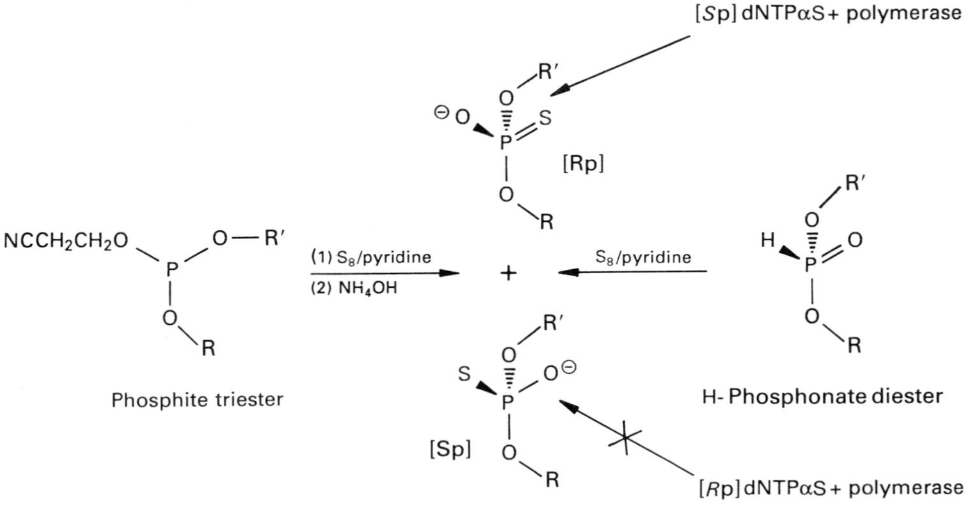

Fig. 3.62 Routes to (R_p) and (S_p) isomers of phosphorothioates.

analogues containing P–S bonds have also been valuable in determining the stereo-chemical pathway of many nucleic acid recognizing enzymes. Phosphorodithioate linkages are showing considerable promise in antisense oligonucleotides and avoid the problem of stereoisomerism at phosphorus.

(ii) **Methylphosphonate** linkages can be introduced by solid-phase synthesis using 3′-*O*-methylphosphonamidite deoxynucleosides under similar conditions to standard phosphoramidite synthesis (Fig. 3.63). Once again a mixture of diastereomers results.

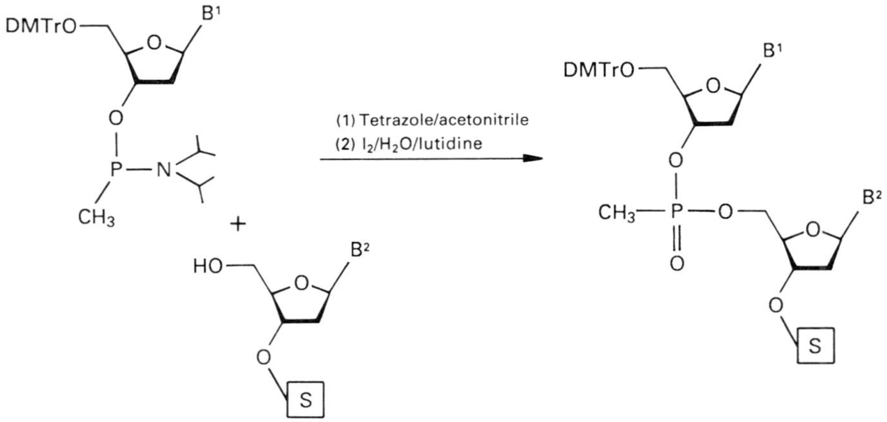

Fig. 3.63 A route to methylphosphonate analogues of oligonucleotides.

Functionalized oligonucleotides

A wide variety of molecules (e.g. fluorescent groups, intercalators, biotin, enzymes) can be attached to synthetic oligonucleotides for diagnostic and other purposes via a specific functional group incorporated during synthesis. Suitable sites for introduction of a functional group (commonly an amino or thiol group) are the 5'- and 3'-ends, C–5 uracil base, or the exocyclic amino group of cytosine. For example, Fig. 3.64a shows a modified pyrimidine deoxynucleoside phosphoramidite which can substitute for a thymidine phosphoramidite in any cycle of solid-phase synthesis. The trifluoroacetyl protecting group is removed during deprotection of the oligonucleotide with aqueous ammonia. Figure 3.64b shows a linker phosphoramidite designed to be added as the final coupling step in oligonucleotide assembly. Once again, treatment with ammonia removes the trifluoroacetyl protecting group and generates an aminohexyl phosphate at the 5'-end of the oligonucleotide. In both of these structures, the terminal amino group can be reacted specifically with any species containing an activated carboxylic acid group (e.g. fluorescein isothiocyanate or biotin N-hydroxysuccinimide).

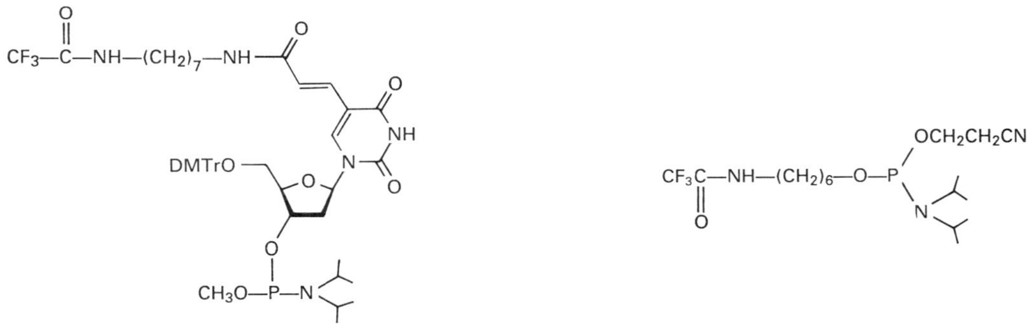

Fig. 3.64 (**a**) A modified pyrimidine deoxynucleoside 3'-O-phosphoramidite. (**b**) Trifluoroacetyl aminohexyl-(2-cyanoethyl)-N,N-diisopropylaminophosphite; a useful linker reagent for modification of 5'-ends of synthetic oligonucleotides.

Summary

Duplex DNA synthesis involves enzymatic phosphorylation of terminal 5'-hydroxyl groups of oligonucleotides and their subsequent joining in a head-to-tail fashion whilst held in a perfect base-paired duplex, each oligonucleotide acting as a splint for joining of two others.

Oligonucleotides containing modified phosphate linkages can be prepared from intermediate phosphite triesters or H-phosphonate diesters. Reactive functional groups can be incorporated into an oligonucleotide using deoxynucleoside derivatives that carry extra protection on the reactive group or by the use of a linker unit added in an extra coupling step.

3.5 Synthesis of oligoribonucleotides

The development of effective chemical methods for synthesis of oligoribonucleotides has been much slower than for oligodeoxyribonucleotides. This is largely because of the added complications resulting from the 2'-hydroxyl group in a ribonucleoside and the difficulty in finding compatible protecting groups for this and other reactive centres. However, it has been possible to prepare short oligoribonucleotides by enzymatic means and somewhat more easily so than their deoxyribocounterparts. At the time of writing, new and more powerful enzymatic procedures are emerging (Section 3.5.3) simultaneously with new and improved techniques for solid-phase oligoribonucleotide chemical synthesis (Section 3.5.2). Although neither process is yet routine, it seems likely that the two procedures will shortly meet all needs. Therefore, this section will concentrate on these new strategies rather than on more classical solution-phase synthetic methods.

3.5.1 Protected ribonucleoside units

Heterocyclic bases

Once again, acyl protecting groups are required for the exocyclic amino functions of adenine, cytosine, and guanine. By and large, the same or similar protecting groups are used as in synthesis of oligodeoxyribonucleotides (Section 3.4.2). However, extra protection for the O^6-position of guanine and the O^4-position of uracil is necessary if phosphotriester couplings are to be used, since these sites are particularly prone to phosphorylation or reaction with coupling agents. The need for extra protection has not been established for either phosphoramidite or H-phosphonate chemistry, but may still be useful sometimes in order to increase the lipophilicity of a nucleoside as an aid to purification.

Hydroxyl groups

Of paramount importance is the choice of permanent protecting group for the 2'-hydroxyl function. The choice is governed by the sensitivity of RNA to hydrolysis under alkaline conditions (Section 3.2.2). The cleavage mechanism involves generation of a 2'-oxyanion which subsequently attacks the neighbouring internucleotide linkage leading either to migration of the phosphate to the 2'-position or to chain cleavage. This attack is particularly rapid if the phosphate is in triester form. Thus the 2'-protecting group should be retained intact throughout chain assembly and be removed only after deprotection of the internucleotide phosphate. Moreover, any protecting group which upon removal gives rise to oxyanion formation should be avoided.

No consensus has yet emerged as to the best 2'-protecting group, but five likely candidates are listed in Table 3.2. The particular choice markedly influences the selection of a temporary 5'-protecting group. For example, tetrahydropyran-1-yl (THP) and 4-methoxytetrahydropyran-4-yl (MTHP), which are ideal in meeting the criteria outlined above, are not fully compatible with a 5′-DMTr group since they are not totally stable to the acidic conditions used for removal of a DMTr group. Instead, new 5′-protecting groups are being sought which are removable under non-acidic but mild conditions. By contrast, 2-nitrobenzyl (NB) and *t*-butyl-dimethylsilyl (TBDMS) protecting groups are stable to acid and can be used in combination with DMTr (or with MMTr, since stronger acids can be used safely in the case of ribonucleosides). 1-(2-Chloro-4-methyl)phenyl-4-methoxypiperidin-4-yl (CTMP), or the corresponding 2-fluorophenyl-derivative, also appears to be compatible with DMTr. At pH 1 or in the presence of a non-aqueous strong acid (e.g. trichloroacetic acid in dichloromethane) it is protonated and thus stable, but at pH 3 it is unprotonated and cleavable.

Table 3.2 Some 2′-hydroxyl protecting groups

Compound	Name	Removal
	Tetrahydropyran-1-yl [THP]	0.01 M HCl
	4-Methoxytetrahydropyran-4-yl [MTHP]	0.01 M HCl
	1-[2-Chloro-4-methyl]phenyl-4-methoxypiperidin-4-yl [CTMP]	0.01 M HCl
	t-Butyldimethylsilyl [TBDMS]	$nBu_4N^+F^-$
	2-Nitrobenzyl [NB]	Photolysis pH 3.5

A typical route to 2′-*O*-THP uridine is shown in Fig. 3.65. Reaction of uridine with trimethyl orthoacetate gives the 2′,3′-methoxyethylidene derivative. This can be used as a 3′-terminal unit in solution-phase synthesis. More importantly, 5′-*O*-acetylation followed by partial acidic hydrolysis affords a mixture of 2′,5′- and 3′,5′-diacetates, of which the latter can be fractionally crystallized in good yield. Tetrahydropyranylation and ammonolysis then give 2′-*O*-THP uridine.

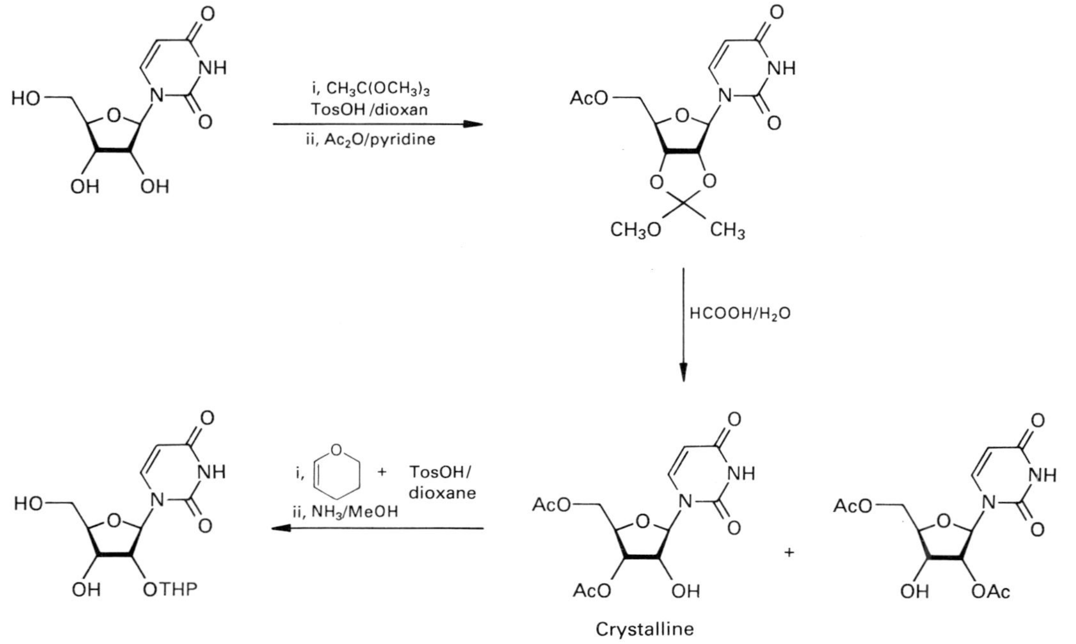

Fig. 3.65 2'-O-Tetrahydropyranylation of uridine.

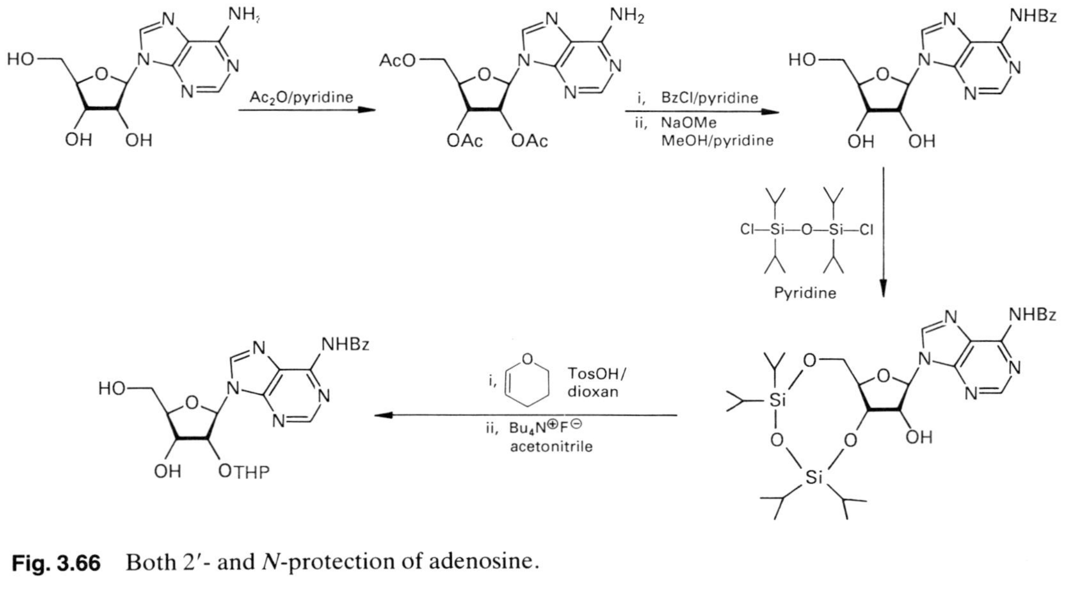

Fig. 3.66 Both 2'- and N-protection of adenosine.

Unfortunately, fractional crystallization of 3′,5′-diacetates is not possible for the other three nucleosides. Here a remarkable bifunctional silylating reagent, 1,3-dichloro-1,1,3,3-tetraisopropyldisiloxane, known as the 'Markiewicz reagent', has significantly improved synthetic routes. For example, N^6-benzoyladenosine (obtained in three steps from adenosine as shown or more simply by the transient protection route (Section 3.4.2)) reacts with the Markiewicz reagent to protect simultaneously 3′- and 5′-hydroxyl groups (Fig. 3.66). Subsequent introduction of THP is followed by removal of the cyclic silyl group with tetrabutylammonium fluoride to give 2′-O-tetrahydropyranyl adenosine. Cytidine is similarly protected except that N^4-benzoylation is more simply carried out by the per-acylation route (Section 3.4.2).

Protection of guanosine presents some practical problems. In contrast to cytidine and adenosine, guanosine compounds protected only on the exocyclic amino group are still relatively polar and are difficult to purify. One option is to silylate first with

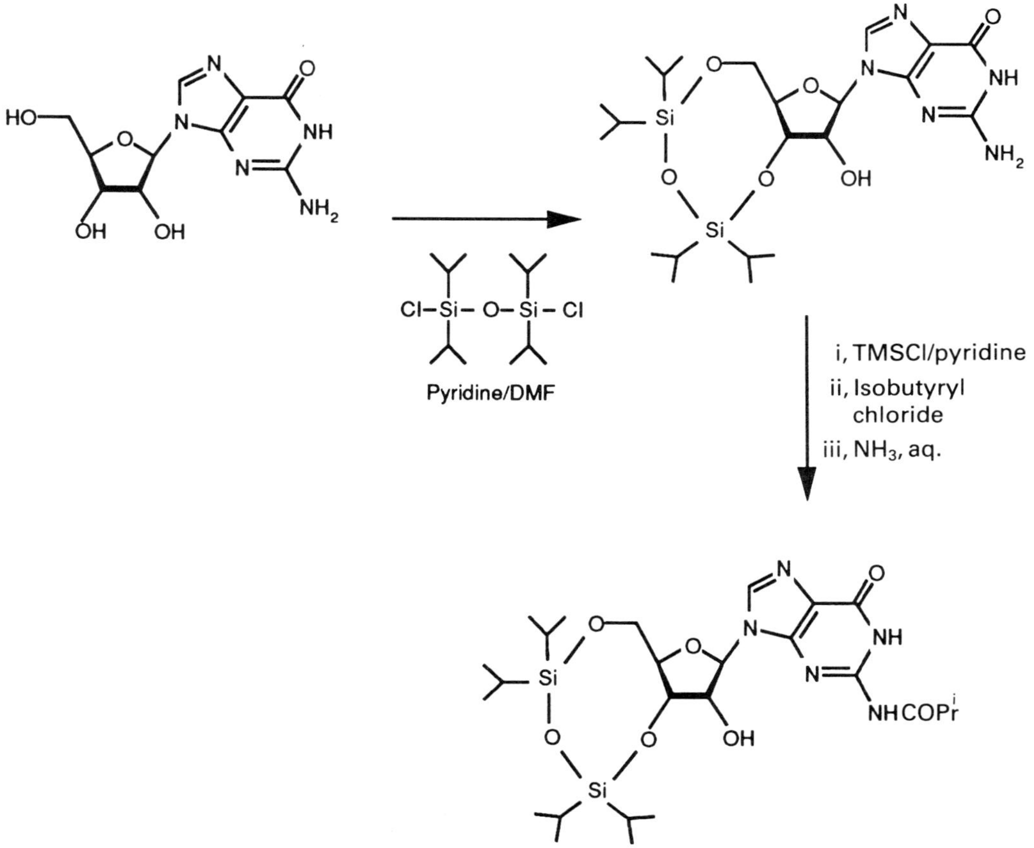

Fig. 3.67 Protection of guanosine.

the Markiewicz reagent and then protect the N^2-position, e.g. with the isobutyryl group by the transient protection route (Fig. 3.67). The 2'-position is now available for protection by an acid-labile group (THP, MTHP, or CTMP) and the cyclic silyl group removed with tetrabutylammonium fluoride.

The Markiewicz reagent cannot be used in the case of the 2'-O-TBDMS group since both are removed by treatment with fluoride ion. Fortunately, selective 2'-silylation of 5'-O-DMTr protected nucleosides can be achieved using TBDMS chloride in the presence of silver salts (Fig. 3.68). Care must be taken during work-up to avoid the migration of the TBDMS group to the 3'-position which can occur under mildly basic conditions.

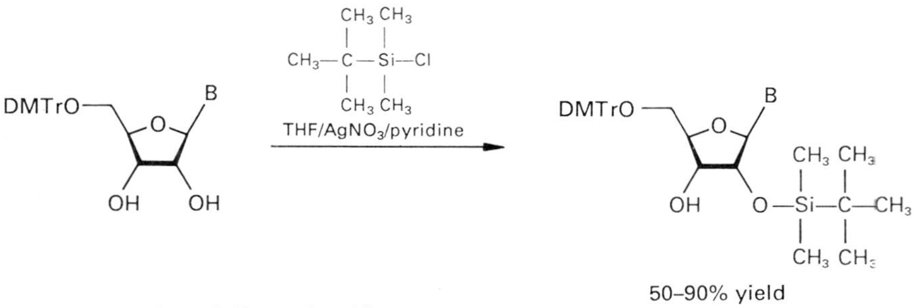

Fig. 3.68 2'-O-Silylation of ribonucleosides.

Techniques for 5'-protection and 3'-phosphorylation or phosphitylation are analogous to those previously described (Section 3.4.2).

Summary

Ribonucleoside building blocks for solid-phase synthesis require a permanent protecting group for the 2'-hydroxyl that is removed at the end of chain assembly ideally by a mechanism that does not give rise to formation of an oxyanion. It should also be stable to the conditions of removal of the 5'-hydroxyl protecting group. Introduction of a 2'-protecting group generally requires prior protection of 3'- and 5'-hydroxyl groups and the exocyclic amino groups of the heterocycles.

3.5.2 Assembly of oligoribonucleotides

In general, the assembly of oligoribonucleotides is entirely analogous to that of oligodeoxyribonucleotides. The only significant difference is the slower rate of some coupling reactions which necessitates either an increase in the time of coupling or use of more powerful activators and coupling agents. This is particularly so when bulky 2'-protecting groups (such as THP and TBDMS) are used.

A typical coupling reaction using a 3'-O-phosphoramidite is shown (Fig. 3.69). Note that when R^1 is 2-cyanoethyl the more active 4-nitrophenyltetrazole is preferred, since the phosphoramidite is less reactive than when R^1 is methyl. Extension of the chain is effected by treatment with acid and further couplings carried out as before (Section 3.4.4). Protecting groups for the 5'-terminus, phosphate, and heterocyclic bases are also removed as before (Section 3.4.4).

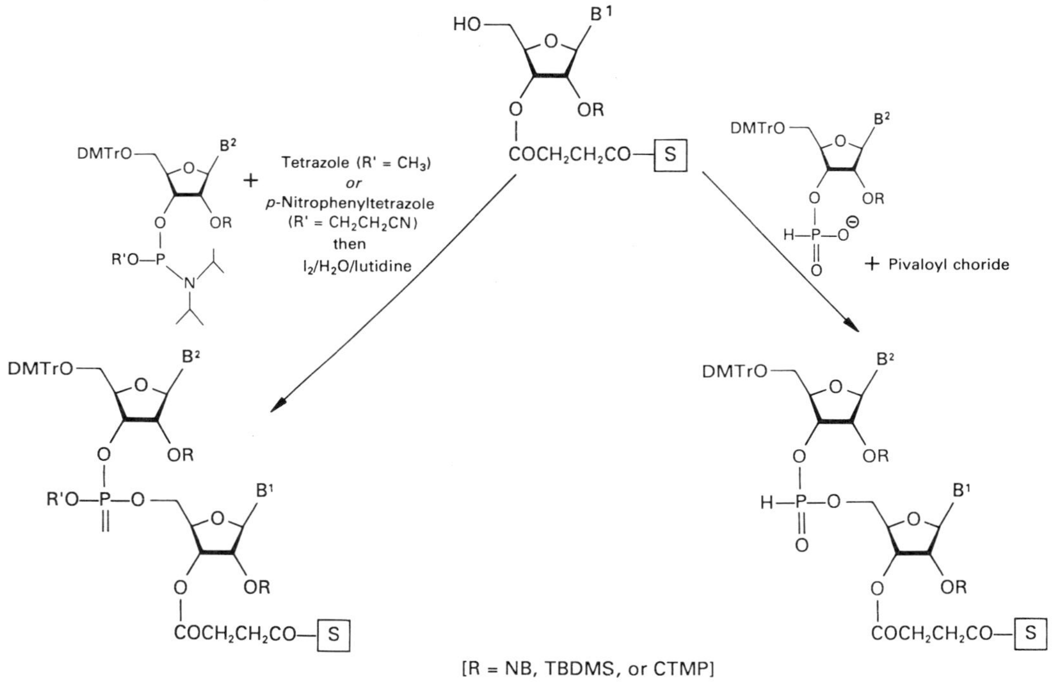

Fig. 3.69 Assembly of oligoribonucleotides by solid-phase phosphoramidite and H-phosphonate synthesis.

In H-phosphonate synthesis (Fig. 3.69) pivaloyl chloride is once again used as activator. In this case the rate of coupling appears to be less affected by the nature of the 2'-protecting group. After assembly, the phosphonate linkages are oxidized to phosphodiesters using iodine. Protecting groups for 5'-terminus and bases are removed as before (Section 3.4.4).

While removal of 2'-protecting groups is accomplished using the reagents specified in Table 3.2, it is often useful to purify and store oligoribonucleotides in their 2'-protected form. This prevents the possibility of inadvertent degradation by ribonucleases. Polyacrylamide gel electrophoresis and HPLC are the purification methods of choice. After removal of 2'-protecting groups, oligonucleotides can be recovered by gel filtration.

Summary

Coupling of ribonucleoside building blocks requires more powerful activators or coupling agents to achieve the same rate as that of deoxyribonucleosides. Assembly of an oligoribonucleotide is otherwise analogous to that of an oligodeoxyribonucleotide.

3.5.3 Enzymatic synthesis of oligoribonucleotides

T4 RNA ligase

Until very recently the best way to make short oligoribonucleotides in small quantities was by use of the enzyme T4 RNA ligase. The enzyme catalyses the joining of a 5′-phosphate group of a **donor** molecule (minimum structure pNp) to a 3′-hydroxyl group of an **acceptor** oligonucleotide (minimum structure NpNpN) (Fig. 3.70). The enzyme exhibits a high degree of preference for particular nucleotide sequences, favouring purines in the acceptor and a pyrimidine at the 5′-terminus of the donor, although there are substantial variations depending on the exact sequences of each. In order to prevent other possible joining reactions, the acceptor carries no terminal phosphate whereas the donor is phosphorylated at both ends. The 3′-phosphate of the donor is in essence a protecting group. After joining it can be removed by treatment with alkaline phosphatase to generate a free 3′-hydroxyl group and thus a new potential acceptor.

Fig. 3.70

$$- - - N_1pN_2pN_3 + pN_4p - - - \xrightleftharpoons{\quad ATP \quad AMP + PP_i \quad} - - - N_1pN_2pN_3pN_4p - - -$$

Trinucleoside diphosphate acceptors can be prepared with widely differing facility. A common starting point is the use of polynucleotide phosphorylase (from *E. coli* or *Micrococcus luteus*) to prepare homopolyribonucleotides and mixed copolymers from nucleoside diphosphates. For example, ApApG is obtained from poly(AG) by digestion with ribonuclease T1, followed by phosphatase treatment to remove the 3′-phosphate. Others require multi-step processes involving polymerases and nucleases. Some donors can also be made this way. In this case, the 3′-phosphate is retained and the 5′-end phosphorylated using T4 polynucleotide kinase.

More simply, nucleoside 3′,5′-bisphosphates can be added sequentially using RNA ligase to catalyse the joining reactions and alkaline phosphatase to remove 3′-terminal phosphates (Fig. 3.71). The products of each joining reaction are separated by HPLC. Nucleoside bisphosphates can be prepared by chemical phosphorylation of the nucleoside using pyrophosphoryl chloride.

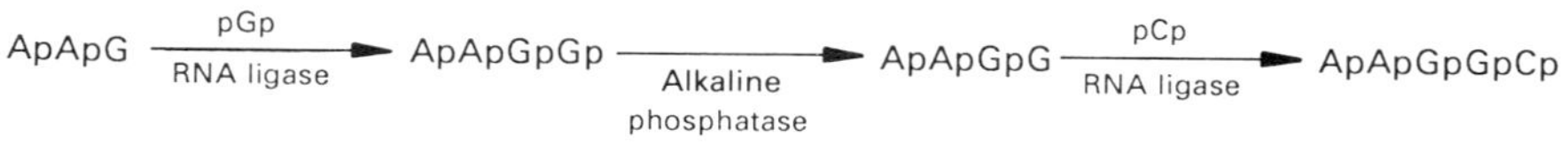

ApApG $\xrightarrow[\text{RNA ligase}]{\text{pGp}}$ ApApGpGp $\xrightarrow[\substack{\text{Alkaline}\\\text{phosphatase}}]{}$ ApApGpG $\xrightarrow[\text{RNA ligase}]{\text{pCp}}$ ApApGpGpCp

Fig. 3.71

T7 RNA polymerase

A new and powerful method of oligoribonucleotide synthesis has been recently described that makes use of the RNA polymerase from bacteriophage T7 to copy a synthetic DNA template. The template is prepared from two chemically synthesized oligodeoxyribonucleotides. Upon annealing, a duplex is formed corresponding to base pairs -17 to $+1$ of the T7 promoter sequence. Position $+1$ is the site of initiation of transcription which in natural DNA would be in a fully base-paired duplex. Here instead, the bottom strand carries a single-stranded 5'-extension corresponding to the complement of the desired oligoribonucleotide. Transcription of this template *in vitro* using T7 RNA polymerase and nucleoside triphosphates gives up to 40 µmol of oligoribonucleotide per µmol of template (Fig. 3.72). Since the enzyme can be prepared in large quantities, the scale can be increased to give milligram quantities of oligoribonucleotides.

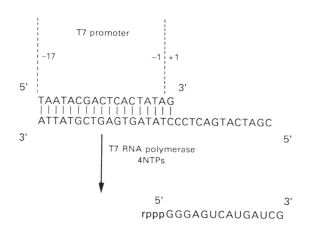

Fig. 3.72 Use of T7 RNA Polymerase to transcribe synthetic DNA templates.

There are two limitations to the method. The exact sequence from $+1$ to $+5$ influences the yield of product considerably. In cases where the sequence is substantially different from the optimum sequence, the enzyme yields a larger proportion of short, abortively initiated chains. Secondly, in some cases a non-template encoded nucleotide is added to the oligonucleotide or it may be one nucleotide shorter than expected. These by-products may need to be resolved by chromatography or gel electrophoresis.

Summary

The joining of two oligoribonucleotides can be effected using T4 RNA ligase without need of a template. The efficiencies of joining reactions are greatly influenced by the composition and sequence of each reactant. The simplest reaction is the joining of a nucleoside 3′,5′-bisphosphate to a trinucleoside diphosphate acceptor. The RNA polymerase from bacteriophage T7 can be used to copy single-stranded oligodeoxyribonucleotide templates with high efficiency.

Further reading

3.1

Hall, R. H. (1971). *Modified nucleosides in nucleic acids*. Columbia University Press, New York.

Hall, R. H. and Dunn, D. B. (1970). Natural occurrence of the modified nucleosides. In *Handbook of biochemistry. Selected data for molecular biology* (ed. H. A. Sober), 2nd edn. CRC Press, Cleveland, Ohio, pp. G99–G105.

Hanessian, S. and Pernet, A. G. (1976). Synthesis of naturally occurring *C*-nucleosides, their analogues, and functionalised *C*-glycosyl precursors. *Adv. Carb. Chem. Biochem.*, **33**, 111–88.

Hobbs, J. B. (1978–1987). In *Organo-phosphorus chemistry. Specialist periodical reports*, Vols. 9–18. Royal Society of Chemistry, London.

Hutchinson, D. W. (1979). *Comprehensive organic chemistry*, Vol. 5. Pergamon Press, Oxford and London, pp. 105–45.

Scheit, K. H. (1983). *Modified nucleotides*. Verlag Chemie, Weinheim, FRG.

Walker, R. T. (1979). *Comprehensive organic chemistry*, Vol. 5. Pergamon Press, Oxford and London, pp. 57–104.

3.2

Cohn, M. (1982). Some properties of the phosphorothioate analogues of adenosine triphosphate as substrates of enzymic reactions. *Acc. Chem. Res.*, **15**, 326–32.

Cox, J. R. and Ramsey, O. B. (1964). Mechanisms of nucleophilic substitution in phosphate esters. *Chem. Revs*, **64**, 317–52.

Eckstein, F. (1979). Phosphorothioate analogues of nucleotides. *Acc. Chem. Res.*, **12**, 204–210.

Eckstein, F. (1983). Phosphorothioate analogues of nucleotides—Tools for the investigation of biochemical processes. *Angew. Chem. Int. Ed.*, **22**, 423–39.

Frey, P. A. (1982). Stereochemistry of enzymatic reactions of phosphates. *Tetrahedron*, **38**, 1541–67.

Goldwhite, H. (1981). *Introduction to phosphorus chemistry*. Cambridge University Press, Cambridge and London.

Knowles, J. R. (1980). Enzyme-catalyzed phosphoryl transfer reactions. *Ann. Rev. Biochem.*, **49**, 877–919.

Lowe, G. (1983). Chiral [^{16}O, ^{17}O, ^{18}O] phosphate esters. *Acc. Chem. Res.*, **16**, 244–51.

3.3

Khorana, H. G. (1961). *Some recent developments in the chemistry of phosphate esters of biological interest*. Wiley, New York.

Wood, H. S. C. (1979). *Comprehensive organic chemistry*, Vol. 5. Pergamon Press, Oxford and London, pp. 489–548.

3.4

Caruthers, M. H. (1985). Gene synthesis machines: DNA chemistry and its uses. *Science*, **230**, 281–85.

Eckstein, F. (1985). Nucleoside phosphorothioates. *Ann. Rev. Biochem.*, **54**, 367–402.

Eckstein, F. and Gish, G. (1989). Phosphorothioates in molecular biology. *Trends Biol. Sci.*, **14**, 97–100.

Gait, M. J. (1984). *Oligonucleotide synthesis: a practical approach*. IRL Press, Oxford.

Itakura, K., Rossi, J. J., and Wallace, R. B. (1984). Synthesis and use of synthetic oligonucleotides. *Ann. Rev. Biochem*, **53**, 323–56.

Miller, P. S., Reddy, P. M., Murakami, A., Blake, K. R., Lin, S-B., and Agris, C. H. (1986). Solid-phase synthesis of oligodeoxyribonucleoside methylphosphonates. *Biochemistry*, **25**, 5092–7.

3.5

Middleton, T., Herlihy, W. C., Schimmel, P., and Munro, H. N. (1985). Synthesis and purification of oligoribonucleotides using T4 RNA ligase and reverse-phase chromatography. *Anal. Biochem.*, **144**, 110–17.

Milligan, J. F., Groebe, D. R., Witherell, G. W., and Uhlenbeck, O. C. (1987). Oligoribonucleotide synthesis using T7 RNA polymerase and synthetic DNA templates. *Nucleic Acids Res.*, **15**, 8783–98.

Reese, C. B. (1987). The problem of 2′-protection in rapid oligoribonucleotide synthesis. *Nucleosides and Nucleotides*, **6**, 121–29

Biosynthesis of nucleotides

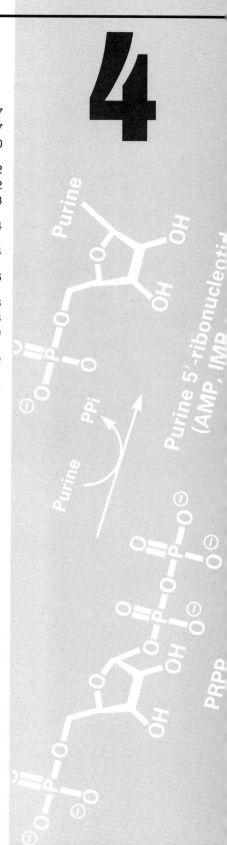

Nucleotides play a key role as the precursors of DNA and RNA, as activated intermediates in many biosynthetic processes, and as metabolic regulators. One particular nucleotide, adenosine 5′-triphosphate (ATP), is an important energy source. For example, a human being turns over 40 kg of ATP per day and during exercise can require 0.5 kg per minute. The biosynthesis of nucleotides involves both constructive (anabolic) and destructive (catabolic) pathways, but the importance of a particular pathway varies substantially between organisms. Moreover, specialized cells sometimes have their own unique metabolic pathways. In this chapter we will concentrate on only the general principles of nucleotide and nucleic acid metabolism and then show how certain steps are prime targets for biosynthetic interference, especially for the design of anti-cancer and anti-viral agents.

4.1 Biosynthesis of purine nucleotides

4.1.1 *De novo* pathways

The key intermediate in the biosynthesis of both pyrimidines and purines is α-D-5-phosphoribosyl-1-pyrophosphate (PRPP) which is formed from α-D-ribose 5-phosphate by a reaction catalysed by the enzyme ribose phosphate pyrophosphokinase (Fig. 4.1). Adenosine 5′-triphosphate acts as the donor of pyrophosphate while ribose 5-phosphate comes mainly from the pentose phosphate pathway.

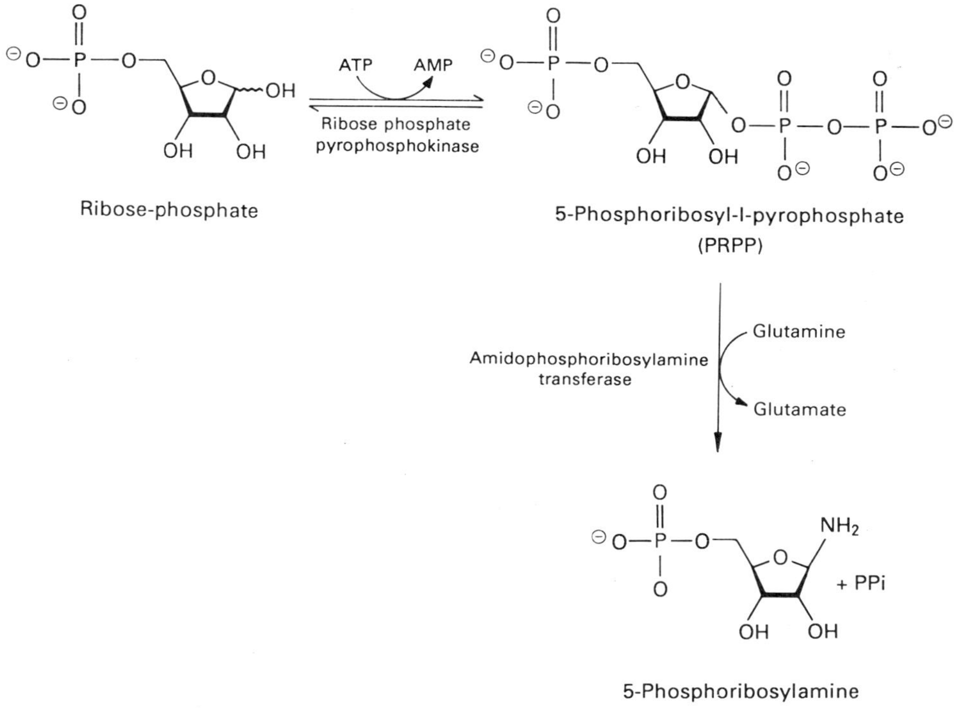

Fig. 4.1   Biosynthesis of 5-phosphoribosylamine.

In contrast to pyrimidine nucleotide biosynthesis, where a preformed heterocycle is incorporated intact (Section 4.2), in purine nucleotide biosynthesis the purine ring is constructed gradually. The first irreversible step (**the committed step**) is the displacement of pyrophosphate at C-1 of PRPP by the side chain amino group of glutamine to give β-D-5-phosphoribosylamine (Fig. 4.1). There is an inversion at C-1 such that the glycosidic bond is now in the β-configuration, the stereochemistry characteristic of all naturally occurring nucleotides. The equilibrium in this reaction is displaced towards the phosphoribosylamine by the hydrolysis of the pyrophosphate co-product.

The five carbon atoms and the remaining three nitrogen atoms of the purine skeleton are derived from no less than six different precursor sources and assembled by nine successive steps (Fig. 4.2). These steps are:

(1) reaction of PRPP with glycine to give glycinamide ribonucleotide;

(2) formylation of the α-amino terminus of the glycine moiety by N^5,N^{10}-methylenetetrahydrofolate to give α-N-formylglycinamide ribonucleotide;

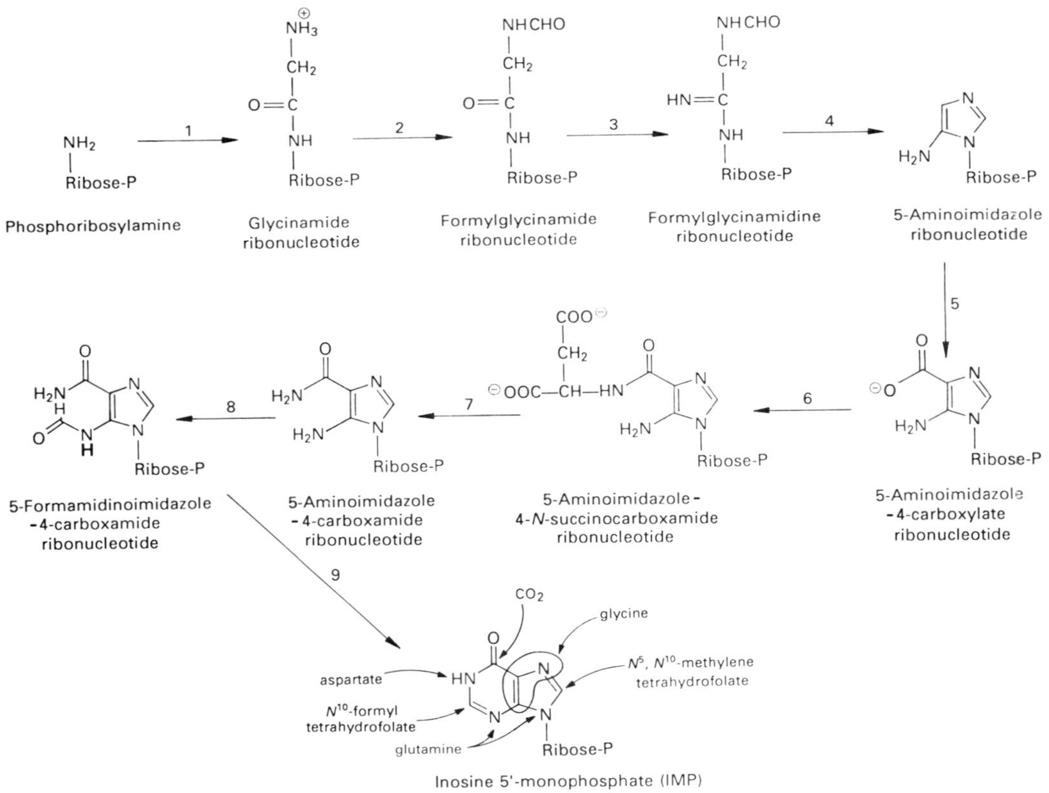

Fig. 4.2 Formation of the purine ring; biosynthesis of IMP.

(3) conversion into the corresponding glycinamidine with a new nitrogen atom derived from glutamine;

(4) ring closure to give 5-aminoimidazole ribonucleotide;

(5) carboxylation of the imidazole C-4 (the carbon atom coming from CO_2);

(6) addition of aspartate;

(7) removal of the carbon skeleton of aspartate (as fumarate), leaving behind its amino group to give 5-aminoimidazole-4-carboxamide ribonucleotide;

(8) formylation of the amino group by N^{10}-formyltetrahydrofolate; and

(9) dehydration and ring closure to form inosine 5'-monophosphate (IMP).

Inosine is a nucleoside rarely found in nucleic acids except in the 'wobble' position of some tRNAs (Section 6.6.8). In such cases, the inosine does not come from IMP nor from deamination of adenosine by an adenosine deaminase, but instead the adenine base is removed from adenosine in the preformed tRNA and replaced by hypoxanthine.

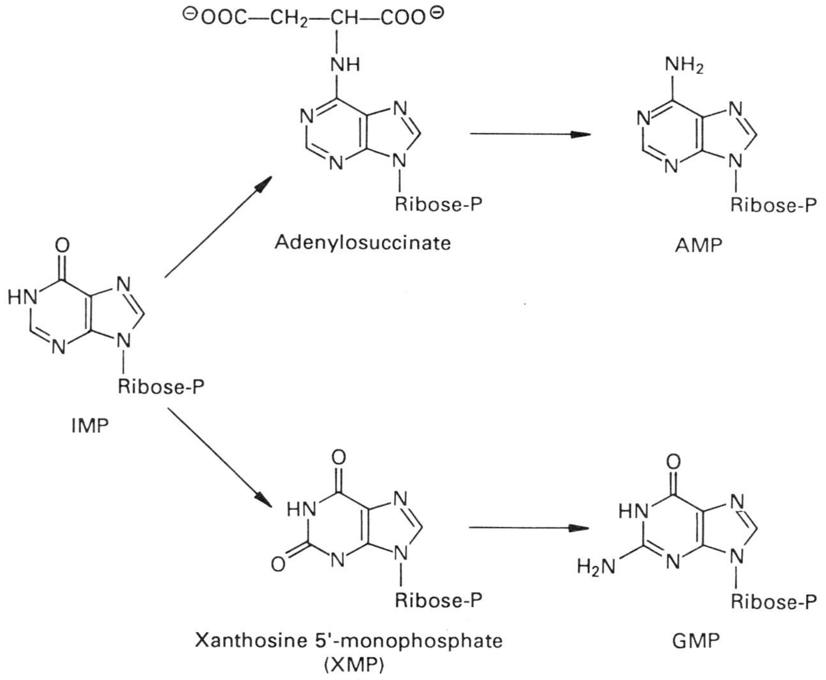

Fig. 4.3 Formation of AMP and GMP from IMP.

IMP is used entirely for the production of the natural purine nucleotides, adenosine 5'-monophosphate (AMP) and guanosine 5'-monophosphate (GMP) (Fig. 4.3). AMP receives its amino group at C–6 from aspartate. GMP is derived in two steps from xanthosine 5'-monophosphate (XMP) with the final amino group being donated by glutamine. In both these pathways, a carbonyl group of an amide is replaced by an amino group to give an amidine. This is a common type of mechanism whereby the amide is phosphorylated by ATP or GTP to its imido-*O*-phosphoryl ester and then the phosphoryl ester displaced by an amine. The leaving group can be inorganic phosphate, pyrophosphate, or even AMP, while the displacing nucleophile is ammonia, the side-chain amide of glutamine, or the α-amino group of aspartate (Fig. 4.4).

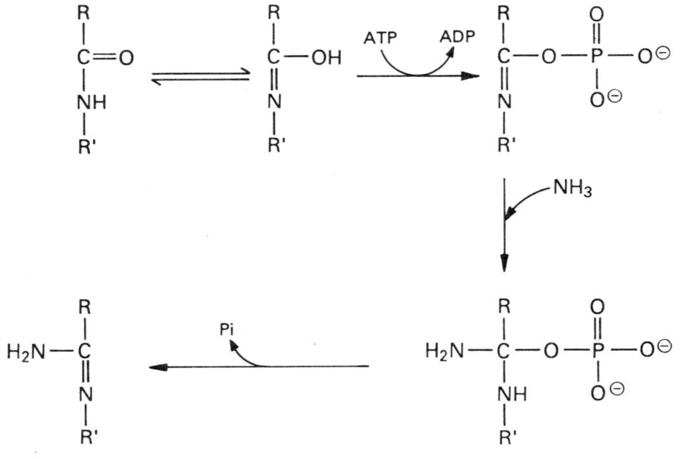

Fig. 4.4 General mechanism for biosynthetic formation of an amidine from an amide.

Steps in the biosynthesis of purine nucleotides furnish good examples of a common control mechanism, namely **feedback inhibition**, where an enzyme catalysing an early step in the pathway is inhibited by the final product of the pathway. For example, the enzyme ribose phosphate pyrophosphokinase (Fig. 4.1) is inhibited by AMP, GMP, and IMP and this inhibition regulates the level of PRPP. Similarly, the enzyme amidophosphoribosyl transferase, which is responsible for catalysing the committed step, is inhibited by a number of purine ribonucleotides including AMP and GMP, which act synergistically. AMP and GMP also inhibit the conversion of IMP into their own immediate precursors, adenylosuccinate and XMP. Another control feature is that GTP is required in the synthesis of AMP, while ATP is required in the synthesis of GMP.

4.1.2 Salvage pathways

Most organisms also use a second pathway of biosynthesis known as **salvage**. This is advantageous since degradation products of nucleic acids can be recycled rather than destroyed, which is much less costly than the energy-demanding reactions of the *de novo* pathways. In some cancer cells or virally infected cells, extra synthesis capacity is required. Here salvage may become the dominant pathway and hence becomes a target for chemotherapeutic inhibitors.

Purine bases, which arise by hydrolytic degradation of nucleotides and nucleic acids, react with PRPP to give the corresponding purine ribonucleotide and pyrophosphate is eliminated (Fig. 4.5). One enzyme, adenine phosphoribosyl transferase, is specific for the reaction with adenine, whereas another enzyme, hypoxanthine-guanine phosphoribosyl transferase (HGPRT) catalyses the formation of IMP and GMP. A deficiency of HGPRT is responsible for the serious disease 'Lesch–Nyhan syndrome', which is often characterized by self-mutilation, mental deficiency, and spasticity. Here, elevated concentrations of PRPP give rise to an increase in *de novo* purine nucleotide synthesis and degradation to uric acid (Section 4.5). The purine analogue 6-mercaptopurine is used in cancer chemotherapy and particularly against lymphoblastic leukaemia. It is utilized by HGPRT to form the corresponding ribonucleotide, which causes feedback inhibition of amidophosphoribosyl transferase in the synthesis of 5-phosphoribosylamine from PRPP (see Fig. 4.1) and also prevents IMP being converted into XMP and into adenylosuccinate (see Fig. 4.3).

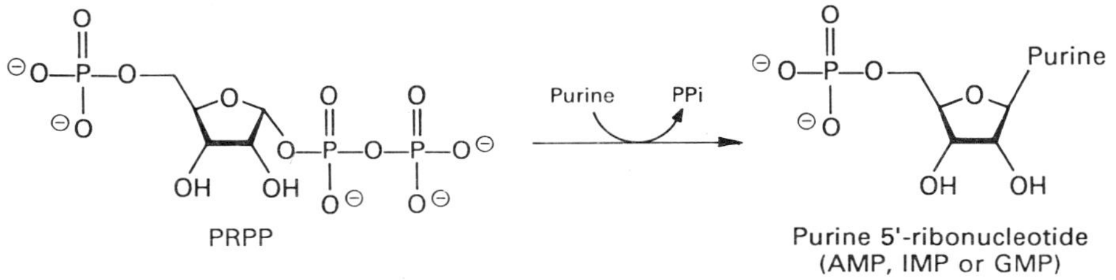

Fig. 4.5   Salvage biosynthesis of purine ribonucleotides.

Another salvage route involves the reaction of a purine (or purine analogue) with ribose 1-phosphate. The reaction is catalysed by a nucleoside phosphorylase (nucleoside phosphotransferase) and the resultant ribonucleoside is then converted into its corresponding 5′-nucleotide by a cellular kinase. Similarly a nucleoside phosphotransferase produces deoxyribonucleosides from purines and 2-deoxyribose 1-phosphate.

Summary

In biosynthesis of purine nucleotides, the heterocyclic ring is gradually built up from β-D-phosphoribosylamine using 6 different precursors to give IMP. IMP is then used for the production of AMP and GMP. Several enzymes in the biosynthetic pathway are **inhibited by feedback** in order to regulate the production of purine nucleotides. Purine bases arise by degradation of nucleotides and can be converted into nucleosides by means of **salvage pathways**. Nucleotides are formed from these nucleosides through the action of kinases.

4.2 Biosynthesis of pyrimidine nucleotides

4.2.1 *De novo* pathways

Carbamoyl phosphate is an important intermediate in pyrimidine biosynthesis and is also used in the biosynthesis of urea. Carbamoyl phosphate is formed from glutamine and bicarbonate in a reaction catalysed by one of two carbamoyl phosphate synthetases (a different one is used in the urea pathway). The reaction requires ATP as an energy source (Fig. 4.6). The committed step is the subsequent formation of *N*-carbamoyl aspartate from carbamoyl phosphate and aspartate. This step is subject to feedback inhibition by CTP which is the final product of the pathway, whereas the synthesis of carbamoyl phosphate is inhibited by UMP. In the next step the pyrimidine ring is formed by cyclization and loss of water followed by a dehydrogenation to give orotate.

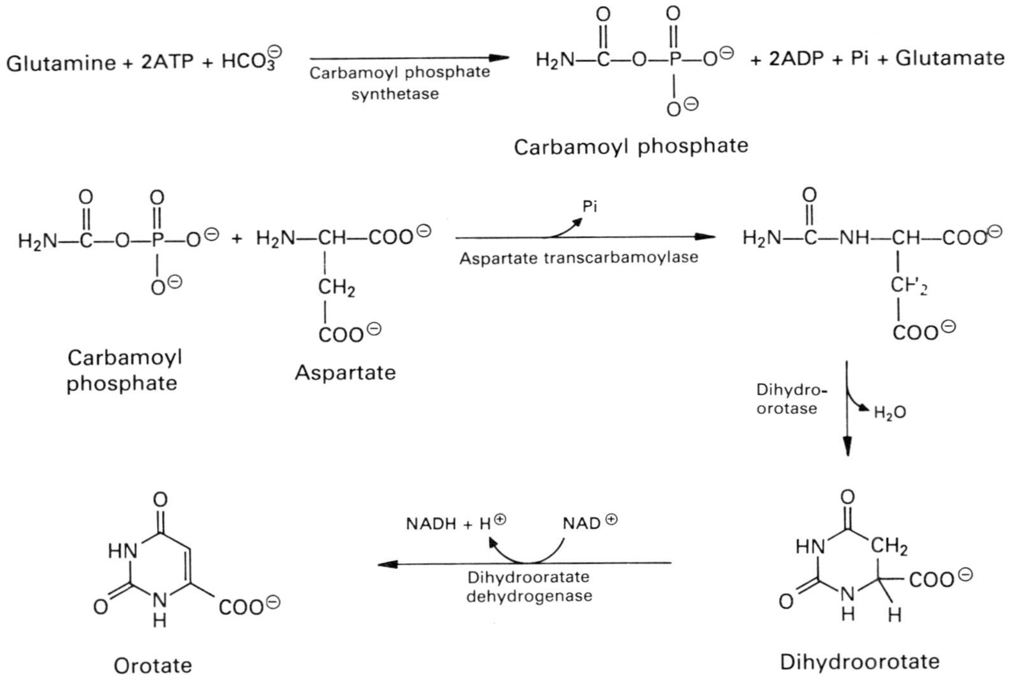

Fig. 4.6 *De novo* biosynthesis of pyrimidines; formation of orotate.

The enzymes involved in the last three steps form a multi-enzyme complex in eukaryotes (but not in prokaryotes) and are located on a single 200 kDa polypeptide chain. A potent inhibitor of the first enzyme, aspartate transcarbamoylase, is *N*-phosphonoacetyl-L-aspartate (PALA) (Fig. 4.7). PALA is an example of a **transition state inhibitor**, which works by mimicking the transition state of a reaction.

PALA binds tightly to aspartate transcarbamoylase and has proved to be useful in the production and isolation of the enzyme complex.

Orotate then reacts with PRPP to give orotidylate (Fig. 4.8). There is an inversion of configuration at C-1 and a β-nucleotide is formed. The equilibrium of the reaction is once again driven forward by hydrolysis of pyrophosphate. Finally, UMP is produced by decarboxylation. The other pyrimidine nucleotides are derived from UMP after its conversion into UTP (Section 4.3).

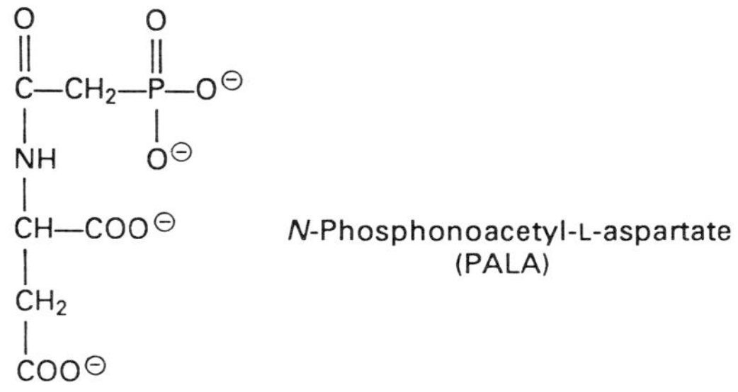

Fig. 4.7

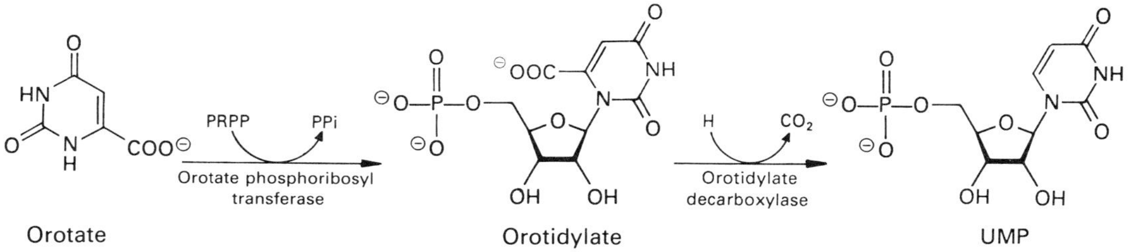

N-Phosphonoacetyl-L-aspartate
(PALA)

Fig. 4.8 Formation of UMP from orotate.

4.2.2 Salvage pathways

The enzyme orotate phosphoribosyl transferase, which is involved in the production of orotidylate from orotate, will also utilize a number of other pyrimidines, produced as a result of hydrolysis of DNA or RNA. In a similar way as for the salvage of purines, phosphorylases will catalyse nucleoside formation from a variety of pyrimidines and either ribose 1-phosphate or 2-deoxyribose 1-phosphate. A cellular kinase is also required to convert the nucleoside into its corresponding 5′-nucleotide. Uridine kinase will accept both uridine and cytosine as substrates. Thymidine kinase will accept deoxyuridine as well as deoxythymidine. The fact that many viral thymidine kinases have a lower specificity for their substrates enables a distinction to be made between normal and virally-infected cells and has led to a strategy for viral interference (Section 4.7.2).

Nucleoside transferases will catalyse base-exchange between nucleosides exclusively in the 2′-deoxy series.

Summary

In pyrimidine nucleotide biosynthesis the heterocyclic ring is preformed as orotate and reacts with PRPP to give orotidylate. UMP is formed by decarboxylation of orotidylate. Salvage pathways operate in a similar way to that for purines.

4.3 Nucleoside di- and triphosphates

The immediate biosynthetic precursors of the nucleic acids are normally the nucleoside triphosphates, whereas in energy conversions diphosphates are also used. Diphosphates are obtained from the corresponding monophosphates by means of a specific nucleoside monophosphate kinase. Adenylate kinase converts AMP into ADP while UMP kinase converts UMP into UDP. Both enzymes utilize ATP as the phosphoryl donor. Nucleoside triphosphates are interconvertible with diphosphates through nucleoside diphosphate kinase, an enzyme that has a broad specificity. Thus Y and Z (Fig. 4.9) can be any of several purine or pyrimidine ribo- or deoxyribonucleosides.

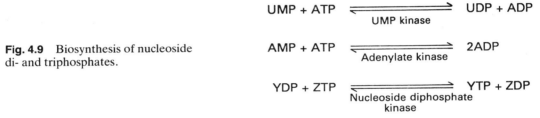

$$UMP + ATP \rightleftharpoons UDP + ADP$$
UMP kinase

Fig. 4.9 Biosynthesis of nucleoside di- and triphosphates.

$$AMP + ATP \rightleftharpoons 2ADP$$
Adenylate kinase

$$YDP + ZTP \rightleftharpoons YTP + ZDP$$
Nucleoside diphosphate kinase

Cytidine triphosphate (CTP) is formed from UTP by replacement of the carbonyl oxygen atom at C-4 by an amino group. In *Escherichia coli* the donor is ammonia whereas in mammals it is the amide group of glutamine. In both cases ATP is required for the reaction.

4.4 Deoxyribonucleotides

Deoxyribonucleotides are formed by reduction of the corresponding ribonucleotides. The 2'-hydroxyl group of the ribose is replaced by a hydrogen atom in a reaction that takes place at the level of the ribonucleoside 5'-diphosphate. The mechanism appears to be simple at first sight but it is in fact extremely complicated. The key enzyme is ribonucleotide reductase (ribonucleoside diphosphate reductase) and the electrons required for reduction of the ribose are transferred from NADPH to sulphydryl groups at the catalytic site of the enzyme. The enzyme from *E. coli* is a prototype for most eukaryotic reductases. A larger subunit ($2 \times 86\,kDa$) binds the NTP substrate, the smaller subunit ($2 \times 43\,kDa$) contains a binuclear iron centre and a tyrosyl radical at residue 122. A mechanism based on all the available

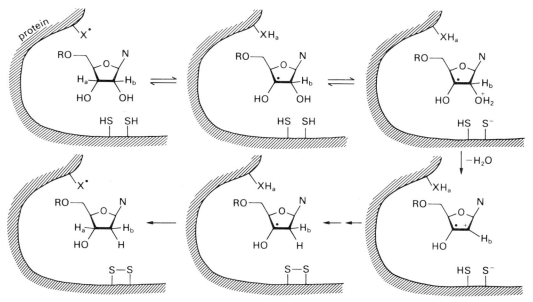

Fig. 4.10 Postulated mechanism for reduction of nucleotides to deoxynucleotides by *E. coli* ribonucleotide reductase ($X^{\cdot} = \text{Tyr}^{122}-O^{\cdot}$).

data is shown (Figure 4.10). The reduction of ribonucleoside diphosphates is precisely controlled by allosteric interactions (an allosteric enzyme is one in which the binding of another substance, usually product, alters its kinetic behaviour). Ribonucleotide reductase has two types of allosteric site that bind a number of nucleoside 5'-triphosphates and lead to a variety of conformations, each with different catalytic properties.

If any dUTP is formed from dUDP, it is rapidly hydrolysed to dUMP by an active dUTPase, which prevents the incorporation of dUTP into DNA. If uracil residues occur in DNA, they are thus likely to have arisen through deamination of cytosines and these mutations are removed by a uracil glycosylase (Section 7.11.2). As a result, deoxythymidine 5'-triphosphate (dTTP) is the dioxopyrimidine nucleotide incorporated into DNA. First, deoxythymidine 5'-monophosphate (dTMP) is biosynthesized from dUMP via the enzyme thymidylate synthetase. The methyl group is derived from N^5,N^{10}-methylenetetrahydrofolate, which also acts as an electron donor (Fig. 4.11) and is oxidized to dihydrofolate. Tetrahydrofolate is regenerated by dihydrofolate reductase using NADPH as the reductant. These two enzymes are excellent targets for cancer chemotherapy because cancer cells have an increased level of DNA synthesis and thus a heavy requirement for dTMP (see 5-fluorouracil and methotrexate in Section 4.7.1).

dTMP is next converted into dTTP in two stages by means of a thymidylate kinase and then a nucleoside diphosphate kinase. In virally-infected cells, the viral thymidine kinase (Section 4.3) often also plays the role of a thymidylate kinase.

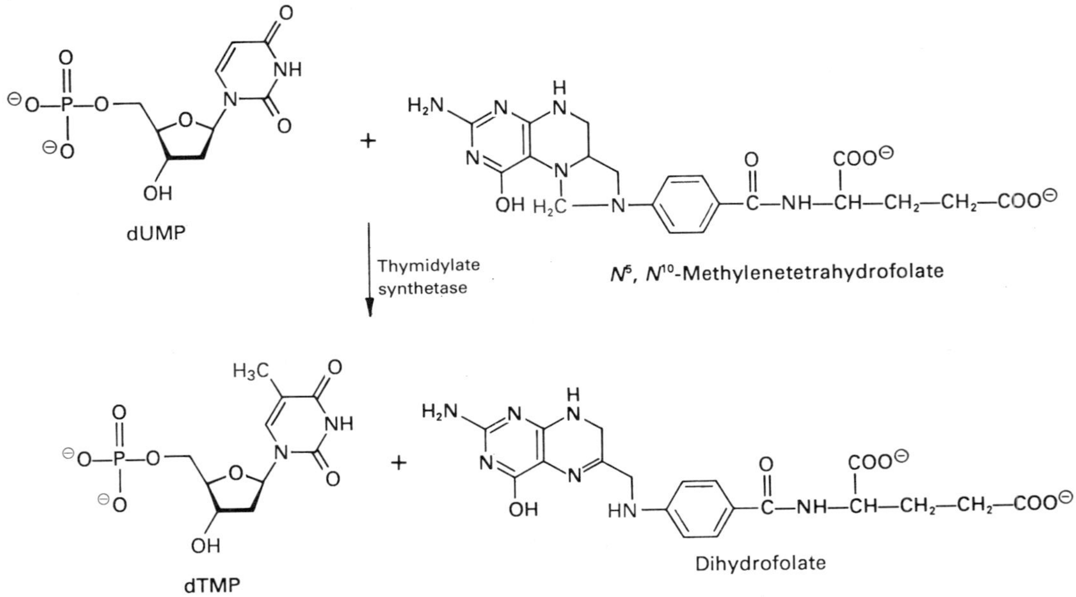

Fig. 4.11 Formation of dTMP from dUMP.

Summary

Specific nucleoside monophosphate kinases catalyse the formation of nucleoside 5'-diphosphates from monophosphates. Triphosphates are interconvertible with diphosphates through the enzyme nucleoside diphosphate kinase. 2-Deoxyribonucleotides are formed by reduction of the corresponding ribonucleotides, the reaction taking place at the diphosphate level. dUTP is hydrolysed to dUMP to prevent its incorporation into DNA. dUMP is converted into dTTP in three steps.

4.5 Catabolism of nucleotides

The degradation of nucleotides is of major importance as a target for drug design. RNA is metabolically much more labile than DNA and is constantly being synthesized and degraded. Degradation occurs initially through the action of ribonucleases and deoxyribonucleases which give oligonucleotides that are further broken down to nucleotides by phosphodiesterases.

Nucleotides are hydrolysed to nucleosides by nucleotidases (and also by phosphatases). Of great importance is the final cleavage of nucleosides by inorganic phosphate to bases and ribose 1-phosphate (or 2-deoxyribose 1-phosphate) catalysed by the widely distributed enzyme purine nucleoside phosphorylase (Fig. 4.12). The ribose phosphate can then be isomerized to ribose 5-phosphate and re-

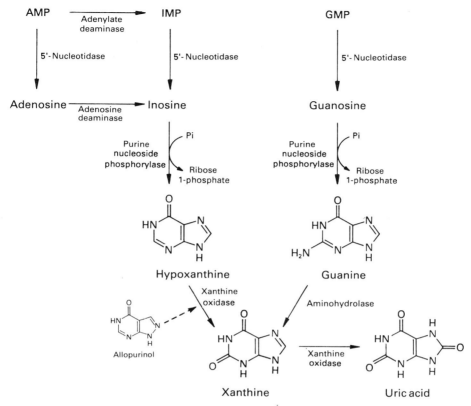

Fig. 4.12 Catabolism of purine nucleotides

used for the synthesis of PRPP. In mammalian tissues, adenosine and deoxyadenosine are resistant to the phosphorylase. AMP is therefore deaminated by adenylate deaminase to IMP and adenosine to inosine by adenosine deaminase, an enzyme which is thought to be present in elevated levels in leukaemic cells. Oxidation of hypoxanthine is catalysed by xanthine oxidase to give xanthine, which is also the deamination product of guanine. Xanthine is further oxidized to uric acid, which in humans is excreted in the urine. Gout is a painful disease caused by excessive production of monosodium urate which is deposited as crystals in the cartilage of joints. Allopurinol, which is an analogue of hypoxanthine, is used to treat gout by acting as a substrate inhibitor of xanthine oxidase. Since the allopurinol becomes tightly bound to the enzyme it is known as a **suicide inhibitor**.

A variety of deaminases can convert cytidine, 2′-deoxycytidine, and dCMP into their corresponding uracil-containing derivatives. All of these products can be broken down to give uracil which, unlike purines, is degraded reductively (Fig. 4.13). Thymine is degraded in an exactly analogous way to uracil.

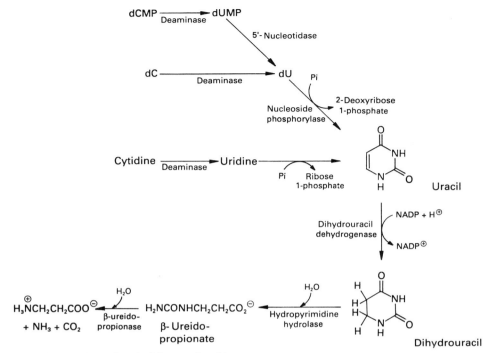

Fig. 4.13 Catabolism of pyrimidine nucleotides.

4.6 Polymerization of nucleotides

While the complex series of reactions involved in the polymerization of nucleotides to form DNA and RNA are described in detail in Chapter 5, we are here concerned primarily with the polymerases as potential targets for chemotherapy since they are the enzymes responsible for polymerization of nucleoside 5′-triphosphates into nucleic acids. In each case there is a requirement for a template strand of nucleic acid and an oligoribo- or oligodeoxyribonucleotide primer.

4.6.1 DNA polymerases

In *E. coli* there are three DNA polymerases. DNA polymerase III is the main replicating enzyme and is composed of at least eight different subunits. DNA polymerase I, which was discovered earlier, works primarily as a repair enzyme, but is still indispensible for chromosome replication. The function of DNA polymerase II is not clearly understood. In eukaryotes there are also three DNA polymerases, designated α, β, and γ. DNA polymerase α is involved in cellular replication and is found in the nucleus whereas DNA polymerase γ is exclusively located in the mitochondria. DNA polymerase β is found both in the nucleus and in the cytoplasm but its function is less clear. All these cellular polymerases use DNA as a template and polymerize in a 5'-to-3' direction. Although these polymerases are potential targets for cancer chemotherapy (e.g. for intercalators, Section 8.4), much greater scope is available for anti-viral therapy since many viruses (e.g. herpes virus) encode their own DNA polymerases and these often have substrate specificities different from those of the cellular enzymes (Section 4.7.2).

One group of RNA-containing viruses, the retroviruses, replicates via a double-stranded DNA intermediate. Retroviruses are important since many cause cancer and one of them, human immunodeficiency virus (HIV), is responsible for the disease AIDS. The RNA is first transcribed into DNA by an RNA-dependent DNA polymerase, also known as reverse transcriptase. In contrast to the cellular polymerases, these reverse transcriptases are unique to retroviruses and they are also tolerant of a wide range of nucleoside triphosphate analogues which identifies them as targets for chemotherapy (Section 4.7.2).

4.6.2 RNA polymerases

DNA is transcribed into RNA by RNA polymerase. In *E. coli* there is a single RNA polymerase having five subunits. Eukaryotes have three RNA polymerases. RNA polymerase I synthesizes a large precursor that is processed to form ribosomal RNA, RNA polymerase II is responsible for synthesis of the precursors of messenger RNA, and RNA polymerase III synthesizes small RNAs such as 5S RNA and precursors of transfer RNA. All are large and complex enzymes. They use a template of double-stranded DNA and synthesize RNA in a 5'-to-3' direction. Several antibiotics are highly potent inhibitors of transcription. Actinomycin D is an intercalator that binds tightly and selectively inhibits ribosomal RNA chain elongation. By contrast, rifampicin interacts directly with one of the subunits of RNA polymerase and inhibits initiation of RNA synthesis. *cis*-Dichlorodiammine-platinum (II) (*cis*platin) has strong anti-tumour activity. It cross-links two adjacent guanines present in the same DNA strand at their N^7-positions and interferes with transcription (Section 7.5.4).

Some viruses encode their own RNA-dependent RNA polymerase. These are also potential targets for chemotherapy, since they are generally specific for viral RNA, but as yet no clinically useful inhibitors have emerged.

Summary

Purine nucleotides are catabolized through the action of deaminases to IMP and inosine and thence to xanthine and finally to uric acid. Cytosine-containing nucleotides are deaminated to the corresponding uracil derivative and further degraded reductively.

Nucleotides are polymerized into DNA and RNA through the action of DNA and RNA polymerases respectively. Viral DNA polymerases are often targets for chemotherapy.

4.7 Drug inhibition of nucleic acid biosynthesis

As the end of the twentieth century approaches, most of the serious bacterial diseases exist only in Third World countries. Developed countries still face, however, the scourge of cancer and many viral and parasitic diseases. In the remainder of this chapter we will examine these problems to show the mode of action of some of the compounds presently used clinically, and to assess the future of rational design for anti-cancer and anti-viral therapy.

Undoubtedly the main reason for the dramatic fall in the spread of bacterial disease is the improved hygiene and better nutrition that have arisen from a higher standard of living. As a result, the natural defences of the body against invasion are given the best possible chance to work. Most bacterial infections are now controlled by vaccination or chemotherapy, but as bacteria are free-living organisms (that is they can exist independently of the organism in which they cause disease), their metabolic processes are usually very different from those of the host and this allows the possibility of selective chemotherapy. It is also much easier to screen compounds for activity or to produce a vaccine against an organism which is under no pressure to change its antigenic determinants and which can easily be grown on a large scale in culture. Thus, bacterial diseases are generally very well controlled.

Many antibacterial agents were discovered by sheer chance (e.g. penicillin). In the case of the cephalosporins and many other compounds, routine screening of extracts from soil samples and other sources gave the first lead. Therefore, it has not proved really necessary in the antibacterial field to have rational design of drugs, since a sufficient number of active compounds have been found in nature and it is relatively easy and cheap to screen for new leads.

By contrast, the anti-cancer and anti-viral fields are very different. As W. H. Prusoff once said 'a cancer cell is a "happy" host cell'. Unfortunately, it is its neighbours which are invaded by the rapidly dividing cancer cells and become unhappy, and our knowledge of the differences in metabolism between the normal and cancer cell is only at a very rudimentary stage. It is thus unlikely that vaccines of the relatively crude type used to combat bacterial infections would be successful, and specific drug-targeting to a cancer cell is probably required. We need to know more about what signal triggers a cell to proliferate and if possible to interfere with this process. Alternatively, we have to discover the differences in metabolism between a normal and a cancer cell and try to exploit them by chemotherapy.

Some leads are being obtained by careful examination of how known anti-cancer compounds work. However, *de novo* design of original compounds in this and in the anti-viral field is altogether another matter, since our knowledge of the criteria necessary to design small molecules which will interact selectively with proteins is very limited. Despite this, anti-viral chemotherapy has made good progress in the past decade and several new anti-viral agents have reached the clinic. Another big success has been vaccines which, for example, have completely eliminated smallpox as well as most of the diseases of childhood, such as tuberculosis, poliomyelitis, and diphtheria, which caused so much suffering and death in previous generations.

Why bother with anti-viral chemotherapy at all if vaccines are so good? There are several answers to this question. First, viruses are usually host and cell-type specific and their metabolic pathways are closely linked to those of their host. For example, Hepatitis B virus currently affects 200 million people out of which 40 million will die of liver cirrhosis. It is only recently that it has been possible to grow the virus in the laboratory and produce a vaccine. A second vaccine effective against Hepatitis B has been constructed by recombinant DNA techniques (Section 10.2) and is now on the market. However, chemotherapy is still required for patients already infected.

Another problem is that of 'latent virus'. Here, after infection, the viral genome is integrated into the host cell DNA and is thereafter associated with it. For example, most people have been infected by one or more types of herpes virus and therefore are always candidates for further attack from 'within', leading to symptoms such as cold sores or the disease, shingles. Chemotherapy is almost certainly required here. Similarly, after HIV infection, the viral genome is integrated into the genome of the host. Thus while the production of a vaccine may prevent future generations from infection, it can do nothing to cure those already infected.

The most common viral diseases are those caused by influenza virus, the rhinoviruses (common cold), and the enteroviruses (diarrhoea). A vaccine is available against some influenza strains, but the antigenic properties of influenza virus change so rapidly, and there are so many serotypes of rhinovirus and enterovirus, that it is not possible to develop a vaccine which would be effective against anything more than a very small percentage of infections.

The need for anti-cancer and anti-viral chemotherapy is very much with us and is likely to remain so for some time. We can now take advantage of the understanding of the metabolic pathways of normal, virus-infected, and cancer cells we have gained earlier in this chapter and use them to investigate the role of anti-viral and anti-cancer drugs.

4.7.1 Anti-cancer chemotherapy

Unfortunately our present knowledge is such that it is very difficult to define unique targets in cancer cells for attack by chemotherapy. Cancer cells appear to arise as a result of multiple insults caused by viruses, chemicals, or radiation. While some genes (**proto-oncogenes**) can give rise to **oncogenes** either following a retroviral infection, translocation, or mutation, for many cancers such an event though necessary, is not a sufficient condition. There is on the one hand a genetic component to carcinogenesis, but in addition it is desirable to avoid known carcinogenic and mutagenic agents—hazards such as those which occur in cigarette smoke and radiation, whether it be ultraviolet, X-ray, or arising from radioactive decay (Section 7.6, 7.8, 7.9).

No cancer cell arising from any of these causes has a qualitatively different metabolism. Many tumours occur with great frequency, but normally pose no problems as they are benign and localized, for example warts. The problems start when such cells spread throughout the body and set up secondary areas of invasive growth. Malignant cells have this ability to **metastasize**.

The main difference between the metabolism of a normal and a cancer cell is usually one of rate of replication. Most successful selectivities in chemotherapy are largely based on this phenomenon. Consequently, one side effect of almost all cancer chemotherapy is depression of the immune system, because the cells responsible for the maintenance of this defence against invasion are also rapidly dividing. The selectivities shown for cancer cells in the examples which follow may well be based on slightly different transport properties between cell types, or perhaps a salvage pathway is being used more, or the cell may have a different pH or oxygen tension due to its rapid metabolism.

5-Fluorouracil

There is no complete accord on the mechanism of action of 5-fluorouracil in killing cancer cells. It is possible that it can be incorporated into DNA. It certainly can be incorporated into RNA. However, it is widely agreed that 5-fluoro-2'-deoxyuridine 5'-monophosphate (dFUrdMP) is the active metabolite (Fig. 4.14). This nucleotide is a potent inhibitor of thymidylate synthetase and hence prevents the *de novo* synthesis of dTMP, dTTP, and DNA. The base 5-fluorouracil and its 2'-deoxynucleoside have similar biological effects and since the base is cheaper, it is usually used clinically. 5-Fluorouracil is presumed to be converted into the 5'-nucleotide via the pyrimidine salvage pathway. The mechanism of action and co-factor requirement of thymidylate synthetase have already been discussed (Section 4.4) and decomposition of the ternary complex formed between substrate, co-factor, and enzyme requires the abstraction of a hydrogen atom from C-5 of the uracil ring. The presence of fluorine, whilst small enough to allow complex formation, prevents this reaction due to the stability of the adduct. Much evidence for the presence of such an intermediate has been obtained and a cysteine residue in the enzyme has been identified as the nucleophile which attacks the 6-position of the pyrimidine ring (Fig. 4.15).

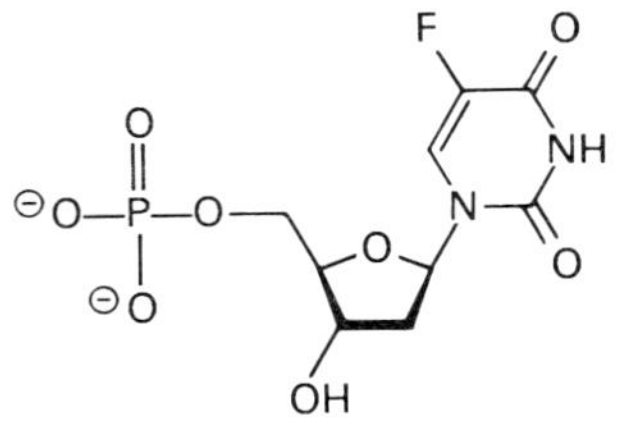

5-Fluoro-2'-deoxyuridine
5'-phosphate

Fig. 4.14

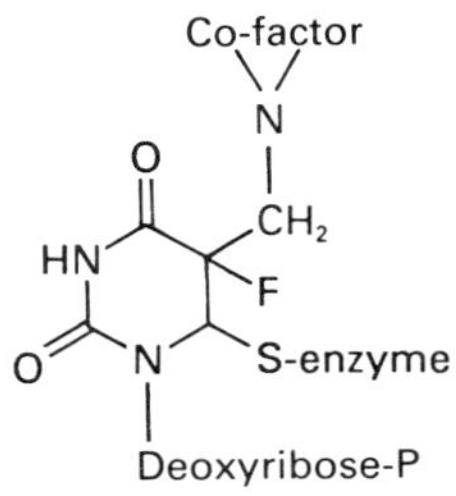

Fig. 4.15 Suicide action of 5-dFUrdMP.

Although 5-fluorouracil, and some pro-drugs of it, are widely used in the treatment of common solid tumours, there is no particular reason why it should show any selectivity. It is therefore toxic and causes suppression of the immune system.

Methotrexate

Another way of preventing the *de novo* synthesis of dTTP is to inhibit biosynthesis of the tetrahydrofolate required as the co-factor and donor of the methyl group in the reaction catalysed by thymidylate synthetase (Section 4.4). The dihydrofolate produced in this reaction is normally reduced to tetrahydrofolate by NADPH using the enzyme dihydrofolate reductase in a process which can be inhibited by folic acid analogues such as **aminopterin** and **methotrexate** (Fig. 4.16).

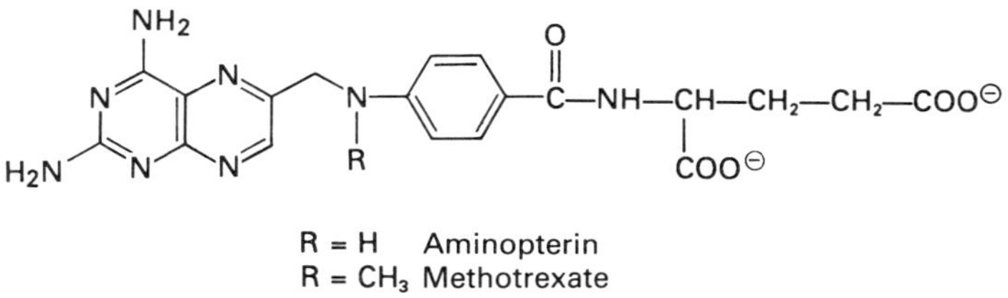

Fig. 4.16

R = H   Aminopterin
R = CH₃ Methotrexate

Methotrexate was introduced as an anti-leukaemia agent ten years before its mode of action was identified, but although many thousands of analogues have been prepared none has been found to have better properties. This is typical of discoveries in this and in the anti-viral field, even though the original discovery may have been based on only limited or no knowledge of the mode of action of the drug. The selectivity which methotrexate undoubtedly shows for cancer cells appears to be due to preferential uptake of the drug. This highlights a major problem in drug design: even the best drug in the world is of no use if it cannot be delivered intact to the correct site and transported into the target cell.

Cyclophosphamide

Many anti-cancer alkylating agents are also mutagenic and carcinogenic, but are still used in chemotherapy under certain circumstances. One example, *cis*platin, is thought to cross-link DNA (Section 7.5.4) and its specificity for tumour cells is not understood. Cyclophosphamide (Fig. 4.17) is a nitrogen mustard and is one of the most clinically useful drugs. It is thought to cross-link DNA and interferes with DNA replication. While the original synthesis was the product of rational drug design, it is now known that the design was based on a false premise and its activity is mediated through an entirely fortuitous mechanism.

Cyclophosphamide itself is inactive as an alkylating agent and it requires enzymatic oxidation in the liver to give the active species (Fig. 4.17). This hydroxylated compound can degrade either enzymatically, leading to relatively inert compounds, or it can undergo an elimination reaction to yield two potent cytotoxic agents: the alkylating agents acrolein and a phosphoramide mustard. Cyclophosphamide is useful because of its selectivity for tumour cells, which may be due to more efficient transport of the active metabolite into or a lack of enzymatic degradation within tumour cells.

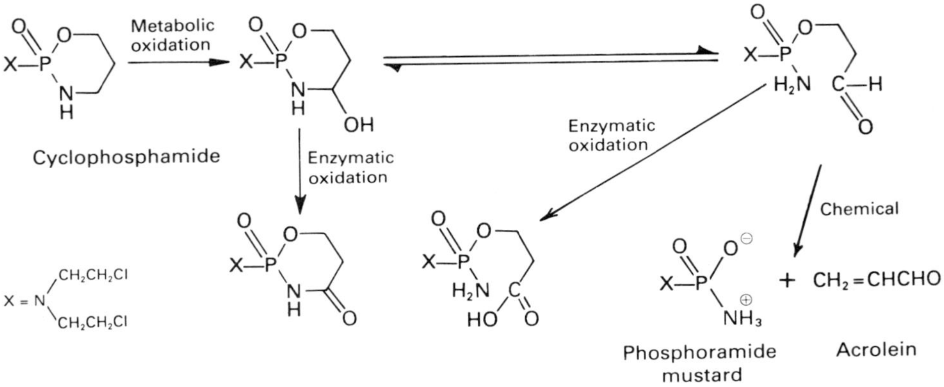

Fig. 4.17 Metabolic activation and deactivation of cyclophosphamide.

In all the above compounds, the eventual target is inhibition of DNA replication, whether by preventing synthesis of precursors or by alkylating the DNA itself. Unfortunately most of the effective drugs are also toxic and patients receiving cancer chemotherapy are unusually susceptible to viral and to bacterial infection. There has been little successful rational design so far and also too few targets at which to aim. As a result, pharmaceutical chemists are often reduced to the synthesis of analogues of previously recognized active compounds rather than aiming to design entirely novel classes of potentially active compounds.

Summary

Unique targets for chemotherapy directed against cancer cells are almost impossible to define. Most anti-cancer drugs discovered so far have toxic side-effects. Their selectivity for a cancer cell is small and other rapidly proliferating normal cells are also killed. 5-Fluorouracil and methotrexate both probably work by preventing synthesis of dTTP. The former is a thymidylate synthetase inhibitor whilst the latter inhibits *de novo* synthesis of dTTP by inhibition of the enzyme dihydrofolate reductase. Cyclophosphamide requires oxidation by liver enzymes to give a hydroxylated derivative which breaks down to yield a DNA-alkylating agent.

4.7.2 Anti-viral chemotherapy

As our knowledge concerning the methods of viral replication has increased in the past few years, it has become clear that rational targets for anti-viral drugs do indeed exist. The establishment of acyclovir (Fig. 4.18) as an anti-viral agent active against some herpes viruses has shown what is possible in situations where the development of a vaccine poses major problems. However, the facts that most viruses are not related to each other and that the replication cycle of each class is unique appear to leave little chance for the discovery of a wide-activity anti-viral agent comparable to broad spectrum antibiotics such as the β-lactams.

Acyclovir   Fig. 4.18

An equally severe problem is the fact that any drug which affects more than a single viral species is likely to impair a fundamental biochemical pathway. This has the almost inevitable result of host cell toxicity. Because such toxicity is unacceptable, extensive screening is required, even though *in vitro* and animal test systems may be insufficiently related to the human situation to provide really accurate models. As a result, an anti-viral drug can take many years to come to market.

The advent of life-threatening viral infections, particularly AIDS and genital herpes has brought a dramatic change. When time-consuming screening has to be set against the reality of death as the likely outcome of HIV infection, action can be swift. One remarkable response has been the marketing of an anti-AIDS drug, AZT, within a year of the report of its *in vitro* properties.

There are very few anti-viral drugs licensed for use (Table 4.1). Of these, all except two are nucleosides or their analogues and most are effective against herpes virus. In the following section, we will take a closer look at a few examples of anti-viral drugs which act against a number of specific viral targets.

Table 4.1. Drugs approved for use against viral infection in humans in 1989

Drug	Virus
Amantadine	Influenza A
Rimantadine	Influenza A (USSR)
5-Iodo-2'-deoxyuridine	HSV
5-Trifluoromethyl-2'-deoxyuridine	HSV
Adenine arabinoside	HSV
Acyclovir	HSV, VZV
Gancyclovir	Cytomegalovirus
5-Ethyl-2'-deoxyuridine	HSV (Federal German Republic)
5-Iodo-2'-deoxycytidine	HSV (France)
Ribavirin	Respiratory syncytial virus
3'-Azido-2',3'-dideoxythymidine	HIV-1

Acyclovir

Before the advent of AZT, this drug was *the* success of ten years of intensive research. It was discovered by chance, it is effective against herpes viruses, and its metabolic conversion into the active form is so bizarre that, with hindsight, it actually gives hope for the possibility of achieving a rational design in the future.

Herpes viruses are a class of double-stranded DNA viruses which cause a variety of diseases in humans: cold sores, eye infections (keratitis), genital sores, chicken-pox, shingles, and glandular fever (infectious mononucleosis). One property of all herpes viruses is that they exhibit latency. This means that after a cell has been infected, the virus produced can go into a latent state in the nerve endings from where it can be reactivated by various stimuli (stress, UV light, other viral infections etc.). Since it has been impossible so far to destroy the virus (i.e. prevent it from replicating) in the latent state, anti-viral chemotherapy must be directed first against primary infection and then against subsequent recurrent episodes. The symptoms of many herpes infections are relatively severe and cause discomfort on a continuing basis, but acyclovir will only alleviate the symptoms and further episodes of infection are bound to recur.

Why is herpes virus the target for so many licensed anti-viral nucleosides? It is known that herpes viruses code for many enzymes involved in their own replication and metabolism. The virus is thereby vulnerable because the properties of some of the virally encoded enzymes are slightly different from the corresponding ones in the host cell, although it is by chance that these properties have been targetted. In particular, herpes viruses rely largely on the salvage pathway for the production of dTTP for DNA synthesis and so the virus encodes its own thymidine kinase. Possibly because of the rate at which this enzyme has to work, its specificity is not as great as that of the host cell and it can phosphorylate a wide range of nucleoside analogues which, once activated, can subsequently inhibit viral replication.

Herpes viruses also code for their own DNA polymerase. Again, it is not clear why this is necessary, but the polymerase thus encoded has a different specificity from the cellular polymerases and hence presents a target for selective attack.

Although acyclovir is a purine nucleoside analogue with C-2' and C-3' of the sugar ring missing (Section 3.1.3), it is specifically phosphorylated at the position equivalent to the 5'-hydroxyl group by the thymidine kinase of the herpes virus. Not surprisingly, no metabolism occurs at all in a normal uninfected cell. However, in a virally infected cell, phosphorylated acyclovir is now recognized by the host cell guanylate kinase and is taken to the diphosphate, from which a nucleoside diphosphate kinase produces the 5'-triphosphate. This is now a substrate for the herpes virus-encoded DNA polymerase and it is incorporated into viral DNA. Since the analogue contains no equivalent to the 3'-hydroxyl group, it is a chain terminator and thus stops the synthesis of viral DNA (Fig. 4.19).

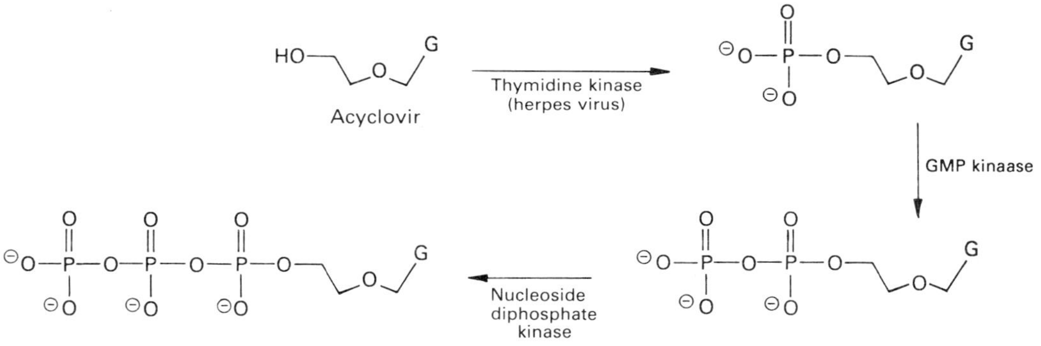

Fig. 4.19   Enzymatic phosphorylation of Acyclovir.

This is a very convincing *post facto* explanation for the efficacy of acyclovir, but the drug was actually discovered by chance before information concerning the replication of herpes virus was available. Since the discovery of acyclovir, it has not been possible to capitalize on this chance discovery by the design of novel compounds which are specific substrates for the viral kinase. However, one successful analogue, 6-deoxyacyclovir (Fig. 4.20), can be considered as a pro-drug of acyclovir

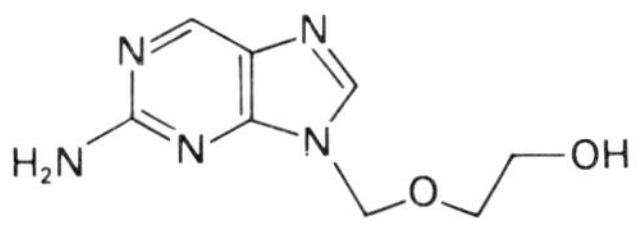

Fig. 4.20   6-Deoxyacylovir.

since it is converted into acyclovir by the enzyme xanthine oxidase (Fig. 4.12), has greater solubility, and so allows higher plasma levels of acyclovir to be achieved.

5-Substituted-pyrimidine 2′-deoxynucleosides

5-Iodo-2′-deoxyuridine (Fig. 4.21a) was discovered by Prusoff in the 1960s and was the first anti-viral nucleoside drug to be marketed. The mode of action of this nucleoside is still not known over 20 years later, although it is a substrate for both cellular and viral thymidine kinases and it is incorporated into cellular and viral DNA. Its anti-viral properties may be a result of this incorporation, of the deoxynucleoside triphosphate being an inhibitor of the viral DNA polymerase, or of some other explanation. It is likely, however, that the toxicity of this drug is a consequence of the fact that both viral and cellular kinases can phosphorylate it and therefore it is further metabolized in infected and non-infected cells.

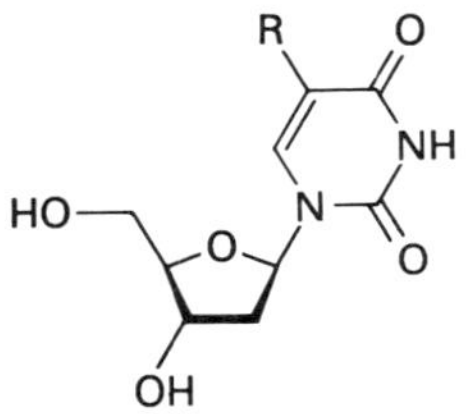

Fig. 4.21 Anti-viral pyrimidine 2′-deoxynucleosides.

(a) R = I
(b) R = CH=CH$_2$
(c) R = CH=CHBr

A breakthrough came when 5-vinyl-2′-deoxyuridine was shown to be much more potent against herpes viruses *in vitro* than the 5-iodo derivative by many orders of magnitude (Fig. 4.21b). While this compound is very toxic in cell culture, presumably because it is also metabolized by cellular enzymes, in animals it is neither toxic nor does it have anti-viral properties. This is because the nucleoside is a very good substrate for nucleoside phosphorylase, an enzyme which is absent in many tissue culture cell lines. The enzyme degrades it to the heterocyclic base, which has no anti-viral properties. From this example we learn that any analogue in this series must be resistant to nucleoside phosphorylase in order to possess anti-viral activity. Unfortunately we do not know the specificity of this enzyme and so it is difficult to design a nucleoside analogue which is not such a substrate.

The next compound in this series to be discovered (again by chance) was (*E*)-5-(2-bromovinyl)-2′-deoxyuridine (BVDU; Fig. 4.21c) which was even more effective (ID$_{50}$: 0.001 mg ml^{-1}) against herpes simplex virus (HSV-1) and varicella zoster virus but less so against HSV-2. After the event, this was shown to be due to the nucleoside acting as a substrate for the thymidine kinase of the virus and not for the kinase of the host cell. This viral thymidine kinase is apparently also a thymidylate kinase and produces the 5'-diphosphate, but only in HSV-1-infected cells, since the diphosphate formation does not occur efficiently with the HSV-2-encoded enzyme. Although BVDU is a substrate for nucleoside phosphorylase, sufficient survives undegraded for it to show useful clinical activity. The base, (E)-5-(2-bromovinyl)uracil is not a substrate for pyrimidine-5,6-dihydroreductase, which is the first enzyme in pyrimidine catabolism. Indeed it is an inhibitor of this enzyme and thus the base can actually be salvaged and the 2'-deoxynucleoside regenerated. The triphosphate of BVDU is a substrate for and an inhibitor of the virally-encoded DNA polymerase. It is still not known which, if either, of these properties is responsible for the anti-viral activity of BVDU.

Ribavirin

The broad-spectrum anti-viral activity of ribavirin (1-β-D-ribofuranosyl-1,2,4-triazole-3-carboxamide; Fig. 4.22a) was first described in 1972. Since then, it is thought to have been studied in more animals and against more viruses than any other anti-viral agent. It is also apparently active in cell culture against about 85 per cent of all virus species studied and to show little or no cellular toxicity. One is then left to speculate why, after fifteen years, this 'wonder drug' has only been approved for use in the USA in aerosol form against respiratory syncytial virus in young children.

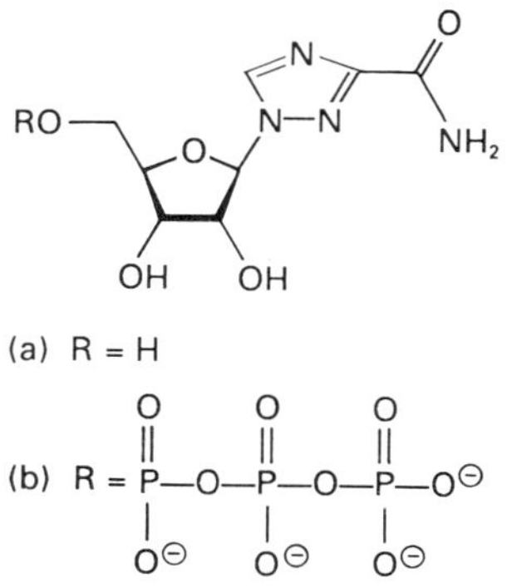

Fig. 4.22 Ribavirin and its 5'-triphosphate.

Ribavirin almost certainly does not act in only one way against all viruses. The most abundant form of ribavirin in cells is the 5'-triphosphate (Fig. 4.22b) and this was originally thought to inhibit inosine monophosphate dehydrogenase, which results in a depletion of cellular GTP pools. This in turn means that ribavirin 5'-triphosphate can now more effectively inhibit competitively the viral-specific RNA polymerase for some viruses. Ribavirin 5'-triphosphate is also known to inhibit the viral-specific mRNA capping enzymes, guanyl transferase, and N^7-methyl transferase, so that viral protein synthesis is interrupted. Ribavirin also is known to be effective clinically against haemorrhagic fever virus, to cure plant virus infections, and has undergone trials against HIV.

Phosphonoformic acid

Phosphonoacetic acid (PAA, Fig. 4.23a) was discovered to have antiherpetic activity *in vitro* following random screening in 1973. Two years later it was shown to be a selective inhibitor of the virally encoded DNA polymerase, and the related phosphonoformic acid (PFA, Fig. 4.23b) was subsequently found to be an even stronger inhibitor of this enzyme. PFA is also widely used for *in vitro* assays of HIV reverse transcriptase. Both these compounds are analogues of pyrophosphate, a product of the polymerase, and presumably bind to the corresponding site on the enzyme thus preventing replication. One problem with compounds of this sort is that they require no prior activation and therefore one has to rely entirely on the difference in affinity between the virus-encoded and host cell polymerase. Any interaction with the latter will presumably cause toxicity and this may be the reason for some of the problems encountered with these agents.

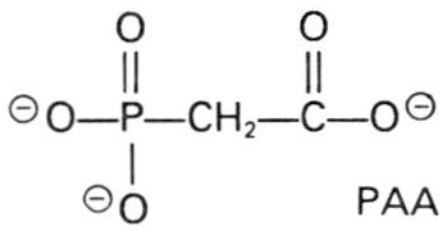

Fig. 4.23 Antiviral analogues of pyrophosphate.

Retrovirus inhibitors

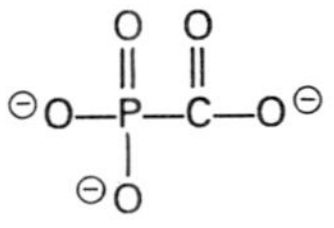

Much effort has been expended on finding a chemotherapeutic agent which can alleviate the symptoms of AIDS. The human immunodeficiency virus is a member of the lentivirus family (a sub-class of retrovirus) and its reverse transcriptase (Section 4.6.1) was an obvious initial target. A number of 2'-deoxynucleoside 5'-triphosphate analogues were quickly found to act as inhibitors or serve as chain terminators of the enzyme. The compound 3'-azido-2',3'-dideoxythymidine (AZT; Fig. 4.24) was found to be the least toxic and is widely used clinically. However, there is a fundamental problem with all compounds of this type. The 5'-triphosphate is necessarily the active species. Since the retrovirus does not encode its own kinase and since it is difficult to get highly anionic nucleotides into cells, one has to rely on the native human cellular thymidine kinase to perform the initial phosphorylation steps.

Fig. 4.24

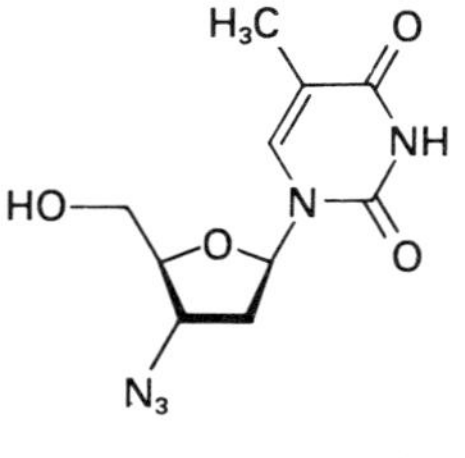

3'-Azido-2',3'-dideoxythymidine (AZT)

As we have already seen (Section 4.7.1), the mammalian cellular kinase is highly selective in its requirements. Thus, although many deoxynucleoside triphosphate analogues may be good inhibitors of the viral reverse transcriptase, as long as it is necessary to rely on a specific kinase of the host cell for activation of the analogue, the choice is greatly restricted. A compound like AZT, having only a small modification, is a suitable substrate for the kinase and thus the triphosphate is a substrate for the reverse transcriptase and can cause chain termination. Unfortunately the high dosage levels of AZT required give rise to considerable toxicity. One suggestion is that this is due to an unexplained depletion in the levels of dTTP and dCTP that have been observed. More plausibly, the triphosphate of AZT could be to some extent a substrate for the DNA polymerase of the host cell. Either situation is likely to contribute to bone marrow suppression, which is one of the observed toxic side-effects of the drug.

Anti-viral drug design

Why should almost all licensed anti-viral drugs and many of these currently under investigation be nucleoside analogues? Although such compounds may well be expected to interfere with viral replication and often do have an observed biological effect, they are also very likely to be metabolized by normal cellular enzymes and thus cause toxicity. There are many other viral targets and there is no particular reason why the inhibitor of a viral enzyme should be a nucleoside analogue. Other steps unique to viral replication are, for example, adhesion of the virus to the host cell and subsequent uptake, uncoating of the nucleic acid, maturation and release of the virion. The problem is to know what sort of compounds will inhibit these processes. Some phenylisoxazoles are one of the few classes of compound apart from nucleoside analogues that are showing promise in the treatment of some RNA-virus infections and amantadine is being used for the treatment of influenza virus infection. The role of virally-encoded sequence-specific proteases is also a target, especially for the HIV aspartyl protease, and anti-viral drug development in this field is under intensive industrial investigation.

It will be very difficult to design novel inhibitors until we know much more about the details of viral replication at a molecular level. We need to have three-dimensional structures of the enzymes involved and much better models of cell and viral surfaces. Even then, it is one thing to inhibit a function *in vitro*, but quite another to get the inhibitor delivered to the infected cell without degradation, at effective concentration and, in the case of a non-lethal infection, in orally active form.

Summary

With many potential targets for anti-viral chemotherapy, most clinically useful compounds have been directed against virally encoded enzymes which are essential for viral replication. Anti-viral agents directed against non-lethal diseases must be non-toxic to the host cell.

Most of the nucleoside analogues possessing anti-herpes virus activity rely upon specific phosphorylation by a herpes virus-encoded thymidine kinase. In the case of acyclovir, the triphosphate subsequently produced is a substrate for the viral DNA polymerase and is a chain terminator.

Since HIV infection is lethal, more toxic anti-virals have been accepted for clinical use. AZT has shown some effect in alleviating the symptoms of AIDS and, following conversion into the 5′-triphosphate, AZT is a chain terminator in the formation of DNA catalysed by reverse transcriptase.

Few anti-viral compounds have so far been designed on rational principles but, with an increasing knowledge of viral replication mechanisms, structures of viral enzymes and other components, it should eventually be possible to design molecules which are effective and non-toxic to the host.

Further reading

4.1–4.6

Nogrady, T. (1985). *Medicinal chemistry—a biochemical approach*. Oxford University Press, Oxford, pp. 343–61.

Stryer, L. (1988). *Biochemistry*, 3rd ed. W. H. Freeman, San Francisco, pp. 601–26.

Stubbe, J. A. (1989). Protein radical involvement in biological catalysis? *Ann. Rev. Biochem.*, **58**, 257–67.

Zubay, G. (1983). *Biochemistry*. Addison Wesley, New York, pp. 697–735.

4.7

Collier, L. H. and Oxford, J. (ed.) (1980). *Developments in antiviral therapy*. Academic Press, London.

Coulson, C. J. (1988). *Molecular mechanisms of drug action*. Taylor and Francis, London, Chapter 2.

DeClercq, E. (ed.) (1987). *Frontiers in microbiology*. Martinus Nijhoff, Dordrecht.

DeClercq, E. and Walker, R. T. (ed) (1984). *Targets for the design of antiviral agents*. Plenum, New York.

DeClercq, E. and Walker, R. T. (1986). Chemotherapeutic agents for herpesvirus infections. In *Progress in medicinal chemistry* (ed. G. P. Ellis and G. B. West), Vol. 23. Elsevier, Amsterdam, pp. 187–218.

DeClercq, E. and Walker, R. T. (ed.) (1988). *Antiviral drug development: a multidisciplinary approach*. Plenum, New York.

Hertzberg, R. P., Caranfa, M. J., and Hecht, S. M. (1989). On the mechanism of topoisomerase I inhibition by camptothechin: evidence for binding to an enzyme: DNA complex. *Biochemistry*, **28**, 4629–38.

Kensler, T. W. and Cooney, D. A. (1989). Inhibitors of the *de novo* pyrimidine pathway. In *Design of enzyme inhibitors as drugs*, (ed. M. Sandler and H. J. Smith), pp. 379–401. Oxford University Press.

Mitsuya, H. and Broder, S. (1987). Strategies for antiviral therapy in AIDS. *Nature*, **325**, 773–8.

Montgomery, J. A. and Bennett, L. I. (1989). Inhibitors of purine biosynthesis. In *Design of enzyme inhibitors as drugs*, (ed. M. Sandler and H. J. Smith), pp. 402–34. Oxford University Press.

Neidle, S. and Waring, M. J. (ed.) (1983). Molecular aspects of anti-cancer drug action. *Topics in molecular structural biology*, Vol. 3. Macmillan, London.

Robins, R. K. (1981). Synthetic antiviral agents. *Chem. Eng. News*, 28–40.

Stuart-Harris, C. H. and Oxford, J. (ed.) (1983). *Problems of antiviral therapy*. Academic Press, London.

DNA sequence information and transmission

5

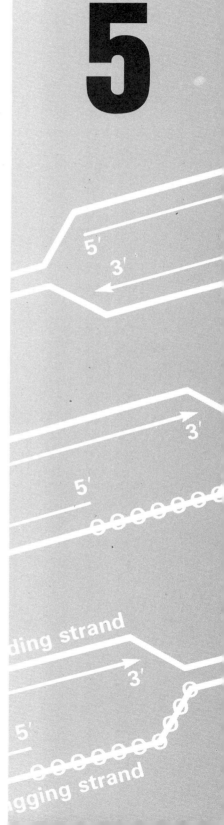

5.1 Isolation of DNA

Although historically the isolation of polymeric DNA from living organisms was problematical (Section 1.2), nowadays the procedure can be carried out easily. This is mainly due to the chemical stability of DNA and the ease with which most deoxyribonucleases can be inactivated. By contrast, the isolation of intact RNA poses a much greater problem (Section 6.1).

DNA can be isolated from prokaryotic or eukaryotic whole cells or from purified nuclei. Usually the use of purified nuclei is unnecessary. DNA isolations can be essentially divided into four steps: (a) isolation of cells; (b) cell lysis; (c) removal of protein; and (d) isolation of pure DNA.

Two essential requirements during DNA isolation are the exclusion of free Mg^{2+} ions (and any other divalent cation if possible) and the absence of rough physical treatment. Mg^{2+} is a cofactor of all common deoxyribonucleases and can normally be excluded by addition of 10 mM EDTA as a chelating agent. To ensure that no magnesium-independent nucleases cause any degradation, an anionic detergent such as sodium dodecyl sulphate is usually also added. Sodium dodecyl sulphate also serves to lyse eukaryotic cells and nuclear membranes. To lyse prokaryotic cells, lysozyme can be used in most cases but sometimes other methods, such as a French press, may need to be employed. The second important point to note is that genomic DNA, when released from any cell, is of extremely high molecular weight and is acutely sensitive to physical shearing. Over-zealous pipetting, vortexing, or shaking should be avoided if DNA larger than 10 000 base-pairs is required. The general procedures are as follows.

(a) *Isolation of cells.* For prokaryotes, monocellular eukaryotes, and tissue culture cells, this simply involves harvesting by low speed centrifugation at 4°C followed by resuspension of the cells in isotonic neutral buffered solution. Tissues will need to be disrupted by physical or enzymatic means.

(b) *Cell lysis.* To avoid clumping, lysis of eukaryotic cells is best achieved by addition of a concentrated lysis buffer to a suspension of cells. A typical lysis buffer is: 1 per cent sodium dodecyl sulphate, 10mM Tris-HCl (pH 7.5), 10 mM EDTA. Prokaryotic cells can be pre-treated with lysozyme and then suspended in a lysis buffer containing sodium dodecyl sulphate. Alternatively they can be physically disrupted in a French press or by use of a sonicator.

(c) *Removal of protein.* A general protease, such as fungal Proteinase K, is added to lysed cells to degrade protein, thereby allowing easy removal by repeated extraction from aqueous solution into phenol. Typically, a 30 minute digestion with 100 µg ml^{-1} Proteinase K is followed by four successive extractions with redistilled phenol (phenol oxidizes readily and must be redistilled before use). Prior to use it should be equilibrated to pH 7–8 with 1 M Tris base). After deproteinization, nucleic acids and polysaccharides are precipitated by addition of sodium acetate (pH 7) to 0.3 M followed by two volumes of 95 per cent ethanol. Genomic DNA can be spooled from the water–ethanol interface using a glass pipette or precipitated by gentle vortexing. Low molecular weight plasmid DNA forms a colloidal suspension and is not separable from RNA by this method.

(d) *Isolation of pure DNA by CsCl density ultracentrifugation*. The best way to remove contaminating RNA and polysaccharides from crude DNA preparations is by equilibrium density gradient ultracentrifugation. This method makes use of the characteristic density of DNA in aqueous solution (1.7 g ml^{-1}). Caesium chloride (CsCl) is added to the crude preparation to a final concentration of 1.7 g ml^{-1} and the solution is centrifuged until a density gradient of CsCl is formed in the tube (typically 48 h at 80 000 *g* in an angle rotor or overnight in a vertical rotor). If plasmid DNA is to be isolated, ethidium bromide is added prior to centrifugation. This compound intercalates between the nucleotide pairs in any DNA duplex without denaturing it (Section 8.4). This intercalation causes a slight unwinding of the helix and a reduction in the density of the DNA. Because plasmids are closed circles this unwinding is compensated by the formation of supercoils (Section 2.3.4). Eventually, the increased free energy thus generated prevents further intercalation at an ethidium:DNA ratio which is significantly below that for a topologically relaxed duplex. The net result is that supercoiled plasmid DNA is less dense than nicked plasmid or genomic DNA and therefore migrates to a different region of the CsCl gradient.

Summary

DNA is isolated by harvesting and lysis of cells, removal of protein, and separation from RNA and polysaccharides. Care is necessary to avoid enzymatic or mechanical breakage of the high molecular weight DNA. DNAs are usually purified according to their buoyant density in CsCl solutions by equilibrium ultracentrifugation.

5.2 Determination of DNA sequence

The methods for determination of the sequence of DNA are now so rapid that they have superseded protein sequencing as the major method of determining polypeptide sequences. Rather than attempting the direct sequence analysis of a protein, it is usually simpler to clone the corresponding gene and determine its DNA sequence and hence the amino acid sequence of the protein.

There are two major ways of determining the sequence of a DNA molecule. These methods were developed in the laboratories of Gilbert and of Sanger for which each received a Nobel prize. Both are dependent upon the partial fragmentation of DNA at one of the four bases followed by an accurate determination of the size of each fragment. Both methods rely upon sequencing only one strand at a time. For example, let us consider how we can determine the sequence of a 20 nucleotide fragment (Fig. 5.1). If we could somehow create one specific cleavage in the top strand at any of the phosphodiester bonds following a C, we would generate a mixture of shorter sequences (Fig. 5.2).

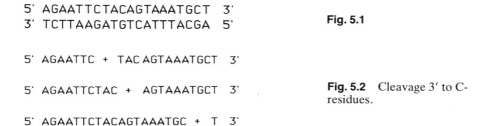

5′ AGAATTCTACAGTAAATGCT 3′
3′ TCTTAAGATGTCATTTACGA 5′   **Fig. 5.1**

5′ AGAATTC + TACAGTAAATGCT 3′

5′ AGAATTCTAC + AGTAAATGCT 3′   **Fig. 5.2** Cleavage 3′ to C-residues.

5′ AGAATTCTACAGTAAATGC + T 3′

Maxam and Gilbert sequencing relies upon radioactive labelling of only one end of the DNA. If in the above example the 5′-end is labelled, then only the left hand sequence of each of the pairs of fragments would be radioactive (Fig. 5.3). These three oligonucleotides can be separated from one another by virtue of their different sizes. In practice, this is achieved by **polyacrylamide gel electrophoresis**. Here, the mixture of oligonucleotides generated by C-cleavage is subjected to an electric field in a polymeric gel matrix such as polyacrylamide. Oligonucleotides migrate towards the anode at a rate which is inversely proportional to their size and thus three radioactive bands are generated in the gel (Fig. 5.4). If the experiment is repeated, but this time with cleavage only after G bases, another set of bands is

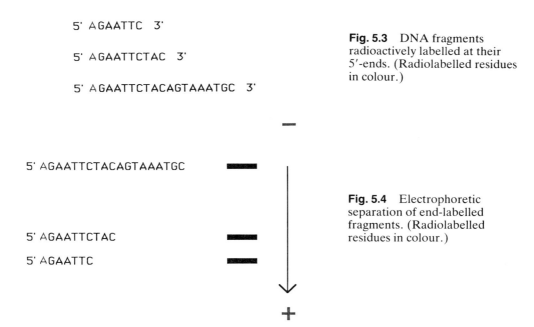

5′ AGAATTC 3′

5′ AGAATTCTAC 3′

5′ AGAATTCTACAGTAAATGC 3′

Fig. 5.3 DNA fragments radioactively labelled at their 5′-ends. (Radiolabelled residues in colour.)

5′ AGAATTCTACAGTAAATGC

5′ AGAATTCTAC

5′ AGAATTC

Fig. 5.4 Electrophoretic separation of end-labelled fragments. (Radiolabelled residues in colour.)

generated (Fig. 5.5). Parallel sets of such reactions can be carried out which are specific for each of the four bases and electrophoresis of these reaction products generates a sequencing **ladder** from which the complete sequence can be read upwards from the bottom of the gel (Fig. 5.6), corresponding to reading 5′→3′ on the DNA.

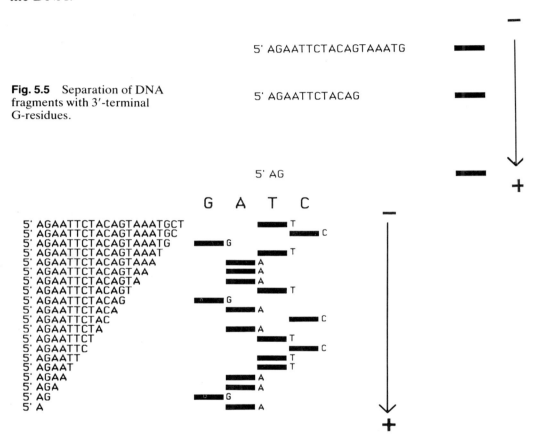

Fig. 5.5 Separation of DNA fragments with 3′-terminal G-residues.

Fig. 5.6 Generation of a DNA sequence ladder.

In practice, the types of partial cleavage most commonly used are shown in Table 5.1. Each is a two-stage process. First, a small fraction of the bases are chemically modified (or removed in the case of G- and A-specific reactions), then the cleavage 3′- to these bases is achieved by alkaline hydrolysis (Sections 7.1, 7.4, and 7.5.3).

Table 5.1. Base-selective cleavages for sequencing DNA

3'-Cleavage adjacent to	Modification	Reagent	Strand breakage
G	Methylation	Dimethyl sulphate	1 M piperidine 90° 30 min
G and A	Depurination	88% Formic acid	1 M piperidine 90° 30 min
T and C	Base ring opening	Hydrazine	1 M piperidine 90° 30 min
C	Base ring opening	Hydrazine, high salt	1 M piperidine 90° 30 min

The **Sanger DNA sequencing** method shares the features of base-specific discontinuities that are resolved by gel electrophoresis, but employs an alternative procedure to generate the ladder. A new strand of DNA is synthesized enzymatically (usually using either the Klenow fragment of DNA polymerase 1 (Section 9.2.4) or the DNA polymerase from bacteriophage T7) with the DNA to be sequenced acting as a template. Radioactive label is incorporated into the growing strand. In the example already considered, a new DNA strand is synthesized on the single-stranded complementary template (Fig. 5.7). Termination of this reaction 3'- to dC residues generates the same three fragments as shown in Fig. 5.3 and termination following each of the four bases in turn gives the same pattern as seen in Fig. 5.6.

5' A G A A T T C T A C A G T A A A T G C T 3'   Newly synthesized strand

3' T C T T A A G A T G T C A T T T A C G A 5'

Fig. 5.7 Dideoxy DNA sequencing involving the synthesis of a new DNA strand.

How is the polymerization terminated at a specific point? 2',3'-Dideoxynucleoside 5'-triphosphates (Fig. 5.8 and Section 7.2) can be incorporated into a growing DNA strand, but, since they possess no 3'-hydroxyl group, they are unable to accept the addition of any extra bases. They are thus **chain terminators**. The addition of a small amount of one of these, together with all four of the normal 2'-deoxynucleoside 5'-triphosphates (one of which is radiolabelled), to a polymerization reaction gives rise to a series of oligonucleotides, each terminated by a dideoxynucleotide. By carrying out four reactions, each with a different dideoxynucleoside triphosphate, and electrophoresing all four in parallel, a sequencing ladder is generated.

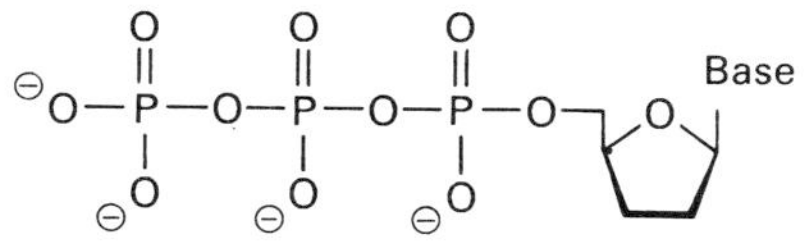

Fig. 5.8 A 2',3'-dideoxynucleoside 5'-triphosphate.

In practice, it is necessary to elongate a short primer which has already been annealed to the template, since DNA polymerases can only elongate existing hybrids (Fig 5.9). For this purpose one usually subclones the DNA fragment to be sequenced into a vector (Section 10.2) with a known sequence flanking the insertion site. Chemically synthesized oligonucleotides (typically 17–25 nucleotides in length) which correspond to one or the other side of the insert are annealed to the subclone of DNA and the dideoxy-sequencing reactions are carried out on these templates. The polymerization reaction can proceed on double-stranded templates, with one strand being displaced by the elongated primer. More usually, single-stranded templates are used, such as the viral DNA from bacteriophage M13-derived recombinants (Section 5.3).

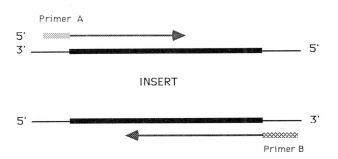

Fig. 5.9 Use of oligonucleotide primers in dideoxy sequence analysis.

Sanger dideoxy-sequencing has several powerful advantages over the Maxam–Gilbert method. First, a set of reactions can be carried out in only an hour or two as opposed to approximately a day for the Maxam–Gilbert reactions. Secondly, the reactions are much cleaner, with few contaminants that might lead to poor quality gel electrophoretograms. This means that 300 nucleotides can be sequenced routinely by this approach for each set of reactions, as opposed to approximately half this number with the Maxam–Gilbert method. Maxam–Gilbert sequencing does have the advantage that one need only purify a restriction fragment in order to sequence it whereas dideoxy-sequencing requires a sub-cloning step into a vector containing the appropriate oligonucleotide primer site. There are also several artefacts generated during both methods that render unambiguous determination difficult. For this reason it is essential to determine the sequences of both strands of a duplex and it is often advantageous to adopt the alternative method when an artefact renders difficult the interpretation of a particularly tricky sequence.

Machines have now been developed to separate and identify the products of dideoxy-sequencing reactions. Here, instead of radiolabelling, a fluorescent molecule is attached to the 5′-end of the primer or is built into an alternative chain terminator (Section 7.5.4). Fluorescently labelled oligonucleotides travelling through a polyacrylamide gel are detected by a laser. Such machines are proving valuable in attempts at large scale sequencing of the human genome.

Summary

There are two major ways of determining the sequence of a DNA molecule. Both rely on the exact measurement of the size of fragments with one end fixed and the other being at one of the four nucleotide bases. Maxam–Gilbert sequence analysis uses base-selective partial chemical cleavage. Sanger dideoxy-sequence analysis relies upon the use of 2',3'-dideoxynucleoside 5'-triphosphates as base-specific chain terminators of the synthesis of a complementary DNA copy by a DNA polymerase.

5.3 Gene structure in prokaryotes and phage

5.3.1 The complexity of genetic information in prokaryotes and phage

DNA molecules are the largest of the macromolecules found in a cell. The *E. coli* **chromosome**, for example, consists of a single molecule of double helical DNA about 4 million base-pairs in length. A more useful and common way used to describe this length is to say that the molecule is about 4000 kilobase (kb) pairs long. The mass of such a molecule is about 2.6×10^6 kDa and is thus about 10^5 times the mass of an average sized polypeptide. These sizes have been arrived at in experiments first done by Cairns where DNA was specifically labelled by growing bacteria in media containing [^{3}H]-thymine. Cells were lysed gently so as not to shear DNA and the chromosomes trapped in photographic emulsion. Under the microscope, tracks of silver grains mark the positions of DNA double helices whose contour lengths can be measured. In the case of *E. coli*, the contour length of the chromosome was 1.4mm and since there is 0.34 μm between successive base pairs, the chromosome was estimated to be about 4100 kb pairs long. A similar size has been deduced recently by the unrelated method of summing the sizes of independent chromosome fragments cloned in cosmids. The *E. coli* chromosome seems to be fairly typical of the length of DNA molecule to be found in other species of bacteria, but it is a tiny molecule compared with those found in the chromosomes of eukaryotes where molecules may be at least 20 to 100 times longer than bacterial DNA molecules (Section 5.5). In phage, DNA molecules are considerably smaller than in bacterial chromosomes. Table 5.2 shows some comparative data.

Before proceeding to consider in detail the nature of DNA in chromosomes, let us stop to consider the implications of DNA size for the variety of DNA sequences possible. At first sight, with only four types of base present, it might seem that there would not be much variety in DNA molecules. However, there are no known restrictions to the order in which bases occur on a polynucleotide chain. Thus, there are 4^2 possible dinucleotides, 4^3 possible trinucleotides and $4^{4\,000\,000}$ possible bacterial chromosomes. As we shall see, it is the sequence of nucleotides in DNA which determines the proteins and enzymes made in a cell and which thus determines the form and behaviour of those cells. Of course, not all of the possible

Table 5.2. The size of some genomes

Source of DNA	Size (kb)	Type
E. coli chromosome	3900	Covalently closed circular DNA
Bacillus subtilis chromosome	2000	Covalently closed circular DNA
F plasmid	95.0	Covalently closed circular DNA
pcolE1	6.4	Covalently closed circular DNA
λ phage	48.5	Double-stranded linear DNA
T7 phage	40.0	Double-stranded linear DNA
φX174 phage	5.4	Single-stranded circular DNA
M13 phage	6.4	Single-stranded circular DNA
MS2 phage	3.6	Single-stranded linear RNA

sequences will give rise to working proteins let alone to viable bacteria, but nevertheless the scope for genetic variation is enormous. Despite the undoubted variety of life on earth, it is tempting to view evolution as having only scratched the surface of the possibilities. What is abundantly clear from DNA sequencing studies is that a great many sequences have been conserved, sometimes strongly so, during the course of evolution over millions of years and we will frequently point out similarities or 'homologies' between sequences, usually with the implication that the homologous sequences serve similar or identical functions.

An unexpected fact to emerge from the experiments of Cairns described above was that the DNA molecules in bacterial chromosomes consist of continuous or 'circular' threads and the same was true of **plasmids**, small extrachromosomal DNA molecules carried by some bacteria. These structures have no free 5'- or 3'-ends and are covalently closed circles of double-stranded DNA, sometimes referred to as cccDNA. **Bacteriophage** (viruses that infect bacteria) show greater variety in their chromosomal structures. In some phage, the chromosome consists of double-stranded linear DNA molecules. In others, such as λ phage, the DNA is linear and largely double-stranded with short single-stranded and self-complementary ends (or **sticky ends**). Other phage such as M13 and φX174 have single-stranded circular DNA as their chromosome and there are some phage where the chromosome consists of linear RNA either single-stranded or double-stranded (Table 5.2). Apparently, single-stranded linear DNA chromosomes and double-stranded circular RNA chromosomes do not exist.

The combined but opposing actions of DNA gyrase and topoisomerase (Section 2.3.4) leave small circular molecules of DNA with negative supercoils, and larger circular molecules would seem to have domains lying between anchorage sequences which have more, sometimes less, negative supercoiling than the average for the whole chromosome.

Finally, supercoiled DNA must be folded backwards and forwards many times in bacteria in order to pack 1.4 mm of DNA into a volume equivalent to a cube of side about 0.25 μm. It has been estimated that adjacent double helices would be about

40 Å apart following this packing and it is difficult to see how molecules like RNA polymerase of about 100 Å diameter can rapidly penetrate the **nucleoid**, as this DNA knot is known in bacteria (Section 2.6).

5.3.2 The functional subdivision of DNA sequences

The astonishing success of early man in breeding plants and domesticated animals indicates a strong belief in the idea that living forms contained heritable information. The idea came more into focus from the mid-nineteenth century onwards and the word **gene** was coined to give some substance to the hypothetical units of heredity. With the discovery that DNA causes bacterial transformation (Section 1.1) the way became open to see that genes were made up of DNA and that the linear arrays of genes postulated by geneticists were in molecular terms linear arrays of DNA sequences. At the same time, geneticists such as Beadle and Ephrussi in the 1930s were concluding that the fundamental role of the gene was to produce enzymes and that deliberate or accidental alteration, or **mutation**, of a gene caused the production of an altered enzyme. Thus was born the '**one gene: one enzyme' hypothesis**.

Finally, workers such as Crick, Brenner, and Yanofsky established that linear sequences of nucleotides in DNA produced, through RNA working copies, primary amino acid sequences in proteins—the '**one gene: one primary sequence' hypothesis**. They were able to show that special DNA sequences signalled the first, N-terminal amino acid in a polypeptide (i.e. the beginning of a gene) and that other sequences signalled the last, C-terminal amino acid (i.e. the end of the gene).

In general, genes are packaged closely along DNA molecules with only a few nucleotides separating the end of one gene from the beginning of the next. In phage T7, for example, the DNA contains 50 genes and more than 92 per cent of the 39 936 nucleotide pairs specify or code for proteins. Overlapping of genes is rare but small overlaps are nevertheless known to occur, particularly in phage **genomes**, as arrays of covalently connected genes are often known. Space for packaging DNA inside phage capsids, the protective protein coat of a virus, is at a premium and the occasional overlap of phage genes may simply reflect the need to reduce the number of nucleotides between genes to a minimum.

The genome of *E. coli* has been more extensively mapped than any other cellular genome and some 1000 genes have been located on it to date. However, while in many parts of the genome the genes are as closely packed as in phage genomes, there exist a few regions, sometimes as long as 200 kb pairs, where no genes have been found. It is possible that the gap is an accident of the types of gene which have so far been investigated, but there is a strong feeling that the DNA in these gaps serves some function other than to code for proteins. The question of apparently redundant DNA is one which will be taken up again when we examine eukaryotic chromosomes.

In keeping with the finding that *E. coli* chromosomal DNA is circular is the finding that the genes also form a circular unbranched array. The location of a few of the known genes is shown (Fig. 5.10). The units of distance on this map are minutes (time), however illogical this may seem to a physical scientist! The unit arises from the experimental way in which the genetic mapping was originally carried out. *E. coli* bacteria, variously known as **Hfr** or **donor** or **male** cells, transfer their chromosomes from a fixed point (0 min), in a regular 'clockwise' direction to **female** or **recipient** or **F⁻** cells. Transfer proceeds fairly uniformly (Section 5.4.6) and takes about 100 minutes to complete. A gene such as the one called *trp* in Fig 5.10 takes 27 minutes to appear first in a recipient and is said therefore to be at 27 min on the chromosome. One minute on the map is equivalent to about 40 kbp. Some day, mapping will have become sufficiently detailed to be able to use kbp as the map unit as is already the case for the maps of many phage and viruses.

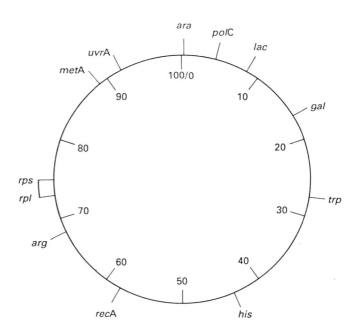

Fig. 5.10   The position of some genes on the *E. coli* chromosome.

Nomenclature in bacteria involves use of three, lower case, italic letters to designate a gene. Frequently the letters give a clue to the gene's function. For example, *trp* indicates a gene coding for an enzyme involved in tryptophan biosynthesis, *lac* indicates a gene involved in lactose utilization, and *uvr* indicates a gene whose product is involved in conferring resistance of the cell to UV light.

Where more than one gene has been found which affects a given cell function, the genes are given a further capital letter to distinguish them. Thus, *trpA*, *trpB*, *trpC*, *trpD*, and *trpE* genes have been identified each of which codes for the production of a separate polypeptide or a separate enzyme needed for different stages in the biosynthesis of tryptophan. Tables listing the known bacterial genes, their function and their map position are available and are frequently updated.

5.3.3 Clustering of functionally related genes

A consistent finding of genetic mapping is that genes coding for different enzymes involved in one metabolic pathway are frequently clustered or even adjacent on the chromosome. The way in which the five *trp* genes mentioned above are organized in *E. coli* and the functions they serve in tryptophan biosynthesis is shown (Fig. 5.11).

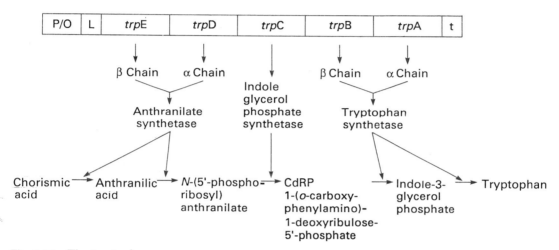

Fig. 5.11 The tryptophan operon.

In phage also there is a frequent clustering of genes which have a related function. In phage λ (Fig. 5.12) there is a group of about 20 genes involved in capsid (or

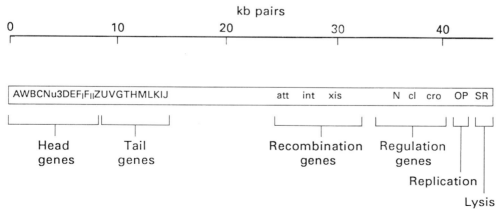

Fig. 5.12 Clustering of gene functions in the phage λ genome.

phage coat) synthesis subdivided into a group involved in forming the head of the capsid and another group in forming the tail. Some phage genes, unlike bacterial genes, are simply denoted by a single capital letter. Further along the λ chromosome is a group of genes involved in phage DNA recombination, in regulation of phage gene expression, and in phage DNA replication. Finally there are two genes, S and R, coding for proteins which lyse open the phage-infected host to release the newly formed infective particles. However, not all genes of related function are clustered. Instead, such genes may be found scattered throughout the genome.

5.3.4 Initiation of RNA synthesis

In the first step in expression of a gene as a polypeptide chain, an mRNA working-copy of the gene has to be made by RNA polymerase. The enzyme binds to DNA and progresses along it, transcribing the gene in a continuous process to produce mRNA, and finally dissociates from the DNA. The process is thought of in terms of 'flow' such that '**upstream**' refers to DNA sequences before the start of the gene and '**downstream**' refers to those after the end of the gene. **Initiation of transcription** is a very precise process and occurs only at specialized DNA sequences known as **promoters** and situated upstream of genes. The function of a promoter sequence is to be recognized by RNA polymerase and is thus different from the function of a gene sequence, which is to be transcribed and in the end to be translated into an amino acid sequence as a polypeptide.

Promoter sequences have been characterized by the technique known as **foot-printing**, illustrated in Fig. 5.13. Purified RNA polymerase is first bound *in vitro* to promoter-containing DNA in which one of the ends has been labelled with ^{32}P. DNA with and without complexed RNA polymerase is next treated with very limiting amounts of an endonuclease, an enzyme which hydrolyses phosphodiester linkages within polynucleotide chains (Section 9.43), to give a **nick** in one of the DNA strands. Hydrolysis is nearly random and different individual DNA molecules will suffer nicking at different distances from the labelled end. The population of molecules as a whole will contain representatives which have been nicked at position-1, position-2, position-3 and so on from the labelled end. Next the DNA is denatured and the single-stranded fragments separated by electrophoresis on gels to produce a ladder of radioactive bands in which adjacent bands contain molecules differing by one nucleotide in length. With free DNA, the ladder will consist of the complete set of bands from one nucleotide in length up to a band of complete length DNA. However, with DNA complexed to RNA polymerase, those phosphodiester linkages covered by RNA polymerase are protected from hydrolysis by the endonuclease, and hence bands corresponding to the protected region do not appear on the ladder. The missing bands are the '**footprint**' left by binding of RNA polymerase. The technique is of general application to other DNA-binding proteins.

Comparison of the footprint (Fig. 5.13 right) and total DNA sequence allows determination of the promoter sequence. Alternatively, complexed and uncom-

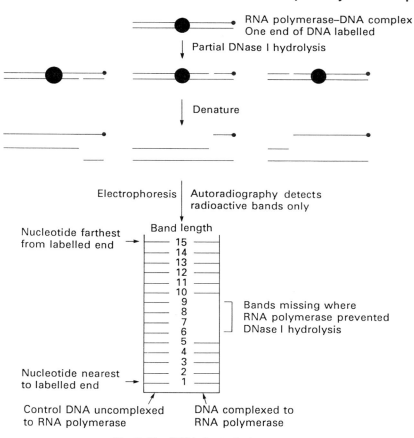

Fig. 5.13 DNA footprinting.

plexed mixtures can be treated to generate four sequencing ladders (Section 5.2) so that the sequence of the promoter footprint can be read off directly.

RNA polymerase footprints for *E. coli* are about 60 nucleotides long and comparison of a large number of different promoter sequences shows that, while great variety is possible, a number of features are common to all promoters. Secondly, promoter mutations have been isolated which either increase or decrease RNA polymerase binding and hence modify transcription from the promoter. Sequencing of these promoter mutations has identified base-pairs which are important for promoter function. The following picture emerges (Fig. 5.14). RNA synthesis begins from a precise nucleotide pair on the DNA, the **transcriptional start**, and is numbered +1 on the sequence. Upstream DNA sequences have negative numbers whereas the gene itself is positively numbered. Most promoters have a sequence TATAAT, sometimes called the **Pribnow box** or the −10 region because it is centred at approximately position −10 on the DNA sequence as numbered above. In different genes, the position of the Pribnow box can vary from the average

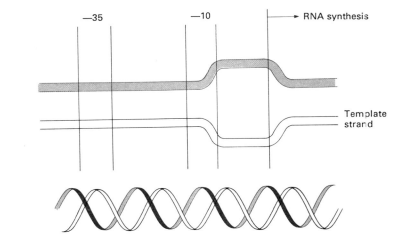

Fig. 5.14 Structure of prokaryotic promoters.

position by several base-pairs. Also the TATAAT sequence is not absolutely conserved from promoter to promoter and one or two changes from this so called **consensus sequence** occur in any given promoter. Conservation of the base at each position varies from 45 per cent to nearly 100 per cent and a summary of the consensus would be: T(80)A(95)T(45)A(60)A(50)T(96), where the numbers in brackets indicate the percentage occurrence of the most frequently found base.

Some 16 to 19 nucleotides further upstream and centred approximately at base-pair −35 is another region where sequence similarities occur for different promoters. The consensus sequence can be represented as: T(32)T(84)G(78)A(65)C(54)A(45) for this **−35 region**. The actual intervening sequence does not seem to be important but the distance separating the −10 and −35 regions does seem to be critical.

Treatment of promoter-RNA polymerase complexes with chemical reagents which react with specific bases have allowed more precise data to be collected on the sites of protein–DNA interaction. This work has shown that RNA polymerase makes contact with one face of the DNA covering two turns of the double helix from about the −35 region to the −10 region with the −35 region serving as a recognition site for the protein. Two turns along the DNA, the −10 region marks the start of a section of 12 to 14 nucleotides from about −9 to +3 positions where strand separation has been shown to occur by treatment of promoter-polymerase complexes with reagents which specifically attack unpaired bases (Section 2.3.3). This is a region which is usually fairly A:T-rich and since A:T pairs require less energy to separate them than G:C pairs, the region will undergo strand separation with a minimum of input energy.

Another factor which may influence the ease with which this strand separation occurs is the degree of DNA supercoiling since less free energy is needed for the melting in highly supercoiled DNA and, as we will see later (Section 5.3.13), some promoters are sensitive to supercoiling both *in vitro* and *in vivo* (Section 2.3.4).

Finally, RNA synthesis is initiated at position +1 within the melted region and synthesis of the RNA takes place from its 5'-end to its 3'-end. Usually the first nucleotide incorporated is ATP or GTP from a dT or dC residue respectively on the **template strand** of the DNA. The other strand of DNA is called the **coding strand** since the sequence is in the same 5'→3' orientation to the resultant RNA and is identical in base-sequence except T in the DNA is replaced by U in the RNA. The 3'→5' template strand is hence sometimes called the **non-coding strand**.

5.3.5 Termination of RNA synthesis

Once RNA polymerase has initiated mRNA synthesis at a promoter, it moves along the DNA of the gene transcribing the sequence information of the template strand into a complementary RNA until it encounters special termination signals. At these signals, RNA synthesis stops and both enzyme and RNA dissociate from the double-stranded DNA template.

Most prokaryotic termination sites show **palindromic sequences** in the DNA just prior to the termination point and the RNA transcribed has short inverted repeats separated by a small number of nucleotides. Hydrogen-bonding between these inverted repeats in the RNA generates **hairpins** with a stem and loop arrangement as shown in Fig. 5.15. Frequently, the palindromic regions are followed by A:T-rich DNA in such a way that the RNA transcribed has a run of residues, usually about U_6 hydrogen-bonded to dA residues in the template DNA strand. It is thought that these hairpin arrangements in the RNA product cause RNA polymerase to stall in its progress along the DNA template and that because the (oligo-rU).(oligo-dA) bonding is very weak, RNA breaks free from its DNA template.

The hairpins of terminators have G:C-rich regions which help to stabilize the structure, and mutations have been found which either reduce the efficiency of termination by disrupting base-pairing in the hairpin or increase the efficiency of termination by increasing the stability of the hairpin. Likewise, mutations which reduce or abolish the stretch of U residues in the RNA also reduce or abolish termination.

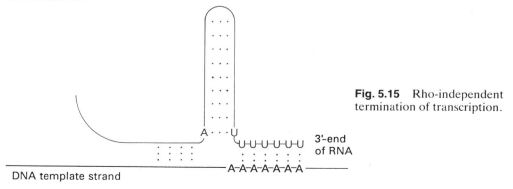

Fig. 5.15 Rho-independent termination of transcription.

DNA template strand

In addition to these simple terminators, or **rho-independent terminators,** are the **rho-dependent terminators.** These terminators also allow the formation of hairpin RNA but the hairpin is not specially G:C-rich and it is not followed by a succession of U residues. Here, termination is thought to be mediated by a special protein called **rho**. One proposal for its mechanism of action is that it binds to the 5'-end of the nascent RNA and travels along behind RNA polymerase (Fig. 5.16). When the polymerase is stalled by the hairpin structure, rho has a chance to catch up. The details of how rho acts to cause termination are not clear but it probably involves the interaction of rho with the β-subunit of RNA polymerase. Possibly the RNA polymerase β-chain is altered in shape by binding of rho such that the enzyme loses its DNA-binding properties. Alternatively, rho may act as a helicase and unwind the RNA from the DNA–RNA complex.

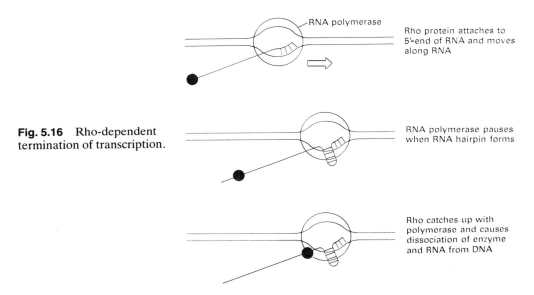

Fig. 5.16 Rho-dependent termination of transcription.

Thus, not only do DNA sequences have to provide information as structural genes, for the synthesis of proteins, they also have to provide the signals for initiation and termination of RNA transcription. We will see shortly that further DNA sequences are involved in regulating gene expression and in DNA replication.

5.3.6 Induction and repression of gene expression

The essential feature of living cells is that almost all their reaction processes are catalysed by enzymes. One way in which a cell can regulate the activities of its metabolic pathways is by alterations in the amounts of the enzymes it contains. Some enzymes are required under all conditions, for example those involved in generating energy, and the genes coding for these enzymes are said to be **constitu-**

tively expressed. Other enzymes are only required when a certain molecule appears in the environment of the cell. Since both protein and the mRNA needed for protein synthesis are very costly to the cell in terms of energy, the efficiency of the cell is vastly improved by synthesizing only those enzymes which are absolutely essential. Thus, some genes are only expressed, or **induced**, when a certain molecule, called the **inducer**, appears in the cell. A well studied case of this which we will look at in some detail is the induction of lactose-metabolizing enzymes when lactose is the only sugar available to provide *E. coli* with essential carbon and energy.

Some genes are normally expressed but are 'switched off' when a compound known as a **repressor** or **co-repressor** appears in the environment of the cell and represses the expression of one or more specific genes. For example, bacteria normally have to synthesize amino acids like tryptophan or histidine from simpler carbon and nitrogen sources through what is often a long series of enzyme-catalysed steps. The presence of preformed amino acids in the environment, however, renders all these enzymes unnecessary and the energy economy of the cell is greatly improved if synthesis of these enzymes is repressed.

Variations in levels of enzymes are also essential in the timing of cell development. This becomes most important in differentiation of eukaryotic cells, but is also found in prokaryotes in periodic processes such as cell division or in the co-ordinated development of phage in infected bacteria. Here, gene expression has to be regulated in a time-related fashion and we find genes divided into **early**, **middle**, or **late genes** according to when their expression is needed in phage development.

In theory, regulation of gene expression could occur at the level of translation of mRNA into enzyme or at the level of transcription of DNA into mRNA, or both. In practice, transcriptional regulation is the most important one in prokaryotes and regulation of promoter activity is the most common way of affecting transcription. However, regulation of transcription at termination sites is also found, particularly in cases of developmentally regulated genes.

Since changes in cell metabolism frequently involve the induction or repression of a whole group of enzymes involved in a pathway rather than induction or repression of one single enzyme, bacteria have evolved ways in which there is co-ordinate induction or repression of all the necessary enzymes. The most important way that bacteria achieve this is to group genes of related function together on the chromosome. When transcription of the group is under the regulation of one single promoter upstream of the first gene in the group and one terminator downstream of the last gene in the group, such a group of genes is called an **operon**. Operon size can vary considerably: operons of two or three genes are common and the operon for histidine biosynthesis in *Salmonella typhimurium* is 11 genes long, while in phage λ there is an operon at least 22 genes in length. Although the mRNA produced from an operon is a single transcript of all the genes which lie between the promoter and the terminator, synthesis of separate polypeptides from this polygenic or **polycistronic messenger** is achieved because of the occurrence within the long mRNA of DNA-coded translational 'start' and 'stop' signals for the ribosome.

The important point to note here is that any process which doubles the frequency with which RNA polymerase initiates at a given promoter will thus double the amount of mRNA for all the genes within the operon and hence will double the amounts of all the proteins coded by all these genes. Therefore we say that there is co-ordinate induction or repression of the genes within the operon.

We must now examine one by one the several mechanisms which regulate gene and operon transcription in bacteria. We must also bear in mind that most transcriptional regulation relies on combinations of these mechanisms.

5.3.7 Negative repression of transcription

The first case we are going to look at is one where initiation of transcription is inhibited when a special protein called a **repressor protein** binds at the promoter and prevents RNA polymerase from binding or prevents it from moving along DNA to transcribe the structural gene. The protein is itself coded by a gene called the **regulator gene**.

The sites where repressor proteins bind to promoters are called **operators** and are designated, for example, as *lacO* and *trpO* in the lactose and tryptophan operons. Operator sequences are about 30 to 40 nucleotide pairs in length and usually overlap with part of the RNA polymerase binding site particularly with the −10 sequences (Fig. 5.17). Their sequences show imperfect dyad symmetry which probably reflects the fact that repressor proteins are symmetrical dimeric or tetrameric proteins.

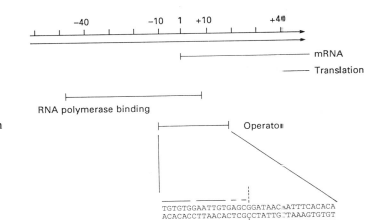

Fig. 5.17 The lactose operon promoter/operator region.

While bound to operator sequences, repressor protein prevents expression of the gene (Fig. 5.18a). However, combination of an inducer molecule with the repressor protein produces a complex which has lost its ability to bind to the operator.

Thus, the promoter is free to bind productively to RNA polymerase and gene expression is induced (Fig. 5.18b).

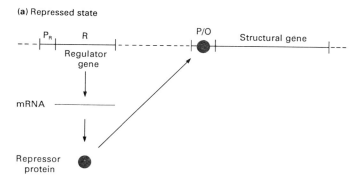

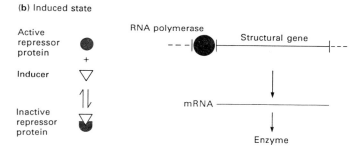

Fig. 5.18 Regulation of gene expression by negative repressors.

5.3.8 Positive regulation of transcription

In the type of regulation just described, the regulator protein acts negatively to inhibit transcription, but in other cases the binding of the regulator protein to the promoter is an essential requirement before transcription can begin. In these positively regulated operons, RNA polymerase cannot bind productively to the

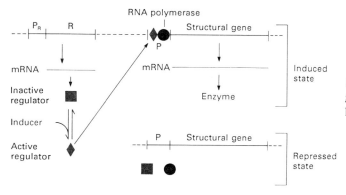

Fig. 5.19 Regulation of gene expression by positive regulators.

promoter unless the positive regulator protein is also bound there. As before, the ability of the positive regulator protein to bind to promoter DNA is affected by the presence or absence of inducer molecules. We illustrate (Fig. 5.19) the situation where the inducer is required to bind to the regulator protein before the latter can bind to the promoter DNA.

5.3.9 Regulation of the lactose operon

The lactose operon consists of three genes (Fig. 5.20). The *lacZ* gene codes for a β-galactosidase which can hydrolyse lactose to glucose and galactose, a *lacY* gene which codes for a lactose permease which transports the sugar across the cell membrane, and a *lacA* gene whose function is obscure, but which seems to cause acetylation and detoxification of noxious β-galactosides. The operon is induced by lactose, but only if there is no glucose available to the bacteria for metabolism. Lack of glucose raises intracellular cyclic-AMP levels and this converts a positive regulator protein called **CRP (cAMP regulator protein)** into a form which binds to the *lac* operon promoter, *lacP*. RNA polymerase is then able to bind to the promoter, but only provided the operator *lacO* is not occupied by *lac* repressor. This condition will be met if lactose is present. Lactose repressor protein is synthesized by the *lacI* gene which happens to be close to, but not part of, the *lacYZA* operon. Note that repressor proteins are freely diffusable within the cell and the genes coding for them need not be close to the promoters to which they bind.

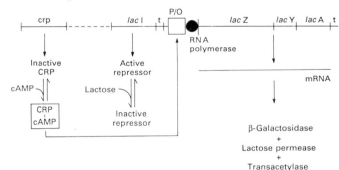

Fig. 5.20 Regulation of *lac* operon expression.

5.3.10 Attenuation

The expression of a number of operons involved in the biosynthesis of amino acids can be subject to aborted mRNA synthesis called **attenuation**. The *trp* operon of *E. coli* is the best characterized example of this process.

The operon (Figs 5.11 and 5.21) consists of five genes and is subject to negative repression as well as to attenuation. When the cell has plenty of tryptophan, the repressor protein product of the *trpR* gene is converted into an active form which largely prevents transcription from the *trp* promoter upstream of the *trpE* gene. Tryptophan starvation inactivates the TrpR protein and stimulates gene expression some 70 fold.

However, once transcription has started, its completion to the terminator down-stream of *trpA* gene is dependent on the severity of the tryptophan starvation and this is sensed when ribosomes try to translate the 5′-end of the newly forming *trp* mRNA. Upstream of the first true gene, *trpE*, lies some DNA called *trpL* (**leader peptide**), which codes for RNA with unusual properties. First, the RNA codes for a 13 amino acid peptide (the leader peptide) containing two successive tryptophan residues. Tryptophan has a fairly low frequency of occurrence in proteins so that it is very unusual for two tryptophans to occur side-by-side. As a result ribosomes attempting to insert these two tryptophan residues into the leader peptide stall when this amino is in short supply, and this has an effect on the second unusual feature of *trpL* mRNA, a change in its secondary structure. The RNA sequence is such that it can adopt alternative secondary structures as shown (Fig. 5.21a). When tryptophan is plentiful, and the ribosome is able to complete synthesis of the leader peptide (Fig. 5.21b), the RNA forms a stem-loop structure (labelled **3** and **4** in Fig. 5.21a) which is rich in G:C pairs and is followed by a run of 8U residues. This is a typical rho-independent terminator (Section 5.3.5) and mRNA from the rest of the operon is not synthesized. However, when tryptophan is scarce (Fig. 5.21c), the ribosome stalls at the position on *trpL* translation where it is expected to put in the two tryptophan residues. This leaves the RNA sequence (labelled **2** and **3** in Fig. 5.21c) to form an alternative hydrogen-bonded stem-loop. Involvement of sequence **3** in pairing with sequence **2** precludes it from taking part in the formation of the termination stem-loop. RNA polymerase thus continues to transcribe the *trpEDCBA* genes, thereby providing sufficient mRNA for the synthesis of much needed stocks of tryptophan.

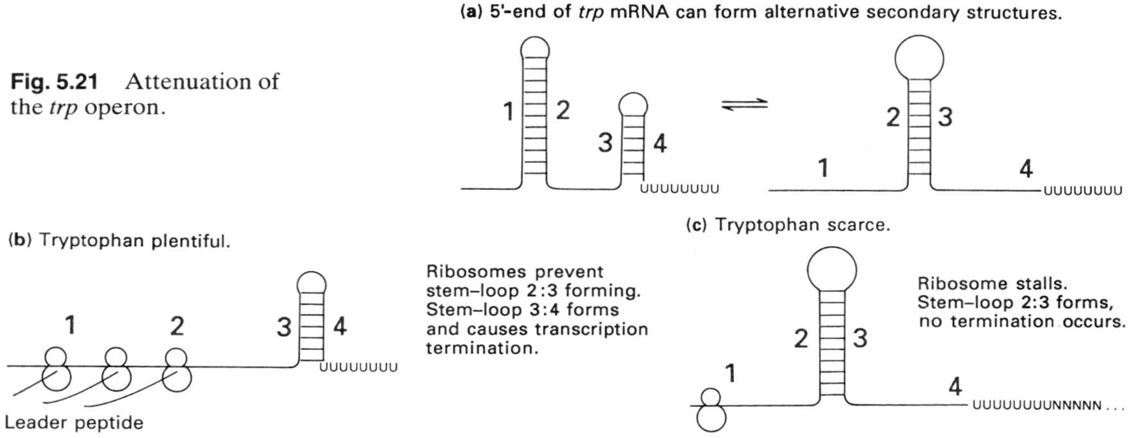

(a) 5′-end of *trp* mRNA can form alternative secondary structures.

Fig. 5.21 Attenuation of the *trp* operon.

(b) Tryptophan plentiful.

Ribosomes prevent stem–loop 2:3 forming. Stem–loop 3:4 forms and causes transcription termination.

Leader peptide

(c) Tryptophan scarce.

Ribosome stalls. Stem–loop 2:3 forms, no termination occurs.

Thus, tryptophan starvation increases *trp* operon expression by inactivating TrpR negative repressor protein which increases transcription about 70-fold. It also prevents normal attenuation, which increases transcription another 10-fold. Over-all, transcription of the operon can be increased some 700-fold. Similar strategies have also been found in the regulation of operons involved in the biosynthesis of histidine, leucine, threonine, and phenylalanine.

5.3.11 Anti-termination in λ phage development

In the previous section we have seen an example where amino acid availability regulates termination of operon transcription. Here, we are going to study a case where a specific protein regulates gene expression by preventing the normal termination of transcription. The case concerns the development of phage λ in *E. coli*

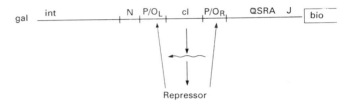

(a) λ Prophage is inserted between *gal* and *bio* operons of *E. col*
cI gene expression negatively represses all other λ gene expre.
(Genome sizes are not to scale.)

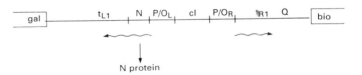

(b) Inactivation of λ repressor allows formation of N protein.

Fig. 5.22 Regulation of gene expression in phage λ.

(c) N protein allows transcription of excision and replication enzymes as well as Q anti-terminator. Late genes cannot be expressed because of termination at $t_{R'}$.

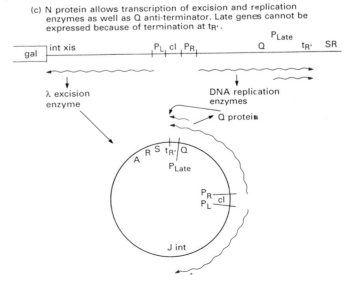

lysogens. These are cells which are carrying λ phage DNA in a quiescent form integrated into the bacterial chromosome. The integrated λ DNA does not seem to affect survival of the bacteria until they are exposed to agents like UV light or mutagens which damage DNA and induce the bacterial DNA repair systems. Then a series of enzymic changes is induced which result in the orderly production of infectious λ particles and the eventual lysis and death of the host bacterium. The details of this ordered development are complex. Here, we wish to concentrate on how **anti-terminator proteins** can regulate gene expression in a time-related manner.

In the normal lysogen (Fig. 5.22a), λ genes are prevented from being expressed by a negative repressor protein (product of the λcI gene) binding at the two promoters P_L (left promoter) and P_R (right promoter). DNA damage, however, causes inactivation of λcI repressor and transcription of λ lysogen DNA begins from the two promoters as shown (Fig. 5.22b). It stops again promptly at the two rho-dependent termination sites t_{L1} and t_{R1}, but not before the key gene N has been transcribed. N protein is an anti-terminator protein and, after a time delay in which concentrations of N protein build up, it binds to RNA polymerase and thereby prevents it from terminating. Thus after a time delay, transcription can proceed past the two terminators t_{L1} and t_{R1} and express genes such as *int* and *xis* involved in excising λ DNA out of the circular bacterial chromosome and in synthesizing multiple copies of the phage DNA.

However, transcription of an even later set of genes involved in making the phage coat is prevented by another terminator $t_{R'}$. Another anti-terminator, product of the λ Q gene, gradually builds up in concentration and can eventually allow RNA polymerase to transcribe past $t_{R'}$ and express the 'late' genes coding for phage structural proteins and for enzymes which lyse open the host bacterium. Thus λ phage development can occur in three sequential stages through the action of two anti-terminator proteins.

5.3.12 Regulation of gene expression by DNA rearrangement

The regulation of transcription of some genes is accomplished by the expedient of removal of its promoter to another part of the chromosome when expression is not required and replacement of it when gene expression has to be switched on. One case of this is the synthesis in *Salmonella* of flagellin, the protein subunit of which flagella are made. Which of the two flagellin genes is expressed is determined by the orientation of a particular DNA segment (Section 10.1.5). Similar types of DNA rearrangements have been found to regulate gene expression in phage P1 and phage Mu of *E. coli*.

5.3.13 Regulation of gene expression by changes in DNA topology

When we compare the frequency of transcriptional initiation at a promoter *in vitro* when it is part of a supercoiled, covalently closed circle of DNA and when it is part of a fully relaxed DNA molecule, we sometimes find that initiation occurs much more frequently from the supercoiled DNA. The same is sometimes found *in vivo* when a gene is taken out of its chromosomal context and cloned into a plasmid DNA molecule where its expression varies according to the degree of supercoiling of the plasmid. Presumably, the stress of supercoiling is relieved by separation of the DNA strands. Such an essential step of promoter function must considerably aid gene expression.

In bacteria, plasmid DNA is normally negatively supercoiled due to the opposing actions of DNA gyrase, which introduces negative supercoils, and topoisomerase which relaxes the DNA (Section 2.3.4). What happens in the chromosome is much more difficult to assess, but it is believed that here too gyrase and topoisomerase affect the extent of supercoiling. In the cases of the genes *proU* and *tonB*, clear evidence has been found that in their normal chromosomal locations their expression is sensitive to changes in supercoiling brought about by mutations in the genes coding for gyrase, or topoisomerase, or brought about by inhibitors of DNA gyrase. *ProU* codes for a betaine transport enzyme whose action protects cells from osmotic stress and *tonB* codes for an iron chelate transport protein.

Summary

Bacterial genomes consist of about 4000 kb pairs of covalently closed circular DNA, but phage have more varied genomes. DNA sequences are grouped into functional units of heredity, **the genes**, with structural genes coding for enzymes, structural proteins, and regulatory proteins. Genes are expressed by transcription into mRNA of working copies of DNA sequences. This involves binding of RNA polymerase at promoter sequences in DNA and dissociation from the DNA at terminator sequences. Genes with related function are frequently grouped as **operons**.

Regulation of gene expression in prokaryotes is principally by regulation of transcription, which may be altered by positive regulators, negative regulators, attenuation, DNA rearrangements, changes in DNA supercoiling, or by combinations of these processes. There is co-ordinate regulation of the genes which make up an operon.

5.4 DNA replication in prokaryotes and phage

5.4.1 Replicons

When a cell divides successfully it must already have created an exactly duplicated set of chromosomes so that both daughter cells carry a set of genes identical to the parental cell. Sometimes, in addition to their chromosome, bacteria possess one or more autonomous circular pieces of DNA called **plasmids** which also have to be replicated prior to cell division. Once again, both daughter cells inherit these plasmids. In addition, cells may be infected by phage DNA which replicates independently of other types of DNA, leading to the production of new infective phage particles, each containing DNA identical to the original infective DNA.

It is useful to be able to refer to each of these independently replicating pieces of DNA as a **replicon**. All replicons have an origin, or **ori** sequence, where replication begins and where replication is controlled by interaction with regulators. Once initiated, DNA duplication proceeds to the end of the replicon without the intervention of other regulators. Each regulator is specific for a specific *ori* sequence, so that different replicons can be replicated independently of one another within one cell.

If DNA molecules which lack *ori* sequences, or have *ori* sequences suited only to another species, find their way into a cell, perhaps by phage transduction or by laboratory transformation processes, they cannot be replicated unless they can be incorporated into a suitable replicon by recombination. Even if such molecules are not degraded by deoxyribonucleases, they will eventually be lost from the population of cells when that cell which carries the extra piece of DNA finally dies. Thus, when in genetic engineering we attempt to propagate the DNA of one species in cells of another species, we must arrange that the foreign DNA is provided with *ori* sequences suited to the cell type in which we wish it to be propagated.

We might imagine that in a linear or circular replicon, DNA synthesis could begin at any base-pair at random and that this could lead to the formation of a replication eye with the movement outwards of one or both **replication forks** until all the DNA has been duplicated. More than one replication eye might be formed and these, as they grow, might merge until duplication is complete (Fig. 5.23).

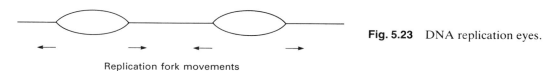

Fig. 5.23 DNA replication eyes.

Replication fork movements

However, in prokaryotes there is overwhelming evidence that: (i) there is only one origin per replicon; (ii) it is a unique sequence of base-pairs in DNA; and (iii) replication forks move outwards from that origin—most commonly bidirectionally or occasionally unidirectionally. For example, examination of a population of replicating plasmid DNA molecules in the electron microscope reveals a collection of theta structures, θ, with a whole range of replication eye lengths (Fig. 5.24). If these molecules are linearized at a unique site, as for example by treatment with a restriction endonuclease which can cut the DNA at only a single site, we find that the linear molecules can be put into a set such as that shown (Fig. 5.24). This analysis shows that all DNA molecules began replication at a site which was at a fixed distance from the unique end. It is also possible to tell whether one or both replication forks are moving with respect to that unique end.

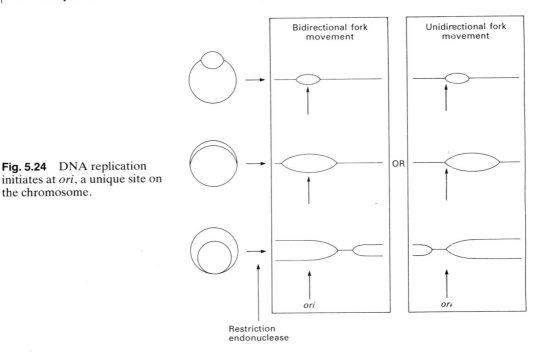

Fig. 5.24 DNA replication initiates at *ori*, a unique site on the chromosome.

Meselson and Stahl employed heavier ^{15}N-bases to achieve differential labelling of the parental DNA from newly synthesized DNA. They then used density centrifugation to examine the course of replication. After the first round of replication, they found that both old strands of a double-helix are conserved intact—one in each daughter duplex where it is paired with a new strand synthesized to complement the old one. After two rounds of replication, two of the four duplexes each contain one intact 'heavy' strand and two duplexes have only 'light' strands. This type of replication is known as **semi-conservative**.

5.4.2 DNA chain elongation

Bacteria such as *E. coli* and *S. typhimurium* have three distinct DNA polymerases, labelled polymerase I, II, and III, but it is DNA polymerase III which is responsible for chain elongation at the replication forks. Structurally, it is very complex and possesses three enzymic activities (a) 5'→3' chain elongation, (b) 3'→5' exonuclease activity, and (c) 5'→3' exonuclease activity (Section 9.2.4).

Chain elongation (Fig. 5.25) requires a template DNA to be provided. It also requires the four deoxynucleoside 5'-triphosphates and it requires a primer with a 3'-hydroxyl group on to which the polymerase can attach new nucleotide residues according to the hydrogen-bonding rules of double-stranded DNA.

It should be noted that the direction of elongation is only from 5'-to-3', that the product shows semi-conservative replication (one new strand is laid down against

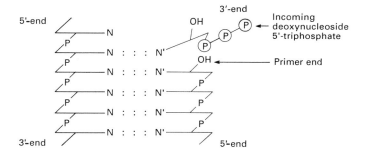

Fig. 5.25 $5' \rightarrow 3'$ Chain elongation by DNA polymerase.

an old strand), and that the sequence of the new strand is complementary to the sequence of the old strand. Thus, there is exact duplication both in amount and in sequence.

No enzyme has ever been found which can synthesize new DNA in the 3'-to-5' direction and so there is a problem in understanding how a single enzyme with only 5'-to-3' activity can replicate both of the anti-parallel strands of DNA simultaneously, as was so clearly shown in the experiments of Cairns (Section 5.3.1). The solution to the problem is that only one strand, the **leading strand**, is synthesized continuously in the 5'-to-3' direction (Fig. 5.26). The other, or **lagging strand**, is

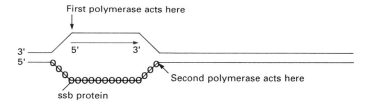

Fig. 5.26 Leading and lagging strand synthesis.

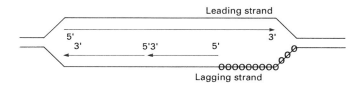

synthesized discontinuously. As elongation of the leading strand proceeds, more and more of the lagging strand template becomes single-stranded and short complementary strands are made by DNA polymerase. These small (1 kb to 2 kb) nascent DNA fragments are called **Okazaki fragments** after their discoverer.

Since there must always be a short section of single-stranded DNA on the lagging strand near the replication fork, and since single-stranded DNA is very susceptible to hydrolysis by nucleases, these sections are protected by **single-strand binding proteins** (product of the *ssb* gene). Ssb proteins are later displaced as DNA polymerase moves in to make the new complementary lagging strand.

5.4.3 The priming of DNA chain synthesis

Another problem with understanding the action of DNA polymerase is that the enzyme cannot begin DNA chain synthesis *de novo*. There is an absolute requirements for a $3'$-hydroxyl priming group paired to a template strand at which the polymerase can add new DNA (Fig. 5.25). In this respect it differs profoundly from RNA polymerase which can initiate new chains *de novo*, for example at promoters. Thus there is a need for a **primer** when leading strand synthesis is initiated at *ori* and also every time synthesis of an Okazaki fragment is initiated on the lagging strand.

The solution to the problem turns out to be that a short section of RNA is first laid down and that the $3'$-end of the RNA serves as the primer to which DNA polymerase covalently attaches the first deoxynucleotide. We will return to the priming event at *ori* later (Section 5.4.4) and consider here only the priming of Okazaki fragments. This is carried out by a form of RNA polymerase called DNA primase which is the product of the *dnaG* gene in *E. coli*. Like RNA polymerase, this enzyme begins RNA chain synthesis *de novo* and makes a primer 20 to 30 nucleotides long and complementary in sequence to the template strand. It differs from RNA polymerase, however, in accepting deoxy- as well as ribonucleotides as substrates, so that the primer is a mixture of DNA and RNA.

Once an Okazaki fragment has been primed by DNA primase, DNA polymerase moves along the lagging strand template laying down new DNA until it runs into the $5'$-end of the RNA which primed the previous Okazaki fragment (Fig. 5.27). Two things have to be done before lagging strand replication is complete: (a) RNA primer has to be removed and replaced by DNA and (b) individual Okazaki fragments have to be polymerized into a covalently continuous new strand.

The first task falls to DNA polymerase I. This is structurally a much simpler enzyme than DNA polymerase III, but nevertheless it has the same range of enzymic capabilities, namely a $5' \rightarrow 3'$ polymerase activity, a $3' \rightarrow 5'$ exonuclease 'proofreading' activity, and a $5' \rightarrow 3'$ exonuclease activity. DNA polymerase I is much more abundant than PolIII and is able to find and bind to nicks in DNA, such as the nick remaining between the $3'$-DNA end of one Okazaki fragment and the $5'$-RNA of the next one (Fig. 5.27a).

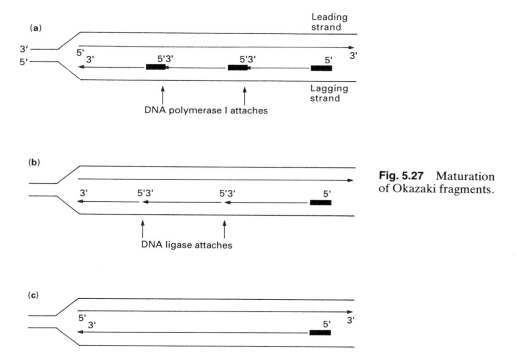

Fig. 5.27 Maturation of Okazaki fragments.

DNA polymerase I uses its $5' \to 3'$ exonuclease activity to remove the RNA primer and, as it moves along, fill in the gap with new DNA by means of its polymerase activity. Eventually, DNA PolI replaces all the RNA primer by DNA, but it cannot join up the 3'-hydroxyl of the inserted DNA to the 5'-end of the DNA laid down previously by DNA PolIII. This task is fulfilled by DNA ligase (Fig. 5.27c), the activity of which is described later (Section 9.5). Thus, Okazaki fragments are polymerized and new lagging strand DNA becomes covalently continuous just like the new leading strand DNA.

The sequence of events occurring in DNA replication clearly has even greater complexity since many more genes are known to be essential for replication than are accounted for in this relatively simple scheme. For example, there are persistent reports that cellular membranes may be involved in some as yet unknown way.

5.4.4 Chain initiation at origins of replication

The most thoroughly studied origin of DNA replication is that belonging to the *E. coli* plasmid pcolE1, described below. However, comparison of the sequence of other origins with that of pcolE1 suggests that similar events may occur at all origins (Fig. 5.28).

In pcolE1, replication begins with the formation of an RNA primer 555 nucleotides upstream from the origin (Fig. 5.28a). RNA polymerase, rather than DNA primase, is the enzyme responsible for synthesis of this primer. Transcription actually passes right through the origin with the formation of a DNA–RNA double-stranded structure over the origin. This serves as a substrate for RNase H which hydrolyses RNA when it forms part of a DNA–RNA hybrid and a 3'-hydroxyl priming group is created at *ori* ready for DNA replication to begin (Fig. 5.28b). Without this processing by RNase H, DNA synthesis cannot be initiated and regulation of the initiation, at least in pcolE1, seems to depend on regulation of the processing event.

The regulation is carried out in the expected way, not by means of a regulator protein but by means of a regulator RNA. A second species of RNA, called RNA I, is synthesized in the same region as the primer RNA but it is coded from the opposite DNA strand to that used for the primer RNA and it is only 108 nucleotides long. The 3'-hydroxyl group of RNA I lies close to the 5'-end of the primer RNA and the two RNA sequences are, of course, complementary to one another over the 108 nucleotides. They are therefore capable of forming a double-stranded RNA complex at the 5'-end of the primer RNA. This event is thought to affect the secondary structure of the rest of the primer, so that it can no longer serve as a substrate for RNase H.

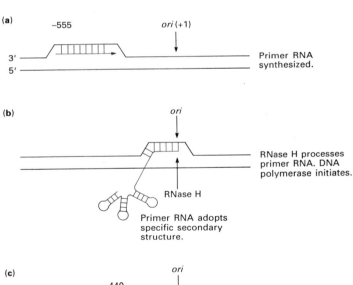

Fig. 5.28 pcolE1 origin of replication.

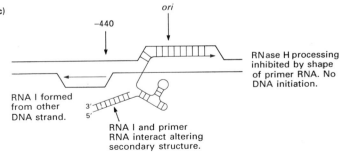

Thus in the presence of RNA I, the primer is not processed and initiation of DNA replication cannot occur (Fig. 5.28c). As the concentration of RNA I falls, for example as cell volume increases prior to cell devision, so there is less RNA I available to complex to the primer RNA and new rounds of plasmid DNA replication are initiated in order to build up plasmid numbers in preparation for cell division.

5.4.5 Accuracy of DNA replication

The accuracy of DNA replication relies on correct hydrogen-bonding interaction between an incoming base and a base on the template strand. When we consider that bases can exist (albeit fleetingly) in tautomeric forms other than the normal ones we see that, for example, instead of a normal dG:dC pair (Fig. 5.29a), a dT in the template strand could hydrogen-bond to the tautomeric form of dG (Fig. 5.29b). Insertion of a dG would be a replication error which would lead to mutation. Based on the expected lifetimes of the tautomeric forms of nucleotides, errors would be expected to occur once every 10^4 or 10^5 nucleotides replicated. In fact, measurements of spontaneous mutation rates show that they are four or five orders of magnitude lower than this expectation: one error occurs roughly every 1000 cycles of bacterial replication, i.e. once in the replication of 10^9 nucleotides.

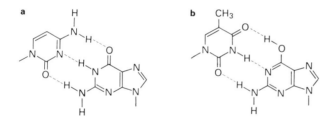

Fig. 5.29 Errors in DNA replication due to hydrogen-bonding using rare tautomeric forms of bases: (a) normal hydrogen-bonding of dG with dC, (b) minor tautomeric form of dG bonding with dT.

The reason for this discrepancy seems to be that DNA polymerase exercises a 'proof-reading' function. Having inserted a new nucleotide, the enzyme moves forward one nucleotide along the template and, before inserting the next new nucleotide, checks that the last base pair formed is correct. If an incorrect pair is found, this induces a change in shape of the enzyme such that its polymerizing activity is inactivated and its hitherto quiescent $3' \rightarrow 5'$ exonuclease activity is activated. This results in the hydrolytic removal of the offending nucleotide and a return of the enzyme to its original polymerizing form with a second attempt being made to insert the correct nucleotide. The probability that the base is in its rare tautomeric form both at the time of insertion and at the time of proof-reading is about 10^{-8} to 10^{-10} which is more in accord with observed mutation rates.

Evidence that the $3' \rightarrow 5'$ exonuclease activity is involved in proof-reading also comes from the isolation of **mutator** strains of bacteria which have abnormally high rates of mutation and prove to have reduced 3'→5'exonuclease activity in their isolated polymerase.

5.4.6 Rolling circle replication

We have seen above how circular replicons can form θ structures during DNA replication, but this mode is not appropriate to the duplication of linear replicons such as linear phage DNA molecules, nor is it appropriate where the duplication of a circular plasmid results in the copy being transferred in a linear fashion from one cell to another. This happens when the *E. coli* **F plasmid** is transferred from an F$^+$ cell to an F$^-$ cell via a sex pilus coded in the donor cell by F plasmid DNA. Rolling circle synthesis of plasmid DNA accompanies this transfer.

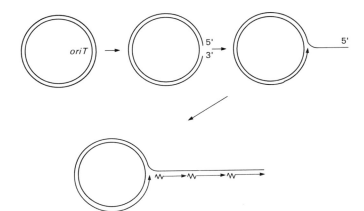

Fig. 5.30 Rolling circle replication during F plasmid transfer synthesis.

First of all, the 95 kb circular plasmid is nicked at a precise location, *oriT* (origin of transfer synthesis), by a specific endonuclease coded by the *traYZ* genes of F plasmid (Fig. 5.30). Next, the free 5'-end of single-stranded DNA passes down the narrow protein tube of the sex pilus driven by replacement of the lost strand by DNA synthesis in the donor cell. As the 5'-end appears in the F$^-$ recipient cell, a new second strand is synthesized in a series of Okazaki fragments. The template strand in the F$^+$ donor can be regarded as a revolving or rolling circle from which an old strand is peeled off and replaced with a new one. Once a circle has completed one revolution, the endonuclease cuts the old strand free and its 3'-end passes into the F$^-$ cell where it acquires a complementary strand and the linear molecule is circularized. Thus, the F$^-$ cell acquires an F plasmid (and becomes F$^+$) and the donor cell retains a copy of the plasmid.

This form of plasmid replication is to be distinguished from that which occurs when F plasmids are duplicated prior to cell division. Here, conventional θ replication takes place which is initiated at an origin quite distinct from the origin of transfer synthesis.

Sometimes F plasmid integrates into the *E. coli* chromosome and, when in this case DNA transfer is initiated from the *oriT* sequence, not only is the F DNA transferred but donor cell chromosomal DNA is also transferred to the F⁻ recipient (Fig. 5.31). The donor cell replaces its chromosome by rolling circle synthesis and the recipient cell acquires a copy of the genes of the donor cell. Donor cells behaving this way have been known as **Hfr** cells because they transfer genetic information with high frequency and they have been extremely important in genetic mapping of *E. coli* (Section 5.3.2).

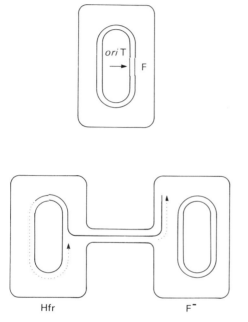

Fig. 5.31   Chromosome transfer during mating of Hfr bacteria with F⁻ bacteria.

5.4.7   Rolling circle DNA synthesis during λ replication

λ Phage DNA is linear with short complementary single-stranded ends which are self-cohesive and referred to as *cos* sequences. Within a host cell, this molecule quickly circularizes via its *cos* sequences and DNA ligase of the host cell makes it covalently closed. This circular λ DNA next undergoes a few rounds of θ replication from an *ori* sequence within gene O, but the major DNA replication occurs through rolling circle mechanisms. Each circle turns many times and many tandem repeats of λ sequence are produced, known as **concatameric DNA** (Fig. 5.32). So that this DNA can be packaged within the phage heads, an enzyme first makes staggered cuts in the two chains at the *cos* sequences. This recreates the single-stranded ends and then the DNA between the two *cos* sequences is packaged into the phage. The only demands made by the packaging system are that *cos* sequences must be present and that they are about 49 kb pairs apart on concatameric DNA, a feature

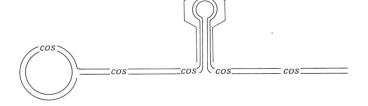

Fig. 5.32 DNA replication in phage λ.

which has been exploited in packaging foreign DNA into λ heads *in vitro* and for subsequent efficient entry into *E. coli*.

5.4.8 Replication of single-stranded phage DNA

DNA polymerases can only catalyse the synthesis of sequences complementary to the template they are given. Therefore, sequences identical to the template can be made only by synthesis of a complement of the complement. Thus, in order for copies of single-stranded circular DNA of phage to be made (Fig. 5.33), the infective phage strand (the + strand) must be used as a template for the synthesis of a complementary, (−) strand.

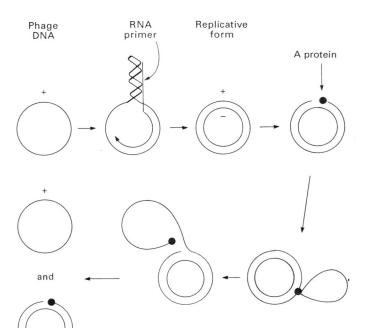

Fig. 5.33 Replication of single-stranded DNA involving (+−) RF DNA.

The first step is the synthesis of a double-stranded (+−) circular DNA molecule, referred to as the replicative form or **RF DNA**. (+−) RF DNA then undergoes many rounds of rolling circle replication to produce new (+) DNA for packaging into phage coats. In contrast to the type of rolling circle replication described above (Section 5.4.7 and Fig. 5.30), only single-stranded (+)DNA is formed here. As well as serving to make more (+) DNA, the RF DNA is also used for transcribing phage genes into the mRNA required for building the protein parts of the new phage, since only double-stranded DNA can be transcribed into mRNA by RNA polymerase.

Just as in the case of initiation of the synthesis of any new DNA strand, formation of the (−) strand on the infecting phage (+) strand requires priming by RNA to produce the double-stranded RF DNA. This occurs at a specialized region of the (+) DNA where the sequence allows for formation of a stem-loop hydrogen-bonded structure. RNA polymerase (in the case of M13) or DNA primase (in the case of φX174) binds to this region of DNA and an RNA primer is produced. The DNA polymerase of the host cell then catalyses the formation of the (−) strand, DNA polymerase I removes the primer, and replaces it with DNA. Finally, DNA ligase seals the nicks and DNA gyrase supercoils the RF DNA.

In the case of φX174, the subsequent rolling circle stage involves a phage-coded A protein that nicks the supercoiled RF DNA at a particular sequence and binds covalently to the 5′-phosphate end. The (+) single-strand then peels off while a new (+) strand is synthesized to replace it. After one turn of the circle, the A protein cuts off the (+) strand and joins the 5′-end of the DNA to its 3′-end, thus forming the first single-stranded (+) circle for packaging into new coat proteins. Repeated turns of the RF circle can peel off multiple copies of the single-stranded (+) DNA.

Summary

A **replicon** is an independently replicating unit of DNA. Replication begins from unique DNA sequences, *ori*, in prokaryotic replicons. Replication eyes proceed bidirectionally, unidirectionally or outwards, with the extension reaction being catalysed in the 5′→3′ direction by DNA polymerase III. High accuracy is achieved by the 3′→5′ exonuclease **proofreading** activity of DNA polymerases. Initiation of new DNA chains requires formation of **RNA primers** on to which DNA chains are added. The RNA primer is replaced with DNA by DNA polymerase I and covalent continuity of the DNA strands is effected by DNA ligase.

Circular DNA is most commonly replicated via θ **intermediates**, but **rolling circle** replication occurs during plasmid DNA transfer synthesis, during λ phage replication, and during the production of single-stranded phage DNA molecules from double-stranded circular replicative intermediates.

5.5 Gene structure in eukaryotes

Eukaryotic genomes are more complex than those of prokaryotes, since eukaryotes need to construct a larger, more complicated cell. There is often an additional requirement of building a complete organism composed of many different cell types all of which must have been derived from a single cell with a fixed amount of DNA sequence information. It is no surprise therefore to find that eukaryotic genomes are larger than prokaryotic ones. In general, this size increase follows the eukaryotic 'evolutionary tree' but there are some exceptions (Fig. 5.34).

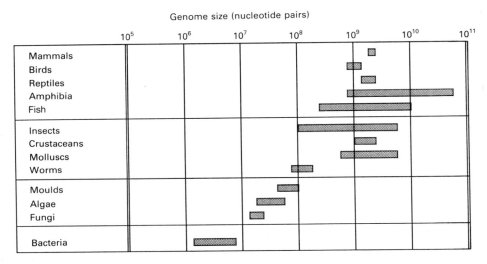

Fig. 5.34   The variation in genome size between different organisms.

The genome size for some types of organism, such as amphibia, is extremely variable with sizes varying over a hundred-fold range. It seems highly implausible that one type of amphibian needs much more genetic information than another. Furthermore, the genome size for *Drosophila melanogaster* (about 2×10^8 base-pairs) seems bigger than is required. If we assume an average size for a gene of four thousand nucleotide pairs, it would mean that *D.melanogaster* contains approximately 50 000 genes (although the calculations are rather approximate). This is roughly ten times the number which is estimated from saturation mutagenesis experiments on the fly. This problem of too much DNA in some organisms has been called the **C value paradox**. In the course of this section we will attempt to find a solution to this parodox by considering what types of DNA sequence are present in eukaryotes and what function they have, if any. It should be mentioned at the outset, however, that much of the DNA found in higher eukaryotes does not encode information which is expressed as protein or stable RNA.

5.5.1 Eukaryotic promoters

We will begin our survey with that proportion of DNA which is expressed as RNA and in particular with the DNA sequences which are implicated in the initiation of transcription. Eukaryotic genes are transcribed by one of three different RNA polymerases (excluding the mitochondrial genes). RNA polymerase I transcribes ribosomal RNA exclusively. Since this is the major RNA species in any eukaryotic cell, this polymerase is the predominant enzyme. RNA polymerase II transcribes those RNAs which will form the messenger RNAs encoding all cellular proteins (and some small non-protein coding RNAs). RNA polymerase III transcribes most small RNAs including tRNAs and 5S RNA. Each RNA polymerase is a multi-subunit complex and we do not know which subunits (if any) are shared by different polymerases.

Let us look at the promoters which are recognized by eukaryotic RNA polymerases to see if we can find consensus 'boxes' in the same way as we found the Pribnow and −35 boxes in prokaryotes (Section 5.3.4). In the case of RNA polymerase I promoters, no strong consensus emerges between the promoters of different species. Mammals share common sequences between −16 and −21 nucleotides and about −33 and −40 nucleotides upstream of the initiation site but these are not conserved in amphibia. However, these regions seem to be important to the mammalian genes, since deletion of them seriously impairs initiation. The only sequence shared between all higher eukaryotes studied so far is a G at 16 bases upstream of the initiation site. A point mutation of this base almost completely abolishes transcription of mouse rRNA.

RNA polymerase III is unique in that its promoter region almost always lies downstream of the transcription initiation site. Regions shown to be essential by deletion analysis of three genes are shown in Fig. 5.35. These internal regions are binding sites not for RNA polymerase but rather for a **transcription factor** called **TFIIIA** which interacts with RNA polymerase III so as to promote transcription.

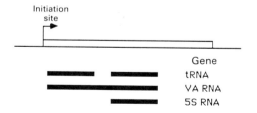

Fig. 5.35 Promoter regions of genes transcribed by RNA polymerase III

RNA polymerase II is the most interesting of the three polymerases, since it is responsible for the synthesis of all mRNA and ultimately of all protein. Since the thousands of different proteins encoded by eukaryotic DNA have a huge variety of requirements for the timing and location of their expression, there has evolved a correspondingly complex set of regulatory controls on the transcription of the genes which encode them. In practice, this means that there are many types of consensus box to be found upstream of different protein-encoding genes with each type donating a particular developmental specificity of transcription. These have been variously named **upstream activator sequences** or **enhancer elements** (Fig. 5.36).

Sequence	Transcription responsive to	Recognized by
5' CTGGAATNTTCTAGA 3' 3' GACCTTANAAGATCT 5'	Heat shock	Heat shock activator protein
5' GGATGTCCATATTAGGACATCT 3' 3' CCTACAGGTATAATCCTGTAGA 5'	Growth factor stimulation	Serum response factor
5' CTGGGTGCAAACCCTTTGCGCCCGG 3' 3' GACCCACGTTTGGGAAACGCGGGCC 5'	Cadmium or copper ions	?
5' CCGCCC 3' 3' GGCGGG 5'	Cell type	Protein Sp1
5' GGCCAATCT 3' 3' CCGGTTAGA 5'	Cell type	CAAT binding transcription factor (CTF)
5' ATGAGTCAG 3' 3' TACTCAGTC 5'	Phorbol esters	TPA-moculated trans-activating factor

Fig. 5.36 Nucleotide sequences of upstream activator/enhancer sequences.

Deletion of these small regions almost abolishes promoter activity. Conversely, addition of, for example, the heat shock consensus box to an upstream region of a gene confers heat shock inducibility upon that gene. Footprinting studies analogous to those used for prokaryotic promoters show that proteins recognize and bind to these consensus boxes. Note also that several of these regions contain dyad symmetry analogous to that seen for restriction enzyme recognition sites (Section 9.3.4). Many upstream consensus boxes (enhancers) have the property of retaining their transcriptional activating potential when they are either moved or inverted with respect to the initiation site (Fig. 5.37). Boxes with dyad symmetry possess the same sequence in either orientation.

Two major models have been proposed for the interaction between upstream boxes and the RNA polymerase at the initiation site. The first suggests that this interaction occurs by looping out of the DNA between them. The second model suggests that the boxes are entry sites for a protein factor which binds to them and then migrates along the DNA in order to interact with the RNA polymerase (Fig. 5.38). Present evidence strongly favours the first explanation. There is also

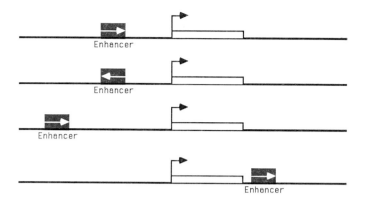

Fig. 5.37 Enhancer function is independent of orientation or distance from the affected gene.

growing evidence that different proteins can compete for binding to the same box and prior binding of one protein to a box can inhibit subsequent binding of another protein to an adjacent or overlapping box.

In addition to the varied upstream activator sequences or enhancer elements which have been mentioned above, another type of consensus sequence is often found at RNA polymerase promoters. The consensus sequence is: TATAAA. This

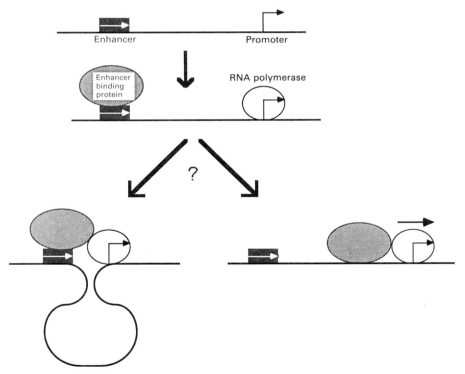

Fig. 5.38 Models for the action of transcriptional enhancers.

sequence is usually called the **TATA box** or **Goldberg/Hogness box** after its discoverers. A TATA box is often found at approximately 25 nucleotides upstream of the transcriptional start site. Although it is similar in sequence to the prokaryotic Pribnow box, its function is different. It is required for accurate positioning of the RNA polymerase initiation site. Thus, promoters which possess TATA boxes begin transcription at a unique spot whereas promoters which lack a TATA box yield RNAs with several different, staggered 5′-ends. TATA boxes, however, do not affect the total level of mRNA production, only the exact structure of the 5′-ends. Lastly, in yeast there are consensus boxes containing the TATA motif which may be more akin to upstream activator sequences than to the Goldberg/Hogness box.

5.5.2 Introns and exons

When we begin to look at the sequences of eukaryotic genes and compare them with those of their prokaryotic counterparts, a major difference immediately becomes apparent. Most eukaryotic genes are discontinuous with respect to the mature RNA or protein encoded by them. A simple example is the β-globin gene of mammals, which was one of the first to be discovered (Fig. 5.39).

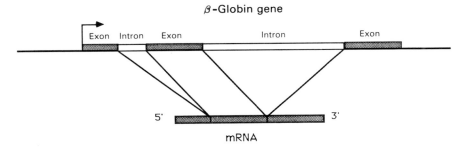

Fig.5.39 Structure of the β-globin gene which contains intervening sequences (introns) that are not present in the mRNA.

The rabbit β-globin gene contains two internal DNA segments which are not found in the mature RNA. These inserts are called **introns** and the regions surrounding them, **exons**. You will see later (Section 6.4.3) that introns are transcribed by the RNA polymerase but they are removed from the nascent transcript by a processing mechanism called **RNA splicing**. Genes which are transcribed by RNA polymerases I and III may also contain introns.

What is encoded by introns? Do they have any direct function? It is not easy to answer these questions definitely, but in general we believe that introns very often have no functions in the day-to-day expression of those genes which contain them. This has been elegantly demonstrated for an actin gene of yeast by the removal of its sole intron (Section 10.4). No discernible effect on the expression of the actin gene could be seen. Our conclusion must therefore be that, at least in this case, an intron is irrelevant to the expression of a gene.

How much of the eukaryotic genome is comprised of introns? We cannot be sure, but there is often more DNA contained in the introns of genes than is found in their exons. Several extreme examples are known such as the *Antennapaedia* gene of *Drosophila*, where the entire gene stretches across at least 100 000 nucleotide pairs, yet the mRNAs encoded by the gene are a mere 5000 and 3500 nucleotides long.

If at least some introns have no direct function, do any have indirect functions? Why are they there at all? The answers to these questions probably lie in the ancestry of genes themselves. We know that introns are at least as old as the prokaryote–eukaryote split because they are occasionally found in prokaryotes. This suggests that they are a relic of the process that originally led to the creation of genes. The genes we see today probably derived from the fusion of many smaller modules of genetic information. As an example, many enzymes with widely different functions share a domain which is capable of binding ATP. It is likely that such a segment only evolved once and the many different genes containing it somehow picked up this sequence module during their evolution. Modules of information (exons) may have been 'shuffled' and dealt in many combinations to yield the genes we see today. This model was first proposed by Gilbert and is supported by two lines of evidence. First, exons often encode discrete units of protein structure (often called **domains**) such at the ATP binding site mentioned above, or the catalytic site of an enzyme. Secondly, removal of introns from RNA can sometimes occur in the absence of any protein. Instead it is an inherent property of the nucleic acid (Section 6.5). This amazing property suggests that RNA splicing evolved before proteins and provides us with a rational explanation for how **exon shuffling** was possible. Presumably, some type of DNA rearrangement must be a prerequisite for a shuffling event, but DNA rearrangements could not be expected to join up sequence modules precisely. There would have to be gaps, many of which would inevitably contain translational stop codons or introduce frame-shifts into the hybrid reading frames. If those gaps had the ability to remove themselves from the RNA containing them by splicing, then these exons would be far more easily shuffled.

Currently, we believe that introns have always occurred in genes simply because they are a part of the way genes evolved in the first place, namely by shuffling of modular units of genetic information (exons) at the DNA level combined with removal of the intervening sequences (introns) at the RNA level.

5.5.3 Gene families

The β-globin gene is not the only globin gene in a mammal. All mammals possess α-globin genes and myoglobin genes. Several genes related to the α- and β-globin genes are only expressed in the fetus or embryo. We call the collection of all these related genes a **gene family**. Often the members of a gene family are all present at the same chromosomal location. It is postulated that each gene family originally evolved from an individual gene. A plausible mechanism which involves recombination, called **unequal crossing-over**, exists to explain how such duplications may occur (Fig. 5.40 and Section 10.1.3). Strong evidence to suggest that this happens in *Homo sapiens* comes from sequence analysis of the human embryo-specific, β-globin-like genes Gγ and Aγ. The DNA sequences of these two genes and of the regions surrounding them are extremely similar, suggesting that a region bigger than the original precursor gene became duplicated relatively recently (in evolutionary terms). Futhermore, the duplicated regions are flanked by pieces of repetitive DNA. The likely mechanism of duplication involved homologous recombination between these repetitive segments. Normally, recombination exchanges a DNA duplex for a closely related sequence. However, if the two DNAs do not align themselves properly, unequal crossing-over transfers a piece of one chromosome to the other, thus generating a duplication (Fig. 5.40).

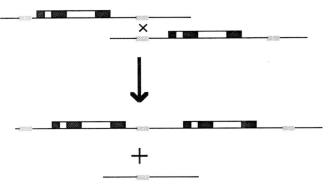

Fig. 5.40 Unequal crossing-over between repetitive sequences generating gene duplications and deletions.

An extreme example of β-globin gene duplication can be found in the domestic goat. Here, at least four consecutive duplications have given rise to the 12 genes which share homology with β-globin (Fig. 5.41).

Thus, genes may be duplicated many times over in the higher eukaryotes. Does this process fulfil a useful function? We believe it does, since in several instances the duplicated genes encode proteins with different specificities. For instance (Fig. 5.41), some of the β-globin-like genes are expressed only during embryonic developmental stages and the protein encoded by them possesses an advantageous oxygen-binding capacity which, in fetal haemoglobin, enables it efficiently to extract oxygen from the maternal haemoglobin. This is a specific example of a much more general phenomenon of great evolutionary significance. If a gene is duplicated, then the new copy is under no selective evolutionary pressure to carry out its old function, since one copy is perfectly capable of doing this. Therefore, a new copy is able to experiment with its sequence by random mutation through the course of time until it finds a novel function which is useful to the host.

Do all duplicated genes eventually find a use? No, some pick up nonsense or missense mutations and become 'dead' genes or **pseudogenes**. Three genes in the

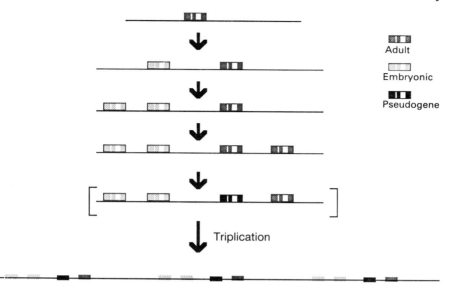

Fig. 5.41   The probable evolution of the β-globin locus of the goat.

goat β-globin-like gene cluster are pseudogenes. Thus, genes have a tendency to multiply in higher eukaryotes, thereby contributing both to the large amounts of DNA seen in these organisms and to the evolution of their hosts by the acquisition of novel functions.

Perhaps the most fascinating gene family discovered thus far in the vertebrates is the immunoglobulin superfamily. A broad class of at least one hundred genes which encode proteins that bind to cell surfaces probably are all derived from a single ancestral gene. Most (and perhaps all) of these proteins are involved in recognition of other cells or invading entities. Presumably, this diversification evolved coincidentally with the structural evolution of higher eukaryotes in order better to define and regulate the activity of different cell types in the organism.

Another consequence of multiplication of genes is the ability to produce large amounts of the gene products. Thus, the genes which encode very abundantly expressed RNAs, such an tRNAs, 5S RNA, ribosomal RNA, and histone RNAs, are repeated many times in higher eukaryotes. The approximate copy number of these genes for several organisms is listed (Table 5.3).

Table 5.3. Approximate copy number of genes encoding abundantly expressed RNA or protein

	tRNA	5S RNA	rRNA	Histone H2A
Saccharomyces cerevisae	250	150	150	2
Drosophila melanogaster	1 000	150	150	100
Xenopus laevis	1 000	25 000	500	50
Homo sapiens	1 000	2 000	300	15

5.5.4 Highly repetitive interspersed DNAs

There are two known classes of DNAs which are present in hundreds of thousands or even millions of copies which are interspersed amongst the rest of the genomes of eukaryotes. Short interspersed elements (or **SINES**) are comprised of short stretches of DNA (a few hundred base-pairs long). These are virtually all pseudo-genes which are derived from genes for various small RNAs. Each organism has its own characteristic family (or families) of SINE elements. The major element in mammals is the **Alu family**, named after the restriction enzyme *Alu* I which generates a characteristic fragment seen in digests of human DNA. It is possible that Alu elements neither fulfil any major function for their host nor do they significantly harm their bearer. These sequences have been termed parasitic or **'selfish' DNAs** in that their only function is their propagation through the genomes they inhabit. At present, we are uncertain of whether such sequences are truly parasitic. Perhaps some as yet undiscovered property which helps the host is awaiting discovery.

How do such sequences become amplified? We believe that they are members of a family of transposable genetic elements (Chapter 10) which employ **reverse transcription** (conversion of RNA into DNA) to create new genomic copies of themselves. A possible mechanism for the creation of a new SINE element is shown (Fig. 5.42).

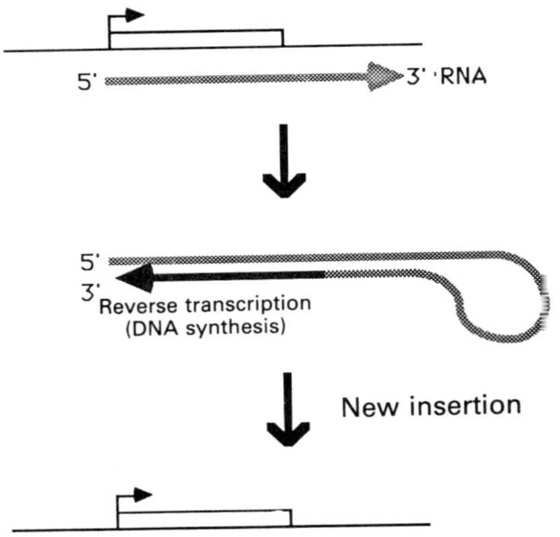

Fig. 5.42 A model for the amplification of SINE elements.

A second class of interspersed repeats which is related to the SINE family is found in higher eukaryotes. These long interspersed elements (or **LINES**) are several thousands of nucleotides long. We believe that they are also of no use to the organisms containing them and that they multiply by a fundamentally similar mechanism to that used by SINES. It is likely, however, that LINES encode the reverse transcriptase which is involved in this process and it is even possible that LINE reverse transcriptase is the enzyme which aids the amplification of SINES (which are far too small to encode such a protein).

5.5.5 Satellite DNA

The last major repetitive component of eukaryotic DNAs which we will consider is satellite DNA. This DNA is composed of many thousands or even millions of repeats of short sequences. Satellite repeats are organized in huge tandem arrays. Because such arrays are so homogenous in sequence, they have a characteristic G–C content. This is sometimes significantly different from the overall average value for the bulk DNA of the organism, with the result that such DNA is of a different buoyant density and will band at a different position in a CsCl density gradient to that of the large part of the bulk DNA (Section 5.1). Such **satellite bands** in CsCl gradients gave satellite DNA its name. The fruit fly *Drosophila virilis* has three related satellite sequences. These are:

5′ ACAAACT 3′ Satellite I
5′ ATAAACT 3′ Satellite II
5′ ACAAATT 3′ Satellite III.

Together, the millions of copies of these three repeats comprise about one third of the entire *D. virilis* genome. This amounts to at least 100 million nucleotide-pairs of satellite sequence. Higher eukaryotes may have longer repeat units. These sequences again encode no RNA or protein and are assumed at best to fulfil a structural role in chromosome organization, since they are found mainly near the centromere.

Summary

Eukaryotic genomes are more complicated than those of prokaryotes. They are in general much bigger than seems necessary for the approximate number of genes they contain. Much of the DNA in eukaryotes is repetitive in sequence, does not encode essential information for the host, and is not expressed as stable protein or RNA.

Eukaryotic genes are transcribed by one of three polymerases. RNA Polymerase I transcribes ribosomal RNA from promoters which are poorly conserved in evolution. RNA polymerase II transcribes mRNA and some small RNAs. RNA polymerase II promoters can be very complex and contain one or more upstream activator sequences or enhancers. RNA polymerase III transcribes small RNAs from promoters located inside the genes.

Many eukaryotic genes contain intervening sequences (introns) which are not present in the mature RNA. The antiquity of introns argues that they are as old or older than recognized genes. They may be a by-product of gene evolution catalysed by DNA rearrangements and RNA self-splicing. Eukaryotic genes are sometimes duplicated by unequal crossing over. This process has given rise to multi-copy gene families encoding the globins and immunoglobulins. SINE and LINE elements account for a significant part of the dispersed repetitive DNA in higher eukaryotes. Satellite DNA is comprised of millions of tandemly repeated copies of simple sequences and is found at centromeres.

Further reading

5.1

Arrand, J. E. (1985). Preparation of nucleic acid probes. In *Nucleic acid hybridisation: A practical approach.* (ed. B. D. Hames and S. J. Higgins. IRL Press, Oxford, pp. 17–44.

Sambrook, J., Fritsch, E. F., and Maniatas, T. (ed.) (1989). *Molecular cloning, a laboratory manual*, 2nd ed. Cold Spring Harbor Laboratory Press, New York.

5.2

Maxam, A. M. and Gilbert, W. (1980). Sequencing end-labelled DNA with base-specific chemical cleavages. *Methods Enzymol.*, **65**, 499–560.

Sanger, F. (1981), Determination of nucleotide sequences in DNA. *Science*, **214**, 1205–10.

Sanger, F. (1988). Sequences, sequences, and sequences. *Ann. Rev. Biochem.*, **57**, 1–28.

5.3

Bachmann, B. J. (1983). Linkage map of *E coli*-K12, Edition 7. *Microbiol. Rev.*, **47**, 180–230.

Cairns, J. (1983). The bacterial chromosome and its manner of replication as seen by autoradiography. *J. Mol. Biol.*, **6**, 208–13.

Lehman, I. R. and Kaguni, L. S. (1989). DNA polymerase α. *J. Biol. Chem.* **264**, 4265–8.

Mclure, W. R. (1985). Mechanism and control of transcription initiation in prokaryotes. *Ann. Rev. Biochem.*, **54**, 171–204.

Miller, J. H. and Reznikoff, W. S. (ed.) (1978). *The operon.* Cold Spring Harbor Laboratory, New York.

Platt, T. (1986). Transcription termination and the regulation of gene expression. *Ann. Rev. Biochem.*, **55**, 339–72.

Ptashne, M. (1986). *A genetic switch: gene control and phage λ.* Cell Press, Massachusetts and Blackwell Scientific Publications, Oxford.

Ptashne, M. (1989). How gene activators work. *Sci. Amer.*, **260** (1), 24–31.

Rosenberg, M. and Court, D. (1979). Regulatory sequences involved in the promotion and termination of RNA transcription. *Ann. Rev. Biochem.*, **13**, 319–53.

Smith, G. R. (1981). DNA supercoiling: another level for regulating gene expression. *Cell*, **24**, 599–600.

5.4

Fersht, A. R. (1980). Enzymatic editing mechanisms in protein synthesis and replication. *Trends Biochem. Sci.*, **5**, 262–5.

Kornberg, A. (1980). *DNA replication.* Freeman, San Francisco.

McHenry, C. S. (1988). DNA polymerase III holoenzyme of *Escherichia coli*. *Ann. Rev. Biochem.*, **57**, 519–50.

Modrich, P. (1989). Methyl-directed DNA mismatch correction. *J. Biol. Chem.*, **264**, 6597–600.

Ogawa, T. and Okazaki, T. (1980). Discontinuous DNA replication. *Ann Rev. Biochem.*, **49**, 421–57.

Radman, M. and Wagner, R. (1988). The high fidelity of DNA replication. *Sci. Amer.*, **259**(2), 24–31.

Tomizawa, J. and Selzer, G. (1979). Initiation of DNA synthesis in *E. coli. Ann. Rev. Biochem.*, **48**, 999–1034.

5.5

Brutlag, D. (1980). Molecular arrangement and evolution of heterochromatic DNA. *Ann. Rev. Genet.* **14**, 121–44.

Lewin, B. (1987). *Genes III*. Wiley, New York, pp. 293–356.

Rogers, J. (1984). The origin and evolution of retroposons. *Int. Rev. Cytol.*, **93**, 188–280.

Watson, J., Hopkins, N., Roberts, J., Steitz, J., and Weiner, A. (ed.) (1987). *Molecular biology of the gene*. Benjamin Cummings, Menlo Park, California, pp. 621–73.

RNA sequence information and transmission

6

6.1 Occurrence and isolation of RNA

We know that genetic information is stored in the form of nucleotide sequences in DNA and that this information is copied into RNA before it can be used. RNAs thus play a central role in the transmission of genetic information and the process of copying the sequence information from DNA into RNA is called **transcription**.

6.1.1 Abundance and size of RNA species

At least three major classes of RNAs have been identified in cells. These are **messenger RNAs** (mRNAs), **transfer RNAs** (tRNAs), and **ribosomal RNAs** (rRNAs). All three classes are essential for protein biosynthesis (Section 6.6). mRNAs contain the information necessary to specify the amino acid sequence of proteins which is present in the form of nucleotide sequences in mRNA. The process by which the information present in mRNA is read into protein is called **translation**.

Approximately, 80–85 per cent of the total RNA in cells is present as rRNA, 10–15 per cent is tRNA, and only 2–5 per cent consists of mRNAs. Ribosomes from bacteria such as *E. coli* contain three different rRNAs. These are the 5S rRNA, 16S rRNA, and the 23S rRNA (the **S** stands for the sedimentation rate often determined by sucrose gradient centrifugation of the RNAs). The 5S rRNA of *E. coli* is 120 nucleotides long whereas the 16S and the 23S rRNAs are 1542 and 2904 nucleotides long respectively. Eukaryotic cells also contain the corresponding three types of rRNAs, in this case the 5S, 18S, and 28S rRNAs, and in addition a fourth, 5.8S rRNA. Although eukaryotic cells contain this 'extra' rRNA in their ribosomes, in bacteria this rRNA species is included as a part of the 23S rRNA. Consequently, the 5.8S rRNA does not endow the eukaryotic ribosome with any particular properties which are lacking in the bacterial ribosome.

Most cells contain 60–70 different tRNAs, the sizes of which range from 74–95 nucleotides except for some mitochondrial tRNAs which are slightly smaller.

The number of different mRNAs in a cell can range from several hundred to a few thousand, depending on the number of proteins made by that cell. Most cells, however, have several hundred different mRNAs. Those few eukaryotic cells which specialize in the production of specific proteins are the only exceptions to this rule. For example, some adult reticulocytes are specialized for production of haemoglobin and contain predominantly two types of mRNAs encoding the α-and β-chains of haemoglobin. Since the size of proteins made in a cell varies over a wide range, the size of mRNAs in a cell also spans a wide range, from a few hundred to several thousand nucleotides.

In bacteria such as *E. coli*, mRNAs are being synthesized, utilized, and degraded continuously which makes mRNAs subject to continual **turnover**. The half life of a typical bacterial mRNA is around two minutes and so the concentration of any particular mRNA in a bacterial cell is low. By contrast, tRNAs and rRNAs are much more stable and, since there are only a few species of tRNAs and rRNAs, the concentrations of these RNAs in a cell are much higher than those of mRNAs.

Eukaryotic cells contain specialized organelles such as mitochondria (for respiration) and chloroplasts in plant cells (for photosynthesis). Such organelles contain DNAs which are different both from each other and from DNA in the nucleus. Their protein synthesizing systems are also distinct from each other and from the one which operates in the eukaryotic cytoplasm. Consequently, eukaryotic cells such as plants have as many as three different sets of mRNAs, tRNAs, and rRNAs which are distinct from each other and which operate in different compartments within the same eukaryotic cell.

Besides mRNAs, tRNAs, rRNAs, there are many other RNAs in a cell with important cellular functions. Several RNAs, collectively called **small nuclear RNAs** (snRNAs), function inside the nucleus of eukaryotic cells and are involved in mRNA biosynthesis, in removal of intervening sequences, and in polyadenylation, etc. (Section 6.4.3). Another RNA, designated 7S RNA, is an essential component of a cytoplasmic ribonucleoprotein (RNP) particle which is involved in the transport of proteins through phospholipid membranes. In addition, there are examples of specific RNA molecules which are essential components of enzymes. In many cases, the RNA alone has the catalytic activity (Section 6.5). Thus, RNAs are a highly versatile class of molecules possessing a wide range of biological activities.

6.1.2 Isolation of RNA

Although RNA is not as long as DNA, care needs to be taken in isolating this class of macromolecule. While the isolation procedures for the different classes vary according to the properties of the RNAs, some generalizations can be made. As in the case of DNA (Section 5.1), the process can be divided into four steps: (a) isolation of cells or organelle (e.g. chloroplasts), (b) lysis, (c) removal of protein, and (d) isolation of the individual class of RNA.

Since RNAs are much more susceptible to degradation by nucleases than DNAs, a major consideration in RNA isolation is the need to eliminate, or at least reduce, the nuclease activity normally present in any cell extract. This is achieved by rapid phenol extraction of the cell lysate, a treatment which denatures most nucleases and other proteins. The aqueous phase contains the nucleic acid derivatives, including nucleotide co-enzymes, while the phenol phase (and the interface) contains most of the protein. In addition, the presence of divalent metal ions during the purification is beneficial, as they are important for preserving the structure and hence the biological properties of RNA.

It is often difficult to isolate biologically active RNA from organs such as the pancreas, because it is difficult to homogenize the tissue. In such cases guanidinium isothiocyanate is used to dissolve the tissue and this process also inactivates any ribonucleases that are present.

The separation of tRNA and ribosomal RNA relies on the different solubilities of the molecules. The aqueous supernatant of the phenol extraction is adjusted to 1 M in salt which precipitates the rRNA. The tRNA remains in solution and is purified by anion exchange chromatography. Usually the 5S ribosomal RNA is also in the tRNA fraction and this can be separated by gel filtration or gel electrophoresis.

Most eukaryotic messenger RNAs possess a stretch of adenylate residues (from 20 to 200 long) at their 3′-ends (Section 6.4.4). The interaction of this **poly A** tail with **oligo-dT cellulose** forms the basis of an affinity method for the purification of mRNA, the development of which was crucial for some of the advances in molecular biology. Oligo-dT cellulose chromatography is based on the principle that oligo-dT (chain length 9–20) attached to cellulose provides a solid matrix on to which poly A-containing RNA can be absorbed (as a result of base-pairing with the oligo-dT stretches) and thereby purified in one step from other RNAs.

Summary

The three major classes of RNA in cells are messenger RNA, transfer RNA, and ribosomal RNA of which rRNAs are the most abundant. Prokaryotic cells contain three species of rRNA whereas eukaryotic cells contain four, but the sets of rRNAs confer similar properties to both prokaryotic and eukaryotic ribosomes. Most cells contain 60–70 different tRNAs, whereas the number of mRNAs can vary from several hundred to a few thousand and these can be of widely varying length. Mitochondria and chloroplasts have their own sets of mRNA, tRNA, and rRNA. Small nuclear RNAs play an important role in processing of mRNA precursors. Some RNA molecules possess catalytic activity.

Isolation of RNA involves isolation of cells or organelles from tissue, lysis, removal of protein (including ribonucleases), and separation of the individual class of RNA. Purification of mRNA is facilitated by use of oligo-dT cellulose chromatography.

6.2 Determination of RNA sequence

Because methods for determining DNA sequences are so rapid, one approach to RNA sequence determination is to clone the gene for an RNA (Section 10.2) and then sequence the cloned DNA (Section 5.2). However, during biosynthesis most RNA molecules are made as longer transcripts (Section 6.4) and such precursor RNAs are then cut by specific nucleases at their 5′- and 3′-ends to produce the mature functional RNA molecules. In addition, there are enzymes which attach special nucleotide sequences to the 5′- and/or 3′-ends of some RNAs. As a result, the DNA sequence alone cannot tell us about the nature of the ends of a mature biologically functional RNA molecule. Also, RNAs often contain modified nucleotides in which the structures of the four common nucleotides, A,C,U, and G have been changed in various ways by enzymes. Since these modified nucleotides are often important in the biological activity of the RNAs, it is necessary to identify and locate them.

6.2.1 Methods involving reverse transcriptase

RNA can be used as a template for the synthesis of DNA using the enzyme reverse transcriptase. This enzyme copies the RNA sequence into a DNA sequence in the presence of an appropriate DNA primer. The resulting DNA (**cDNA**) can then be cloned and sequenced. This method has been used widely for sequence determination of eukaryotic mRNAs.

If the required RNA is abundant within an RNA preparation, it can be sequenced directly by application of the Sanger method of DNA sequencing using chain terminators (Section 5.2) except that RNA is used as the template and reverse transcriptase is used to copy it. In this case, some part of the RNA sequence must already be known so that a complementary oligodeoxyribonucleotide can be designed for use as a primer. This method is particularly useful in sequence determination of viral RNAs where some part of the sequence is highly conserved between species. Neither of these methods identifies modified nucleotides.

6.2.2 Direct methods for sequence determination

The sequence analysis of an RNA species by direct methods predates reverse transcriptase methods and was first achieved for a tRNA by Holley in 1965. Soon after, Sanger and Brownlee developed a more general sequencing method which required the RNA to be radiolabelled with ^{32}P at each phosphate. The RNA was then subjected to complete cleavage by ribonuclease T1, which cuts after every G residue, and separately by pancreatic ribonuclease, which cuts after every pyrimidine residue (Table 6.1). The products of each digestion were separated by a two-dimensional electrophoresis/chromatography procedure to generate characteristic **fingerprints**. Individual radioactive RNA fragments were isolated from each fingerprint and further degraded using phosphodiesterases (Section 9.4.3) to establish their base sequences.

Nowadays, there are three methods that are commonly used for direct sequencing of an RNA. They all require the RNA to be available in pure form.

Use of base-specific ribonucleases

In contrast to the Sanger and Brownlee method, a single ^{32}P-radioactive phosphate is first introduced at either the 5′-end of the RNA using γ-^{32}P-ATP and T4 polynucleotide kinase (Section 9.4.1) or at the 3′-end of the RNA using [5′–^{32}P]-pCp and T4 RNA ligase (Section 9.4.2). This allows the identification of all the oligonucleotide degradation products of the RNA which contain the labelled end.

The second step is to treat the ^{32}P-labelled RNA in separate reactions with different base-specific ribonucleases, which cut the RNA phosphodiester bonds only on the 3'-side of specific nucleotides. Enzymes commonly used for this purpose cut RNAs after G residues, after A residues, after C residues, or after both U and C residues (Table 6.1). In a parallel reaction, the RNA is treated with alkali which cuts randomly after any residue. All of the RNA cleavages, whether by enzymes or by alkali, are carried out to a limited extent, so that each susceptible phosphodiester bond is cut in only a small fraction of the ^{32}P-labelled RNA molecules and not more than once per molecule.

Table 6.1. Enzymes commonly used in RNA sequencing

Ribonuclease	Base specificity
T1	G
A	U or C
U2	A > G
B. cereus (B.c.)	U,C
Chicken liver (C.l.3)	C >> A or U
Physarum M (PhyM)	U,A or G >>> C
T2	U,C,A, or G

For example, consider an RNA with the sequence *pAGUCGAAGUU (*p = ^{32}P-phosphate). If this RNA is completely digested with ribonuclease T1, which cuts after every G residue, the only radioactive oligonucleotide produced will be *pAG. However, if the same RNA is digested only partially, a series of radioactive oligonucleotides *pAG, *pAGUCG, *pAGUCGAAG is obtained. Each of these oligonucleotides contains a G residue at the 3'-end. Similarly, the use of the enzyme ribonuclease U2, which cuts an RNA after A residues, gives a different series of radioactive oligonucleotides, each of which has an A residue at its 3'-end. Use of the other two nucleases gives rise to two other series of ^{32}P-labelled oligonucleotides which either end with C or end with U or C. In the case of alkaline treatment, cleavage occurs at all possible ester sites to produce a complete series of ^{32}P-labelled oligonucleotides. Such a pattern of RNA is called a **ladder**.

Each of the different series of ^{32}P-labelled oligonucleotides can be separated according to their size by use of polyacrylamide gel electrophoresis. Based on the pattern of radioactive bands obtained (Fig. 6.1), one can locate the positions of G, A, U, and C residues relative to the ^{32}P-labelled 5'-end. Thus we can read the sequence of the RNA simply by following the pattern of oligonucleotides from the shortest towards the longest.

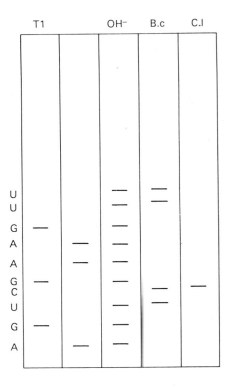

Fig. 6.1 Autoradiographic pattern obtained after polyacrylamide gel electrophoresis of partial digests of 5'-^{32}P-labelled RNA with RNase T1 (T1), RNase U2 (U2), alkali (OH$^-$), *Bacillus cereus* RNase (B.c.) and *Chicken liver* RNase (C.l.). The sequence deduced from the pattern is shown on the left.

Use of base-specific chemicals

This method is very similar to the Maxam–Gilbert method for sequencing DNA (Section 5.2). Four different base-specific chemical reactions are used to modify either G, A, U, or C residues in an RNA. Subsequent treatment of each of the incompletely modified RNAs with aniline results in cleavage of the RNA at the site of modification. Once again, the ^{32}P-labelled oligonucleotides are separated according to their size by polyacrylamide gel electrophoresis and the resulting pattern of radioactive bands determines the positions of G, A, U, and C residues with respect to the labelled end. Reagents that are used for the base-specific reactions are somewhat different from those used in sequence determination of DNA (Table 6.2). Also, aniline cleavage of the RNA at the site of modification has to be carried out under rather mild conditions to prevent cleavage at unmodified sites (cf. DNA which is much more stable to alkali, Section 3.2.2). These conditions result in formation of a 5'-phosphate group at the nucleoside on the 3'-side of the cleavage site, but the nucleoside on the 5'-side is converted into a mixture of species. Thus, the chemical method is more suitable for sequence analysis of 3'-^{32}P-labelled RNA rather than 5'-^{32}P-labelled RNA.

Table 6.2. Chemical reagents used in RNA sequencing

Reagent	Base specificity
Dimethyl sulphate	G
Diethyl pyrocarbonate	A >> G
Hydrazine (aqueous)	U >> C
Hydrazine + NaCl	C >> U

Use of random chemical cleavage

Neither of the above methods permits the direct identification of a modified base naturally present in the RNA. To identify modified bases, one can use a method first described by Stanley and Vassilenko (Fig. 6.2).

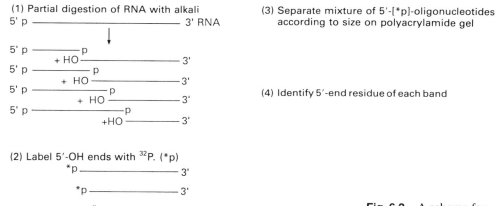

(1) Partial digestion of RNA with alkali

(2) Label 5'-OH ends with ^{32}P. (*p)

(3) Separate mixture of 5'-[*p]-oligonucleotides according to size on polyacrylamide gel

(4) Identify 5'-end residue of each band

Fig. 6.2 A scheme for sequencing RNA by the random chemical cleavage method.

In the first step, the RNA is partially hydrolysed under alkaline conditions to generate a ladder. Under such conditions, one obtains a homologous series of oligonucleotides each of which contains the 3'-terminal end of the original RNA and another series each of which contains the 5'-terminal end of the original RNA. Each of the oligonucleotides that contain the 3'-terminal end of the original RNA has a free 5'-hydroxyl group at its 5'-end. By treatment of the entire mixture of oligonucleotides with the enzyme T4 polynucleotide kinase in the presence of γ-^{32}P-ATP, a radioactive phosphate group is transferred to each of these 5'-hydroxyl groups. All the ^{32}P-labelled oligonucleotides are now separated according to their size by polyacrylamide gel electrophoresis.

In the final step, the 5'-terminal nucleotide of each radioactive band is identified by complete degradation of the oligonucleotide to mononucleotides either by use of alkali or by use of a non-specific ribonuclease. The 5′-terminal nucleotide, which in each case contains all of the ^{32}P-label, is identified by further separation using paper or thin layer chromatography or by electrophoresis. Because the 5′-end of each oligonucleotide is identified directly, it is now possible to determine whether the 5′-residue is one of the four normal nucleotides (A, C, U, or G) or whether it is a modified nucleotide by comparison of its chromatographic and electrophoretic behaviour using marker nucleotides of known structure.

Summary

The sequence of an RNA can be determined indirectly by cloning and sequence analysis of the corresponding DNA or by application of the Sanger method of DNA sequencing to the purified RNA using chain terminators and the enzyme reverse transcriptase.

The sequence of a purified RNA can be directly determined by methods that involve the use of either base-specific ribonucleases or base-specific chemicals to generate fragments of RNA which can be separated by electrophoresis in polyacrylamide gels. The identification of modified bases naturally present in an RNA is possible using the random chemical cleavage method.

6.3 Secondary structure of RNA

The many biological functions of RNA are based on very specific three-dimensional RNA structures. What are these structures and how are they formed and maintained?

In contrast to DNA, RNA occurs mostly as a single polynucleotide chain. However, by virtue of the inherent ability of RNA to make up different conformations (Section 2.4), many RNAs exist in elaborate, defined structures possibly as distinct as those of proteins. RNA molecules usually contain short double-helical regions connected by single-stranded stretches (Fig. 6.3). The helical hairpin regions can form because of the anti-parallel orientation of some complementary sequences found in different parts of the RNA chain. For instance, in the secondary structure of tRNA (Fig. 6.5, Section 6.3.2) the 5′-end of the tRNA forms a helix with nucleotides located near the 3′-terminus. Such double-helical structures contain not only the standard A:U and G:C base-pairs but may contain also a number of energetically less stable G:U base-pairs.

A large RNA, such as rRNA, can have many possible secondary structures based on the frequent occurrence in different parts of the RNA of short sections of complementary bases. Thus to determine the most probable secondary structure of 16S rRNA (Fig. 6.4), it was necessary for supporting evidence to be obtained for the various stem-loop regions in the RNA (see below). Additional pairing between

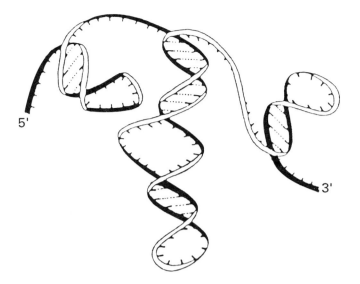

Fig. 6.3 A hypothetical secondary structure of RNA showing short double-helical regions separated by single-stranded regions. Dotted lines indicate base-pairs formed between the strands.

bases distant in the secondary structure generates, through formation of tertiary base-pairs, the structure of 16S rRNA.

The maintenance of RNA structure is dependent on factors such as salt concentration, pH, temperature, and the presence of specific ions (e.g. Mg^{2+} or polyamines). If the temperature is raised sufficiently, then all the base-pairs in the helical regions will dissociate to yield the fully 'denatured' state of the RNA.

6.3.1 The determination of RNA secondary structure

Ultimately, X-ray crystallography will provide us with high resolution structures of RNA molecules either alone or complexed with proteins. At the present time, however, only tRNA species have given crystals which diffract at high resolution. Fortunately, a combination of other techniques may be used which, together with some of the principles of RNA structure learnt from the crystal structure of tRNA, allow meaningful structures to be derived for many RNA molecules.

Biophysical methods, such as NMR spectroscopy, X-ray crystallography, or neutron scattering, can provide structural information when a relatively large amount of an RNA is available. In most other cases where only small amounts of RNA are available, two other methods are especially useful and particularly so when used in combination.

RNA modification by chemical or enzymatic reagents

RNA (or an RNA:protein complex) can be subjected to limited modification by structure-selective probes. These are chemicals or enzymes having selectivity for a particular base in either a single-stranded (i.e. unpaired) environment or in a

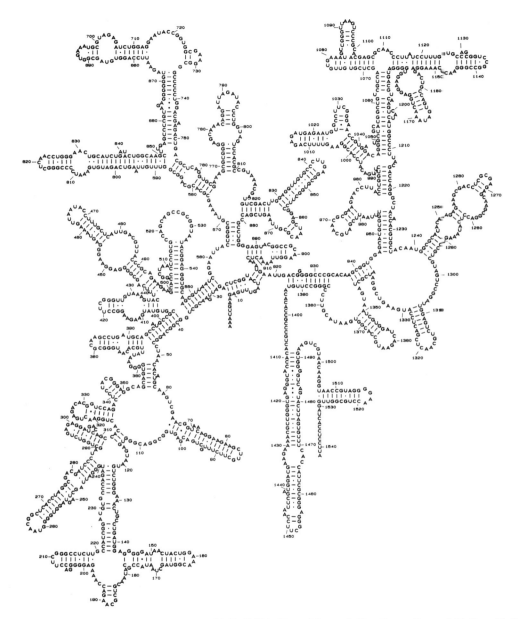

Fig. 6.4 Proposed secondary structure of 16S rRNA (from Moazed, D., Stern, S., and Noller, H. F. (1986). *J. Mol. Biol.*, **187**, 400).

base-pair (secondary or tertiary) (Table 6.3). If such modifications are introduced randomly at a low level and give rise directly or indirectly to chain scission, then gel electrophoretic methods can be used to separate the fragments (Section 6.2.2). If the RNA is end-labelled before modification, then the size of the labelled oligonucleotide fragment identifies the nucleotide in the chain which was modified. This method is powerful and requires little material. For example, it was established by this technique that the bases in the D-loop of tRNA (Section 6.3.2) are in exposed single-stranded regions. Under certain conditions, tertiary base-pair interactions (Section 6.3.2) can also be distinguished. These interactions are less stable than the standard Watson–Crick pairs and are broken under mild denaturing conditions.

Table 6.3.[a] Some probes for RNA secondary structure

Probe	Specificity
RNase T1	Unpaired G
RNase U2	Unpaired A > G
RNase CL3	Unpaired C >> A > U
RNase T2	Any unpaired base
Nuclease S1	Any unpaired base
RNase VI	Any paired or stacked base
Dimethyl sulphate	N^3C, N^1A, N^7G
Diethyl pyrocarbonate	N^7A
CMCT[b]	N^3U, N^1G
Kethoxal	N^1G, N^2G
Bisulphite	Unpaired C ($\rightarrow$ U)
Ethyl nitrosourea	Phosphates

[a] Adapted from Ehresmann et al. (1987). Nucleic Acids Research, **15**, 9109. [b] 1-Cyclohexyl-3-(2-morpholinoethyl) carbodiimide metho-p-toluene sulfonate

In an extension of this technique, called **reverse transcriptase mapping**, chemically modified RNA can be used as a template for DNA synthesis catalysed by reverse transcriptase. Base residues which are accessible to reaction with a modifying reagent can be identified by a block to DNA synthesis at each site of modification, since base-pairing is disrupted between the modified base and the base's normal complement. Only those residues in single-stranded regions of the RNA are modified and hence identified. The method has been particularly useful for the analysis of rRNA and also for the identification of sites of interaction of RNA with antibiotics and proteins.

Phylogenetic comparisons

This approach can be used to refine a structure determined by RNA modification experiments. The method is based on a comparison of the nucleotide sequences of homologous RNA species from closely and more distantly related organisms. Such a comparison, which was crucial in establishing the accepted structure of many RNA molecules (e.g. 16S rRNA, Fig. 6.4), shows that helical regions are usually conserved. Although changes in sequence are often found in double-stranded regions, base-pairing (including G:U pairs) is preserved as a result of compensating mutations in the opposite strand of the duplex. Such phylogenetic comparisons have shown that the secondary structures of some RNA molecules are more conserved than their primary sequences or their lengths (e.g. 16S and 23S rRNA, RNase P RNA).

6.3.2 Transfer RNA: an example of the intricate features of RNA structure

The sequences of several hundred tRNAs from many sources have now been established. Transfer RNA is also the only class of RNA molecule for which X-ray crystallographic structural information is currently available. In addition, since tRNA is of small size, is easily prepared, and plays a central role in protein synthesis, it has been the object of many biochemical and biophysical studies.

All tRNAs contain a common sequence (. . . CCA) at their 3'-termini, the **acceptor end**. During protein synthesis, the ribose of the 3'-terminal A residue becomes attached to an amino acid through an ester linkage. In addition, there are certain nucleotides which are highly conserved between different tRNAs (Fig. 6.5a). Most of these nucleotide residues are involved in tertiary base-pair interactions which stabilize the 3'-dimensional structure of the tRNA.

Although the sizes of tRNAs vary (74–95 bases), they can all be folded into a common secondary structure (**cloverleaf**), in which four or five base-paired regions (**stems**) are separated by single-stranded regions (**loops**) (Fig. 6.5a). The sizes of the **acceptor stem** (7 base-pairs), the **anticodon loop** and stem (7 in the loop and 5 base-pairs in the stem), and the TψC loop and stem (7 in the loop and 5 base-pairs in the stem) are constant in all tRNAs. The variation in size between tRNAs is due to differences in the number of nucleotides in the D loop and stem and/or in the variable loop, leaving a common secondary structure intact. Additional hydrogen-bonds between nucleotides quite distant in the primary sequence cause the cloverleaf to fold into a stable tertiary structure in which two helical segments stack on top of each other to give the appearance of two continuous (yet not covalently joined) double-helices. The D-loop and the TψC-loop provide the hinge in the characteristic L-shaped structure (Fig. 6.5b) in which the anticodon loop and the acceptor end are the furthest apart. The recently completed X-ray crystallographic analysis of a tRNA complexed to a protein shows that the L-shaped structure is retained in the complex (front cover).

Many of the tertiary hydrogen-bonding interactions show structures very differ-

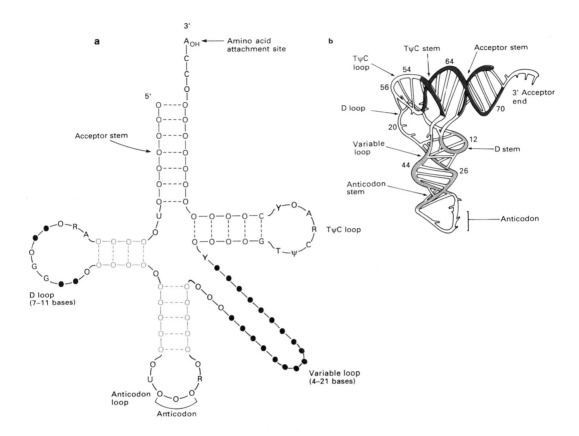

Fig. 6.5 (a) Clover-leaf secondary structure of tRNA and (b) three-dimensional structure of tRNAs. The horizontal part of the L-shaped structure in (b) is formed by colinear stacking of base-paired regions in the acceptor stem and the TψC stem whereas the vertical part has base-pairs in the D-stem stacked over the anticodon stem.

ent from the standard A:U and G:C base-pairs (Fig. 6.6) and involve Hoogsteen and other types of pairs (Section 2.1.2). In addition, the phosphates and 2'-hydroxyl groups of the RNA backbone are also involved in tertiary interactions. Further stability of the tRNA molecule comes from stacking interactions between the heterocyclic bases. As a result, tRNAs are very compact, stable, yet flexible molecules with structures that accommodate all the different tRNA sequences while allowing for slight conformational changes which might be crucial for specific interactions with proteins.

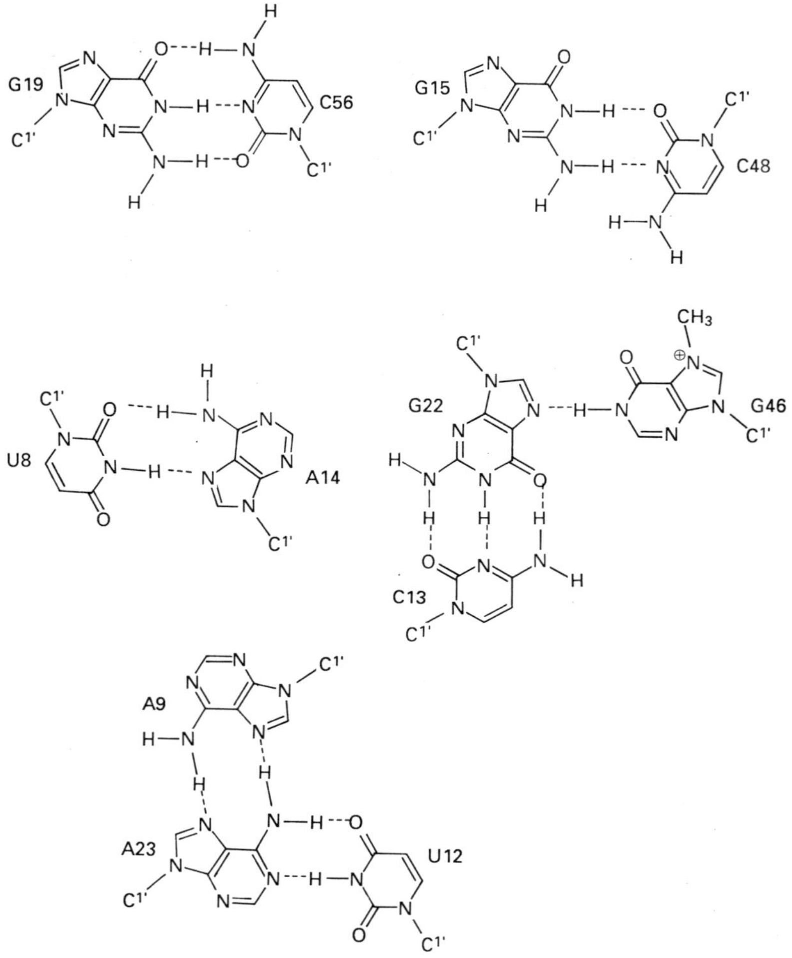

Fig. 6.6 A sample of the tertiary hydrogen-bond interactions found in tRNA. Except for the G19:C56 pair, all of the others involve non-Watson–Crick interactions.

6.3.3 Prediction of higher-order structure in RNA

Some features of secondary structure can be predicted from thermodynamic estimates of helix stabilities based on nucleotide sequence, the number and nature of the different base-pairs, and destabilizing features such as loops, bulges, and base-pair mismatches etc. The contribution of each of these features to the overall stabilization energy must be established empirically (Section 2.4.4).

Progress in the chemical synthesis of oligoribonucleotides (Section 3.5) is making available many model compounds in amounts suitable for the study of their biophysical properties. Such studies should greatly enhance the reliability of computer predictions of structure in the future.

Summary

RNAs adopt specific three-dimensional structures which are important for their interactions with proteins and for their biological functions. Secondary structures of RNAs can be determined by a combination of techniques including the use of chemical and enzymatic reagents selective for bases in either single-stranded environments or in base-pairs and the use of phylogenetic comparisons. Reverse transcriptase mapping depends on the identification of blocks to reverse transcriptase-catalysed DNA synthesis that arise as a result of chemical modifications carried out in bases in RNA.

Transfer RNAs are folded into a cloverleaf secondary structure containing stems and loops that are mostly conserved in size. Additional tertiary interactions occur between nucleotides distant in the primary sequence and involve unusual types of base-pairs. These interactions are crucial for the folding of tRNA into an L-shaped tertiary structure.

Prediction of higher-order structure in RNAs other than tRNAs involve calculations of thermodynamic stabilities of helixes and empirical estimates of their contributions to the overall stabilization energies.

6.4 Biosynthesis of RNA

RNA is produced in cells by transcription of DNA. The initial products of transcription in most cases are longer than the mature products. The additional nucleotides in these longer RNAs, called **precursor RNAs**, may occur in regions **upstream** (5′-leader) or **downstream** (3′-trailer) of the sequence found in the mature RNA or even within the precursor RNA (called **introns** or **intervening sequences, IVS**). During the processing of RNA (**maturation**) these extra sequences are removed (**splicing** in the case of introns) by a variety of different nucleases, each being specific for a certain reaction on a specific class of RNA. Some of these enzymes are proteins, others are ribonucleoprotein particles and yet others are ribonucleoproteins in which the RNA component contains the essential activity. In some cases, the precursor RNA contains its own processing activity

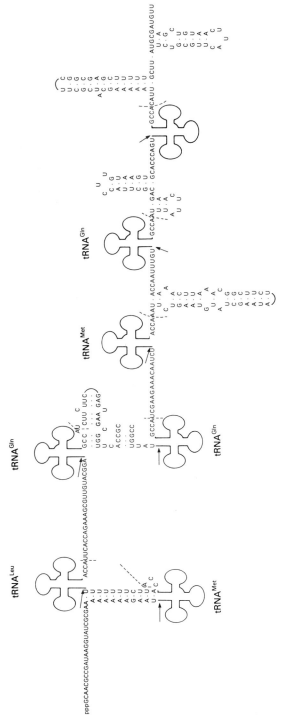

Fig. 6.7 Schematic depiction of a multimeric tRNA gene transcript containing seven tRNA sequences. The tRNAs are indicated in the form of clover-leaf structures. Coloured arrows indicate sites of processing by RNase P. Dashed lines indicate the 3'-end of the mature tRNAs (adapted from Nakajima, N., Ozeki, H., and Shimura, Y. (1981). *Cell.* **23**, 245. Copyright (1981) Cell Press).

(Section 6.5.2)! The maturation process may also involve post-transcriptional addition of nucleotides not encoded in the gene. For example, the 5'-end of a eukaryotic mRNA undergoes a **capping reaction** and the 3'-end of eukaryotic mRNA is extended by a stretch of poly A (Section 6.4.3).

6.4.1 Transfer RNA

The biosynthesis of transfer RNA is a process that is well understood. In bacteria, tRNA transcripts contain either single tRNA sequences (monomeric) or many tRNA sequences (multimeric). In eukaryotes, the transcripts are mostly monomeric, since each tRNA gene possesses its own promoter.

The precursor tRNA sequences fold into tRNA-like structures. This folding appears to be necessary for recognition of the precursors by processing enzymes which then act to remove the leader and trailer sequences. In the majority of cases, the maturation of the 5'-end precedes that of the 3'-end, but only after both ends have been matured are any intervening sequences excised. However, the number of tRNA genes that contain intervening sequences is only a small fraction of all tRNA genes.

The processing of a monomeric or a multimeric tRNA transcript (Fig. 6.7) involves the enzyme RNase P which acts to release the mature 5'-end of the tRNA. This enzyme is a non-covalent complex of RNA and protein in which the RNA component is essential for activity (Section 6.5.2). At the 3'-end, either exo- or endonucleases trim the trailer sequence. In the cases where the 3'-terminal universal CCA sequence is not encoded by the tRNA gene (as in all eukaryotic tRNA genes), the CCA sequence is added by a tRNA nucleotidyl transferase.

Most tRNAs contain a number of modified bases. These are introduced post-transcriptionally, usually by direct enzyme modification of a pre-existing nucleoside. However in some cases, an unmodified base is excised by the action of glycosylases and replaced by a modified one. Some of these modified bases are present in virtually all tRNAs, for example, ribothymidine (T, 5-methyluridine) and pseudouridine (ψ, an isomer of uridine with a C5-to-C1' glycosidic bond) (Section 3.1.1).

6.4.2 Ribosomal RNA

Ribosomal RNA is the most abundant RNA in the cell. Its synthesis is highly regulated because in periods of rapid cell growth the number of ribosomes increases. Biosynthesis is regulated in such a way that the three rRNAs required for assembly of ribosomes are produced in equal amounts. Thus in a bacterial cell, one large RNA transcript contains one copy each of the 5S, 16S, and 23S rRNAs. This transcript also contains some specific tRNAs (Fig. 6.8).

RNase III is an enzyme specific for cleaving double-stranded regions and also appears to be involved in rRNA maturation. It cuts the transcript in the short

Fig. 6.8 Schematic depiction of an *E. coli* rRNA precursor. Arrows indicate sites of processing by RNase III.

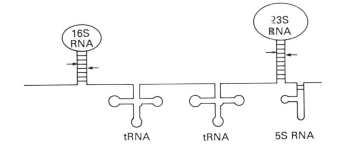

double-helical regions which are found on each side of the 16S and of the 23S rRNA sequence. It is interesting to note that the ribosomal proteins start to assemble with the RNAs soon after the beginning of transcription and well before RNase III cleavages occur.

In eukaryotic cells, the genes for 5.8S, 18S, and 28S rRNAs are linked. These three genes are transcribed as a long precursor by RNA polymerase I, whereas the gene for 5S RNA is transcribed separately by RNA polymerase III. In a few cases, rRNA genes contain intervening sequences (Section 6.5.2).

In both bacterial and in eukaryotic cells, assembly of ribosomal proteins with RNA takes place before the precursors are processed. Since transcription in a eukaryotic cell takes place in the nucleus and translation takes place in the cytoplasm, ribosomal proteins must enter the nucleus and assemble there on the rRNA precursor. The fully assembled ribosomes then leave the nucleus and participate in protein synthesis in the cytoplasm.

6.4.3 Messenger RNA

The formation of mRNA is a process that requires initiation and termination at precise sites on a DNA template (Sections 5.3 and 5.5). All mRNA transcripts begin with either guanosine 5'-triphosphate or adenosine 5'-triphosphate and at the 3'-end there is a free hydroxyl group.

In bacteria and other prokaryotes, there are no processing or modification reactions necessary to produce mature mRNA. In eukaryotic cells, transcription and translation take place in different cellular compartments, and thus mRNA is synthesized in the nucleus and transported to the cytoplasm for use in protein synthesis. Organelles such as mitochondria and chloroplasts in plant cells have independent machinery for mRNA synthesis.

Eukaryotic mRNA transcripts are extensively modified before transport. At the 5'-end, a highly unusual structure called a **cap** is introduced. To achieve this, the 5'-triphosphate of the mRNA transcript is hydrolysed to a diphosphate and a guanosine 5'-phosphate is transferred from GTP to the 5'-end to give a 5'–5' pyrotriphosphate linkage. The N-7 position of the terminal guanine is then methylated using *S*-adenosylmethionine. This resultant cap structure (7MeG(5'–5')-

pppXpY . . .) is further methylated in some cases at the 2′-hydroxyl position of nucleotide X or both nucleotides X and Y. At the 3′-end of the mRNA, a **poly A** tail of up to 200 adenylate residues is added by the enzyme poly A polymerase. The addition of poly A occurs after specific endonucleolytic processing close to the 3′- end of the mRNA transcript. Both the cap and the poly A appear to be import-ant to translation and to the stability of the mRNA, but the poly A may also be necessary for transport.

Another feature of eukaryotic mRNA transcripts is the frequent occurrence of **intervening sequences** (introns) (Section 5.5) which are located between coding sequences (**exons**) and which are removed by a process known as **splicing** (Fig. 6.9). All intervening sequences begin with the sequence G:U and end with A:G.

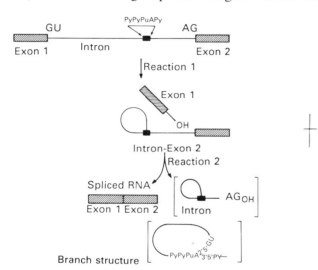

Fig. 6.9 A scheme for processing of introns in mRNA precursors. The two exons are shown as coloured bars and the intron as a thin line. The solid box within the intron contains a consensus sequence and the 2′-hydroxyl group of the A residue within it is involved in the formation of a branch point during splicing (adapted from Brody, E. and Abelson, J. (1985). *Science*, **228**, 963–7. Copyright (1985) AAAS).

Splicing can be considered to be a two step transesterification reaction. In the first step, the 2′-hydroxyl group of an A-residue, found within the consensus sequence PyPyPuAPy in the intron, attacks the phosphodiester bond located at the 5′-exon–intron junction. This results in the cleavage of the phosphodiester bond and the formation of a **branch structure (lariat)** in the intron. In the second step, the newly generated 3′-hydroxyl group in the 5′-exon (exon I of Fig. 6.9) attacks the phosphodiester bond located at the 3′-intron–exon junction. This results in an RNA in which the two exons have been spliced together and the intron originally located between them removed as a lariat structure.

Some pre-mRNAs contain more than one intron which interrupt the coding sequence. In special cases, these introns are spliced out in more than one way (**alternative splicing**) to produce different mRNAs. For example, such mRNAs may code for proteins which differ only slightly in their activities, but which are located in different compartments of a cell. However, when alternative splicing results in a change in the reading frame of the mRNA, the proteins produced will be quite different. Alternative splicing may play a very important role in cell metabolism.

The machinery of splicing is currently under much investigation but it is not yet fully understood. Several different complexes of proteins and RNAs are thought to be involved in splicing. Of these, small nuclear ribonucleoproteins (**snRNPs**) are the most crucial. Each snRNP ('snurp') is characterized by an unique and specific RNA (e.g. U1 RNA, Fig. 6.10). The function of a snRNP in assembly of the splicing complex (**spliceosome**) is to interact with the nuclear pre-mRNA (by complementary base-pairing of the pre-mRNA with the specific RNA component), to recognize the splice sites and position the exons to be joined, and to hold together the resultant spliced pieces.

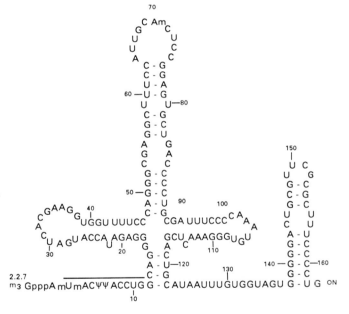

Fig. 6.10 Proposed secondary structure of U1 RNA, one of the components of small nuclear ribonucleoprotein particles (snRNPs) (adapted from Mount, S. M. and Steitz, J. A. (1981). *Nucleic Acids Research*, **9**, 6351–68).

Summary

The initial products of transcription in most cases are RNA precursors which are longer than the mature products.

tRNA precursors fold into tRNA-like structures and both 5'-leader and 3'-trailer sequences are enzymatically removed. In those cases where the 3'-terminal CCA sequence is not encoded by the tRNA gene, the CCA sequence is added by a tRNA nucleotidyl transferase. Modified bases are introduced post-transcriptionally.

Ribosomal RNA transcripts in bacterial cells contain one copy each of 5S, 16S, and 23S rRNA which are released from the transcript after cleavage by RNase III. Ribosomal proteins begin to assemble with the transcripts before RNA processing takes place. In a eukaryotic cell, transcription takes place in the nucleus and ribosomal proteins are imported from the cytoplasm for assembly. Fully assembled ribosomes are exported to the cytoplasm for use in protein synthesis.

The 5'-end of an mRNA transcript contains either GTP or ATP. Bacterial mRNA does not require further processing. In eukaryotic cells, a cap is introduced at the 5'-end and a poly A tail at the 3'-end. Intervening sequences are removed by splicing, a process involving a number of different complexes of proteins and RNAs, the most crucial being small ribonucleoproteins (snRNPs).

6.5 Catalysis by RNA

The catalysis of reactions in biology has long been considered exclusively to be the property of proteins. However, much recent excitement has been generated by the discovery of several cases of catalysis by RNA and crowned by the award of the 1990 Nobel Prize for Chemistry to Sydney Altman and Tom Cech. Classically, an enzyme is defined as a macromolecule which (i) enhances the rate of a reaction by lowering its activation energy, (ii) is specific with regard to substrates and products, and (iii) is a true catalyst (i.e. is not consumed in the reaction). Although proteins have a wider range of functional groups than does RNA, there is clearly enough chemical variety in an RNA to provide specific substrate binding in well defined sites as well as functionalities for general acid-base catalysis. RNA enzymes are sometimes called **ribozymes**.

RNA can act as an enzyme either by itself or when complexed to a protein. To date, all well-characterized reactions catalysed by RNA involve other RNAs as substrates. There have been suggestions that RNA catalysis may be also involved in the degradation of glycogen.

6.5.1 Ribonuclease P

RNase P is an endonuclease which is responsible for producing the 5'-ends of mature tRNA (Section 6.4.1). This ribonucleoprotein contains an RNA which is essential for catalytic activity, since the activity is lost if the RNA component is destroyed by nuclease digestion. There is much evidence to suggest that the enzyme is present in every cell and organism, but the best understood enzyme is that from *E. coli* which consists of an RNA of 377 nucleotides and a protein subunit of 14 kDa. At high Mg^{2+} concentration (non-physiological) *in vitro*, the RNA by itself can carry out accurate cleavage of the precursor tRNAs, albeit at a somewhat slower rate than when the protein is present. It is believed that the high ionic strength overcomes the repulsion between the anions of the RNA enzyme and of the substrates. The protein by itself has no enzyme activity.

The enzymatic reaction does not need energy, but it is dependent on the presence of native RNA structure in both the enzyme and substrate. There is also a specific requirement for divalent metal ions (Mg^{2+} or Mn^{2+}). By use of mutagenesis techniques and by phylogenetic comparisons, a model has been derived of the secondary structure of the RNA component of RNase P from *E. coli* (Fig. 6.11). Although the primary sequence of this RNA varies widely between different organisms, its secondary structure is well conserved. RNase Ps obtained from eukaryotes

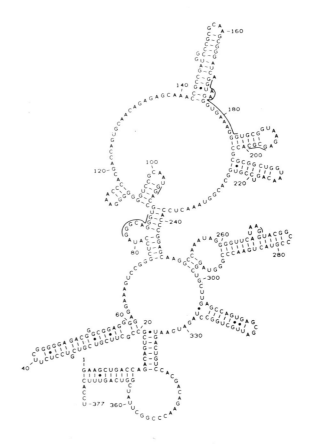

Fig. 6.11 Proposed secondary structure of M1 RNA, the catalytically active component of *E. coli* RNase P. Regions protected by C5 protein from RNase T1 digestion are marked by a solid line next to the sequence (adapted from James, B. D., Olsen, G. J., Liu, J., and Pace, N. R. (1988). *Cell*, **52** 19–26. Copyright (1988) Cell Press).

appear to be more complex, however, and so far none of these RNAs has been found to be catalytically active by itself. While it is accepted that the catalytic activity resides in the RNA, the protein component of the enzyme may be needed for binding to substrate.

6.5.2 The ribosomal RNA of *Tetrahymena*

An exciting story has emerged from the study of the biosynthesis of the nuclear 26S rRNA from the protozoa *Tetrahymena thermophila*. Gene sequencing studies showed that the single rRNA gene had a 413 base-pair intron. Attempts to purify the enzyme responsible for removing the intervening sequence in the rRNA showed that the splicing activity resides totally within the rRNA precursor molecule. This self-splicing is brought about by two consecutive transesterification reactions (Fig. 6.12). The reaction sequence is initiated when the 3'-hydroxyl group of an external (free) guanosine or guanosine 5'-phosphate displaces the nucleotide

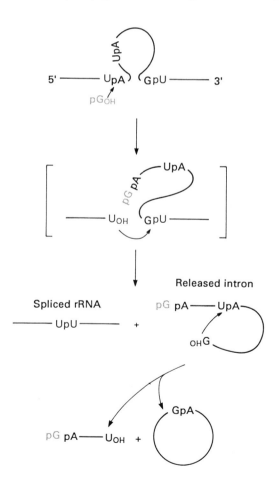

Fig. 6.12 Steps in the self-splicing of *Tetrahymena thermophila* rRNA precursor. Exon sequences are shown in red and the intron sequence in black. The pG$_{OH}$ which initiates the splicing reaction is shown in partial colour (adapted from Zaug, A. J., Grabowski, P. J., and Cech, T. R. (1983). *Nature*, **30**, 578–83. Copyright (1983) Macmillan Magazines Ltd.).

at the 3'-end of the 5'-exon and adds itself via a normal 3'-to-5' phosphodiester bond to the 5'-end of the intervening sequence. The newly created 3'-hydroxyl group of the 5'-exon then attacks the phosphodiester bond preceding the 5'-terminal nucleotide of the 3'-exon, thus generating the spliced RNA. The released intervening sequence (L-IVS) then undergoes a cyclization reaction which is initiated by nucleophilic attack of the newly-generated 3'-hydroxyl group on a phosphodiester bond near the 5'-end of the IVS. This creates a short oligonucleotide (15-mer) which at its 5'-end carries the original G residue which initiated the chain of reactions. A circularized and shortened intervening sequence (C-IVS) is also produced.

What are the requirements of these reactions and where does the energy come from? It turns out that no free energy change is needed for this concomitant cleavage and rejoining of RNA. The reaction seems to be driven by a high concentration of guanosine, or a derivative of guanosine, relative to the RNA. However, it is clear that the RNA must form a particular structure in order to render the specific phosphodiester bonds susceptible to cleavage and to provide the binding site for the 'substrate'. The binding site for the initial RNA substrate contains a **guide sequence** (CUCUC) which positions the scissile bond in the correct orientation by base-pairing to a complementary sequence in the RNA.

Even more remarkably, it has been shown that a shortened form of the intervening sequence (L-19IVS) can act as a true enzyme and can carry out a range of interesting reactions involving nucleotidyl transferase, phosphoryl transferase, phosphomonoesterase, and phosphodiesterase activities. In this respect the molecule 'mimics' the generic activity of a number of different protein enzymes.

6.5.3 Group I introns

Mitochondrial mRNA and rRNA genes often contain introns. Based on a group of conserved sequence elements, several of these introns have been classed as Group I introns. Each intron contains six conserved sequence elements which can form complementary pairs (A to B, 9L to 2, and 9R to 9R') (Fig. 6.13). These elements also occur in the *Tetrahymena* rRNA IVS. These mitochondrial Group I intervening sequences each give rise to a structure which allows them to undergo self-cleavage reactions and some will also self-splice *in vitro*. In the case of *Tetrahymena* rRNA IVS, the reaction requires the addition of guanosine or guanosine 5'-phosphate.

6.5.4 Group II introns

The mitochondrial mRNA precursors contain a second group of self-splicing intervening sequences. These RNAs lack the characteristic group I consensus elements and they splice by a different mechanism which does not require an external nucleotide substrate. Like the splicing of introns in nuclear pre-mRNA (Section 6.4.3), the mechanism proceeds through a branched 'lariat' intermediate structure

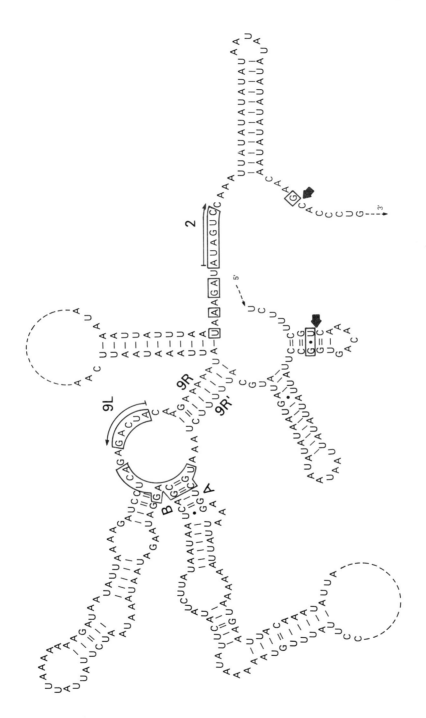

Fig. 6.13 Model for secondary structure of a Group 1 mitochondrial intron. Solid arrows (colour) indicate the splice sites. Each exon ends in a U residue and each intron ends in a G residue. Boxed sequences shown in colour are present in most Group 1 introns (adapted from Michel, J. and Dujon, B. (1983). *EMBO Journal*, **2**, 36).

(Fig. 6.9) which is produced by the attack of a 2'-hydroxyl group of an internal adenylate on the phosphate of the 5'-splice site. However, unlike nuclear pre-mRNA splicing, the splicing activity of Group II introns is an inherent activity of the RNA and can proceed *in vitro* without added proteins or ribonucleoprotein complexes.

6.5.5 RNA self-cleavage

Another type of catalysis reaction involves intramolecular cleavage of RNA that produces 5'-hydroxyl and 2',3'-cyclic phosphate ends. The mechanism is believed to involve a transesterification reaction where the 2'-hydroxyl group plays a crucial role.

The first example was the discovery of sequence-specific cleavage of the D-loop of yeast phenylalanine tRNA by Pb^{2+}. Here, it was demonstrated that the three-dimensional structure of the tRNA influences the reactivity of a particular phosphodiester bond. Analysis of the crystal structure of the complex revealed how Pb^{2+} is bound to a specific pocket in the RNA comprising bases U59 and C60 in such a way that the metal ion is oriented properly towards the polynucleotide chain (Fig. 6.14). The active species is probably $(PbOH)^+$ which has a pK_a of about 7. Removal of the proton from the 2'-hydroxyl group of dihydrouridine at position-17 leads to nucleophilic attack on the phosphodiester bond between D17 and G18 resulting in the specific cleavage.

More recent interest has centred upon the self-cleaving RNAs contained in plant viruses. For example, the satellite RNA of tobacco ringspot virus, which has a size

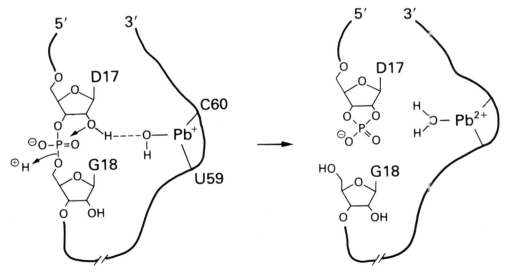

Fig. 6.14 Proposed mechanism for the cleavage of a phosphodiester bond in yeast tRNA^{Phe} by Pb^{2+} bound to a specific site on the tRNA (adapted from Cech, T. R. and Bass, B. L. (1986), *Annual Review Biochemistry*, **55**, 599–629. Copyright (1986) Annual Reviews Inc.).

of only 359 bases, undergoes a self-cleavage reaction that is crucial to the replica-tion of this RNA. Specific phosphodiester cleavage reactions have also been dis-covered in several satellite RNAs of other plant viruses, in viroid RNA, and in RNA from newt and from the hepatitis delta virus. In all cases, the RNAs contain sequences which can be folded into a particular secondary structure called a **ham-merhead**. In some cases, notably viroid and newt RNAs, the active structure prob-ably involves a double hammerhead (Fig. 6.15). Self-cleavage rates are rapid and the only external requirement is for Mg^{2+}. Each hammerhead contains three base-paired sections of which stems I and II do not show sequence conservation, but must be correctly base-paired. However, some nucleotides (shown boxed in Fig. 6.15) are strictly conserved in all hammerheads and it is clear that once again a par-ticular three-dimensional structure is essential. By use of small synthetic oligoribo-nucleotides (Section 3.5) it may be possible to simulate and determine the structure of the active core of the hammerhead by NMR spectroscopy or by X-ray crystallo-graphy and thus give more information about this unusual phenomenon.

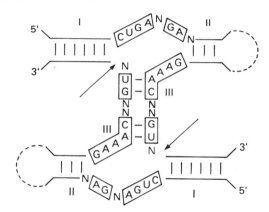

Fig. 6.15 The proposed double hammerhead secondary structure of some self-cleaving RNAs. Arrows indicate the sites of cleavage and boxed residues indicate nucleotides that are strongly conserved. I, II, and III indicate the base-paired stem regions found in each hammerhead.

Based on this knowledge, one should be able to design synthetic RNA molecules having catalytic activity for cleaving a specific sequence in any desired RNA. This may prove to be a powerful approach in understanding the roles of particular RNAs *in vivo* and in the use of catalytic RNAs as possible therapeutic agents against viral and other diseases by inactivation of specific genes.

6.5.6 The universality and role of RNA catalysis

The increasing number of examples of reactions catalysed by RNA suggests that RNA catalysis is a widespread phenomenon. A number of enzymes with essential RNA components have now been identified. In addition, some biochemical reac-tions are catalysed by large ribonucleoproteins or complexes thereof. In these cases, it is possible that the RNA components possess catalytic activity. For example, the peptidyl transferase activity of the ribosome (Section 6.6), which has never been assigned to a particular protein, may well be an RNA-catalysed reaction.

The discovery of enzymatic activity in RNA has also provided a solution to a long-standing problem of how specificity in protein synthesis and replication evolved. The existence of an early 'RNA world' composed of RNA molecules acting as enzymes and carrying out vital functions of life (e.g. self-replication, protein synthesis) is an attractive idea. These early RNA enzymes might not have been particularly specific or efficient and therefore might be expected to have given way to proteins during evolution. If so, one wonders why enzymatic activity has survived in some RNAs. This question can only be answered when more is known about RNA catalysis and especially if new enzymatic activities not involving RNA substrates are discovered.

Summary

RNA can act as an enzyme either by itself or in combination with protein. A catalytic RNA is sometimes called a ribozyme and usually involves RNA as a substrate.

Ribonuclease P contains an RNA subunit essential for catalytic activity. RNase P from *E. coli* can cleave precursor tRNA by itself, but RNase Ps obtained from eukaryotes require the protein component also for catalytic activity.

The ribosomal RNA from *Tetrahymena* carries out self-splicing. The excised intervening sequence has a variety of enzymatic activities.

Some mitochondrial mRNA and rRNA precursors contain intervening sequences that self-splice and are classed as Group I introns. These are analogous to the *Tetrahymena* rRNA. A second class of introns occurs in mitochondrial mRNA precursors. These Group II introns self-splice by a different mechanism which is analogous to the splicing of nuclear mRNA precursors.

Some plant satellite and viroid RNAs undergo self-cleavage reactions. The RNAs can be folded into hammerhead secondary structures.

6.6 Protein synthesis

During protein synthesis, a linear sequence of nucleotides in mRNA is translated by the **ribosome** into a specific sequence of amino acids in protein. Transfer RNAs play a central role by acting as adapter molecules between the nucleotide sequences (**codons**) on the mRNA and the amino acids specified by them. Amino acids destined for incorporation into protein are first attached to tRNA and brought to the ribosome as **aminoacyl-tRNAs**. On the ribosome (Fig. 6.16), codon-1 on the mRNA binds to a specific tRNA, which in turn is attached to the first amino acid. Similarly, codon-2 on the mRNA binds to another tRNA which is attached to the second amino acid. The amino group of the second amino acid then attacks the ester of the first aminoacyl-tRNA to form a peptide bond. The order of amino acids in the protein is thus specified by the sequence of nucleotides in the mRNA. The role of tRNAs in translation depends on two important properties:

(1) specific base-pairing between three nucleotides (**anticodon**) on the tRNA and the mRNA codon; and

(2) attachment of specific amino acids to specific tRNAs.

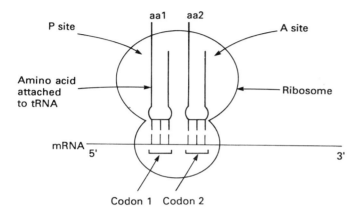

Fig. 6.16 Schematic depiction of a ribosome translating an mRNA into protein. The three vertical lines on the mRNA (red) indicate the nucleotides of the codon and the three vertical lines on the tRNA indicate the complementary anticodon sequence. The A site and P site are the two well-characterized tRNA binding sites on the ribosome.

6.6.1 Aminoacylation

Since there are 20 amino acids in most proteins, at least 20 different tRNA species are needed for protein synthesis. In fact, most cells contain from 60 to 70 different tRNAs. Thus, the same amino acid can be attached to more than one species of tRNA. Such different tRNA species which can be aminoacylated by the same amino acid are called **isoacceptor tRNAs**. Individual tRNA species are usually identified by their amino acid specificity. Thus, tRNA^{Ala} specifies a tRNA which can be aminoacylated with the amino acid alanine. The enzyme which aminoacylates this tRNA is called alanyl-tRNA synthetase (abbreviated as AlaRS).

The accuracy of protein synthesis depends crucially on the covalent attachment of a specific amino acid to a specific tRNA. This reaction, called the **aminoacylation reaction**, is catalysed by a group of enzymes called **aminoacyl-tRNA synthetases**. There are 20 aminoacyl-tRNA synthetases in most cells, each specific for one particular amino acid. Each enzyme is also highly specific for tRNA and can discriminate among tRNAs which have different sequences, but similar three-dimensional structures. Only the correct amino acid becomes attached to the corresponding tRNA. For example, the enzyme alanyl-tRNA synthetase will attach the amino acid alanine to tRNA^{Ala} and not to any other tRNA. The discrimination of tRNAs by aminoacyl-tRNA synthetases involves only a few critical nucleotides in each tRNA. Depending on the individual tRNA species, these positions are localized near the acceptor end, in the anticodon, or are distributed throughout the molecule. The recently determined crystal structure of the complex between *E. coli* glutaminyl-tRNA synthetase and tRNA^{Gln} provides a physical demonstration of protein:RNA contacts in the anticodon and acceptor end regions (see back cover).

Aminoacylation of tRNAs occurs in two steps. First, the synthetase catalyses the reaction of the amino acid and ATP to form an activated aminoacyl-adenylate (Fig. 6.17). Secondly, the enzyme catalyses the transfer of the amino acid to the terminal adenosine of the correct tRNA to form an aminoacyl-tRNA, AMP is released, and the enzyme is free to recycle.

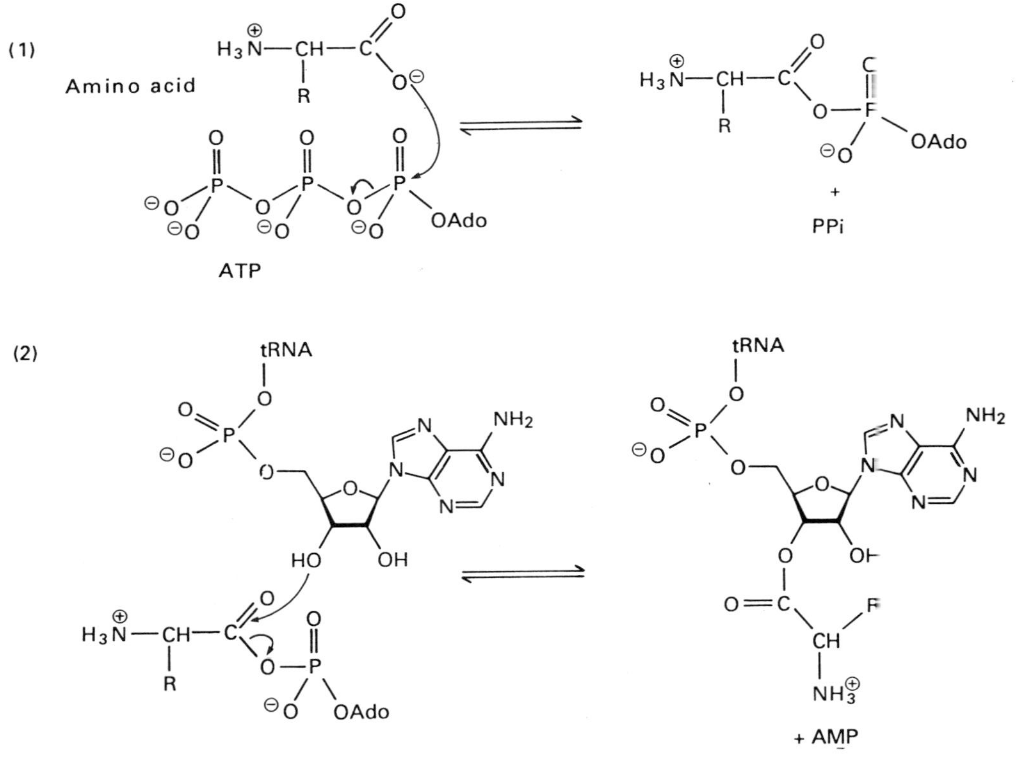

Fig. 6.17 Steps in aminoacylation of tRNAs catalysed by aminoacyl-tRNA synthetases. (1) activation of the amino acid; (2) transfer of the activated amino acid to the correct tRNA.

Following attachment of the amino acid to a tRNA, the amino acid can, in some cases, be further modified by enzymes which are highly specific with respect to the tRNA. For example, methionyl-tRNA is converted into formyl methionyl-tRNA by the enzyme methionyl-tRNA transformylase. This enzyme uses N^{10}-formyltetrahydrofolate to transfer a formyl group onto the methionine tRNA species that is to be used for initiation of protein synthesis. All cells have at least two methionine tRNA species one of which is specially used to initiate protein synthesis.

6.6.2 Ribosomes

The ribosome is a large ribonucleoprotein which is an essential component of the protein synthesizing system. The *E. coli* 70S ribosome contains 51 different proteins and the three rRNAs. The ribosome can be reversibly dissociated into a 30S (small) subunit, containing 21 proteins and the 16S rRNA, and a 50S (large) subunit which contains the remaining 30 proteins and the 5S and 23S rRNAs. The ribosome subunits dissociate and re-associate during their function in protein synthesis. The binding domain for mRNA lies in the 30S subunit whereas the tRNA-binding domain and the catalytic domain for peptide bond formation is largely in the 50S subunit. There are two well-characterized tRNA binding sites on the ribosome. These are adjacent to one another and are called the **A site** (aminoacyl-tRNA site) and the **P site** (peptidyl-tRNA site) (Fig. 6.16). There is also good evidence for a third site, called the **E site** (exit site), thought to be located mostly on the 50S subunit.

6.6.3 Messenger RNAs

In bacteria such as *E. coli*, mRNA synthesis (**transcription**) and utilization (**translation**) are coupled. Even before the synthesis of one mRNA is completed, part of it may be translated by the ribosome. This is possible because bacteria do not contain different cellular compartments.

Most bacterial mRNAs are **polycistronic**, which means that a single long transcript codes for several proteins. Here, ribosomes can usually bind independently to the various translational start sites on the mRNA and initiate protein synthesis. A polycistronic mRNA may code for a number of proteins which are all part of a related biochemical pathway, as in the case of enzymes involved in tryptophan biosynthesis (Section 5.3.3). On the other hand, it may code for proteins which have no obvious relationship to one another. For example, in *E. coli* a single mRNA codes for three proteins: the sigma protein involved in RNA biosynthesis, the *dnaG* protein involved in DNA biosynthesis, and the *trm* protein involved in base modification of tRNA. The relative amounts of proteins which are made from a polycistronic mRNA can also be quite different, which is the case for the mRNA encoding the sigma, *dnaG*, and *trm* proteins.

In contrast to most bacterial mRNAs, mRNA from eukaryotic cells are mostly monocistronic and each mRNA has a single **ribosome binding site** and codes for a single protein.

Several factors determine the relative activities of mRNAs in protein synthesis. These include the secondary and tertiary structure of mRNA and sequences upstream (on 5'-side) of the translational start site in mRNAs. In a polycistronic mRNA with independent translational initiation sites, the affinity of ribosomes for one site may be quite different to that for another site, probably because of differences in accessibility of the mRNA to the ribosome. In such a case, some proteins encoded by the mRNA would be made in significantly different amounts to others.

In the case of the mRNAs coding for most of the ribosomal proteins in *E. coli*, apart from the ribosome binding site nearest to the 5′-end, all the others are inaccessible because of particular mRNA secondary and tertiary structure. As the ribosomes translate this segment and approach the next coding region, the mRNA structure downstream is opened up and the downstream translational start site now becomes accessible. This process is called **sequential translation** and is one of the strategies used by *E. coli* to ensure that proteins which are part of a multisubunit complex, such as the ribosome, are made in equal amounts.

Sequences upstream of the translational start site can also have a major effect on relative efficiencies of mRNA translation. In *E. coli*, base-pairing between a string of pyrimidine nucleotides near the 3′-end of 16S RNA and a purine-rich sequence about ten nucleotides upstream of the translational start site on the mRNA helps to align the initiating AUG codon on the ribosome for translation. The extent of base-pairing here is one of the factors which determines the relative affinity of the ribosome for a translational start site on the mRNA.

In eukaryotic cells, sequences which immediately flank the AUG start site are also important in translation. Since eukaryotic mRNAs are monocistronic, the first AUG near the 5′-end of mRNA is normally used as the start site for translation. Here, the optimal sequence for initiation has been identified as ACC*A*UG*G*, where AUG is the initiator codon (Section 6.6.8). It is not yet established whether a sequence in 18S rRNA of eukaryotic ribosomes, complementary to the ACC sequence, plays a role analogous to that of *E. coli* 16S rRNA in aligning the mRNA on the eukaryotic ribosome.

6.6.4 Protein factors

Several other proteins, called **initiation factors**, **elongation factors**, and **termination factors**, function at specific stages of protein synthesis. Each of these factors associates with the ribosome, carries out a particular function and then dissociates from the ribosome to allow the binding of the next protein factor. GTP acts as an important co-factor for many of these proteins. At each step, GTP is hydrolysed to GDP and inorganic phosphate is released. The hydrolysis of GTP plays a crucial role in recycling of the protein factors from the ribosome.

6.6.5 Steps in protein synthesis

There are three distinct stages in protein synthesis: **initiation**, **elongation**, and **termination**. In initiation, ribosomes bind to the initiation site on the mRNA and associate with fMet-tRNA to form an **initiation complex** (step 1 of Fig. 6.18). In *E. coli*, this reaction requires the participation of three initiation factors IF-1, IF-2, and IF-3. Formation of an initiation complex consists of several steps. First, 70S ribosomes dissociate to form 30S and 50S subunits. IF-3 promotes this dissociation by binding to the 30S subunit to form a 30S·IF-3 complex and by preventing reassociation of the 30S and 50S subunits. The 30S·IF-3 complex then binds both to mRNA at a particular AUG codon and also to formylmethionyl-tRNA (fMet-tRNA). This reaction requires the other two initiation factors, IF-1 and IF-2, as well as GTP. At this point, IF-3 is released allowing the 50S ribosomal subunit to join the 30S·mRNA·fMet-tRNA complex to form the 70S initiation complex. Subsequently, GTP is hydrolysed to GDP, and IF-1 and IF-2 are released from the ribosome. At this stage, the fMet-tRNA is bound to the P site on the ribosome.

The elongation cycle of protein synthesis consists of three distinct processes which are repeated many times (Fig. 6.18, step 2 to step 4). At each elongation cycle, a peptide bond is formed and an amino acid is added to a continuously growing polypeptide chain.

In step 2, an appropriate aminoacyl-tRNA (in this case alanyl-tRNA) binds to the A site (as directed by the particular codon–anticodon interaction) in the form of a ternary complex with elongation factor Tu (EF-Tu) and GTP. Once the correct aminoacyl-tRNA is bound, GTP is hydrolysed to GDP and EF-Tu is released from the ribosome as EF-Tu·GDP complex. In order to regenerate EF-Tu·GTP, the GDP bound to EG-Tu must be exchanged with GTP. This reaction is facilitated by elongation factor EF-Ts which binds to EF-Tu to form EF-Tu·EF-Ts and helps dislodge GDP from EF-Tu. The EF-Tu·EF-Ts complex then combines with GTP to form EF-Tu·GTP and EF-Ts.

In step 3, a protein on the ribosome, peptidyl transferase, catalyses the formation of a peptide bond. This involves attack by the free amino group of aminoacyl-tRNA bound to the A site on the carbonyl group of fMet-tRNA bound to the P site, resulting in formation of a peptide bond at the expense of an ester bond (Fig. 6.18). The peptide chain which was linked to the tRNA on the P site is thus extended by one amino acid and the incoming amino acid is added at the C-terminus of the growing peptide chain.

In step 4, the peptidyl-tRNA on the A site is **translocated** to the P site, the displaced, deacylated tRNA on the P site moves to the E site (not shown in Fig. 6.18) and eventually falls off the ribosome, and the A site is empty once again. The translocation step thus moves the ribosome three nucleotides along with respect to the mRNA and brings the next codon into the reading frame. The translocation step is catalysed by another elongation factor, EF-G, which also uses GTP as a co-factor. In the process of translocation, GTP is hydrolysed to GDP and EF-G is released from the ribosome.

After many cycles of elongation, involving steps 2, 3, and 4, the ribosome eventually reaches a codon on the mRNA, in the case UAA, which specifies termination of protein synthesis. This leads to binding of an appropriate termination factor to the ribosome, more commonly called a release factor (RF) as an RF·GTP complex. Binding of the RF results in hydrolysis of the ester linkage between the completed polypeptide chain and the last tRNA. The polypeptide and tRNA are released from the ribosome and once again GTP is hydrolysed to GDP. RF is then released

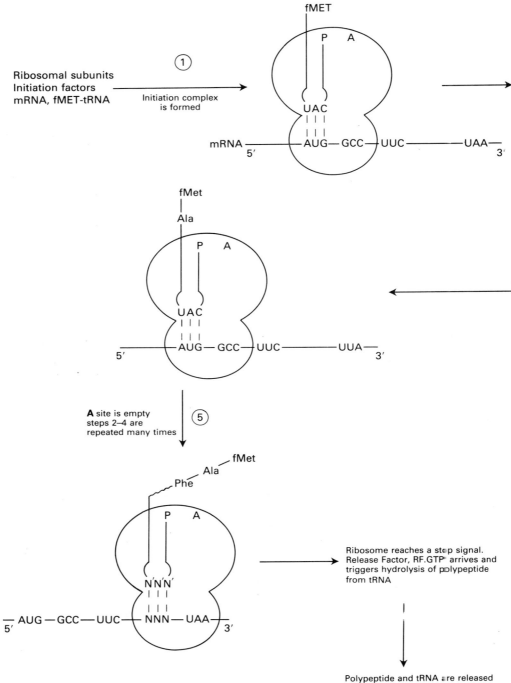

Ribosomal subunits
Initiation factors
mRNA, fMET-tRNA

① Initiation complex
is formed

A site is empty
steps 2–4 are
repeated many times ⑤

Ribosome reaches a stop signal.
Release Factor, RF.GTP arrives and
triggers hydrolysis of polypeptide
from tRNA

Polypeptide and tRNA are released
from the ribosome

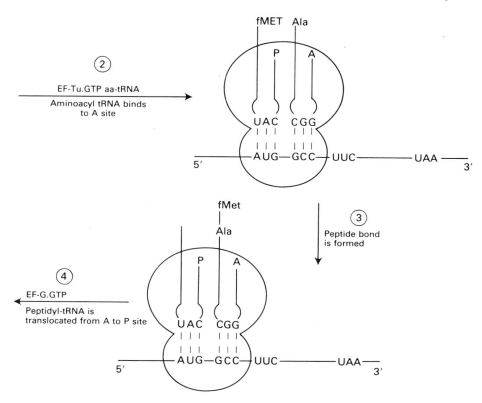

Fig. 6.18 Schematic depiction of the steps in protein synthesis.

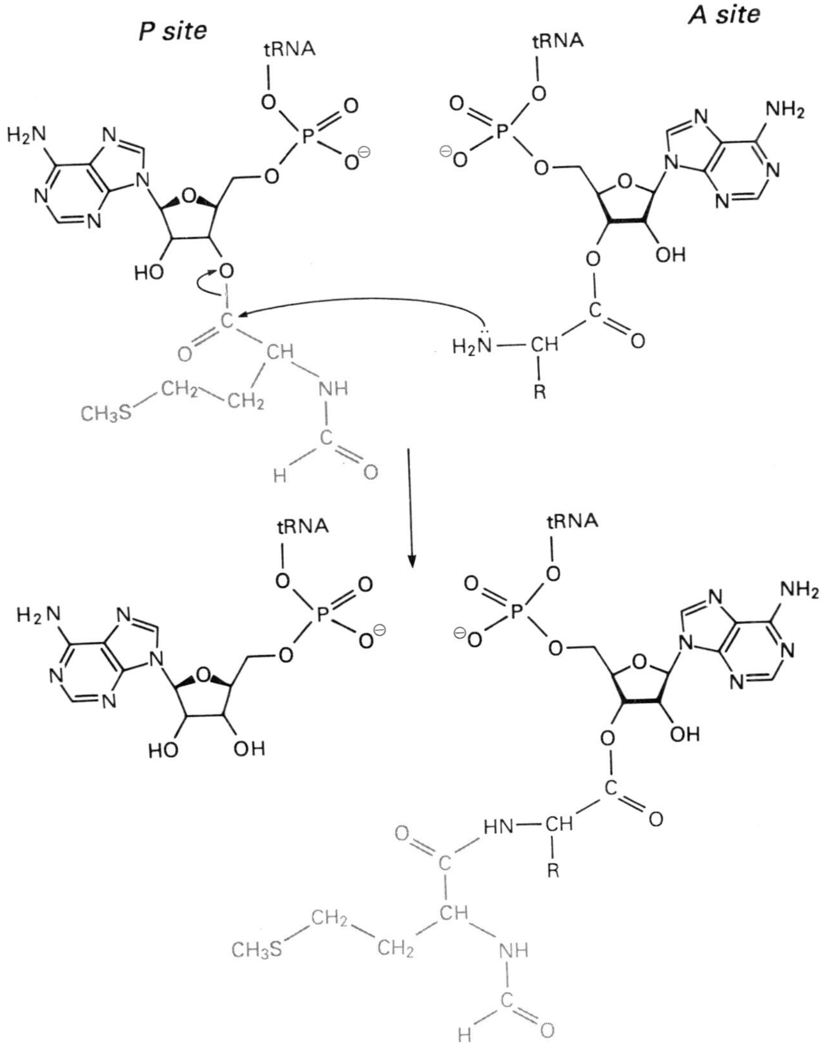

Fig. 6.19 Mechanism of peptide bond-formation. The aminoacyl portion is shown in full colour and the fMet portion is shown in partial colour.

from the ribosome. In the absence of any tRNA bound to the ribosome, the 70S ribosome falls off the mRNA, associates with IF-3 to form a 30S·IF-3 complex, and is now free to participate in another round of protein synthesis.

6.6.6 Polyribosomes

After a ribosome becomes attached and moves along the mRNA, another ribosome can follow and bind at the same initiation site. Thus, a single mRNA is translated by may ribosomes at the same time. An mRNA which is being translated is often found as part of a complex called a polyribosome. In a polyribosome, the individual ribosomes and the nascent polypeptide chains are held together by mRNA (Fig. 6.20). Gentle treatment of such a complex with ribonuclease breaks the mRNA chain at points where it is not protected by bound ribosomes. This results in conversion of a polyribosome into individual ribosomes which contain fragments of mRNA.

Fig. 6.20 A mRNA (red) being translated simultaneously by many ribosomes to give a polyribosome.

6.6.7 Processing and transport of proteins

Virtually every protein is altered in one way or another following its synthesis on the ribosome. Protein synthesis is always initiated with either formylmethionine in case of bacterial, mitochondrial, and chloroplast proteins or with methionine in the case of eukaryotic, cytoplasmic proteins. However, very few proteins contain fMet or Met at their N-termini. The reason for this is that cells contain enzymes that remove the fMet- or Met-residue from the N-terminus, often as the nascent polypeptide chain is being elongated on the ribosome. Besides these changes, there are several other examples of covalent modifications that are introduced into a protein. These include hydroxylation, methylation, acetylation, phosphorylation, addition of sugar residues, and cross-linking of one or more polypeptide chains. A common and important modification involves cleavage of proteins by proteases.

6.6.8 Genetic code

The genetic code was established in 1965 largely through the work of Nirenberg, of Khorana, and of Ochoa (Section 1.5). They showed that the codon for an amino acid consisted of three consecutive nucleotides (**codons**) (Fig. 6.21). Of the 64 possible codons, 61 specify the 20 amino acids and three, UAA, UAG, or UGA, specify termination of protein synthesis. AUG is the codon used to specify initiation, but occasionally GUG and UUG are also used.

A remarkable feature of the genetic code is that, with very few exceptions, it is totally universal. The genetic code in bacteria, fungi, plants, fish, and animals is the same. Except for methionine (AUG) and tryptophan (UGG), all the other amino

Fig. 6.21 The genetic code. The 64 codons are divided into 16 four-codon boxes. The four codons of a codon box differ only in their 3'-terminal nucleotide. Full colour shows where an amino acid is specified by all four codons of a codon box, partial colour shows where an amino acid is specified by two of the four codons, and grey shows where an amino acid is specified by a single codon.

First nucleotide (5'-end)	Middle nucleotide U	C	A	G	Third nucleotide (3'-end)
U	Phe	Ser	Tyr	Cys	U
	Phe	Ser	Tyr	Cys	C
	Leu	Ser	STOP	STOP	A
	Leu	Ser	STOP	Trp	G
C	Leu	Pro	His	Arg	U
	Leu	Pro	His	Arg	C
	Leu	Pro	Gln	Arg	A
	Leu	Pro	Gln	Arg	G
A	Ile	Thr	Asn	Ser	U
	Ile	Thr	Asn	Ser	C
	Ile	Thr	Lys	Arg	A
	Met[a]	Thr	Lys	Arg	G
G	Val	Ala	Asp	Gly	U
	Val	Ala	Asp	Gly	C
	Val	Ala	Glu	Gly	A
	Val	Ala	Glu	Gly	G

[a]AUG is also used as a start signal

acids are specified by more than one codon, most often by two, four or six codons. The codons that specify the same amino acid are often closely related and differ only in third position. The two codons for histidine, CAU and CAC, differ only in that the third nucleotide can be either of the pyrimidines. Similarly, the two codons for glutamine, CAA and CAG, differ only in that the third nucleotide can be either of the purines. Likewise, the four codons for proline, CCU, CCC, CCA, and CCG, differ only in that the third nucleotide can be either U, C, A, or G.

It might be expected that at least 61 different tRNAs would be needed to read the genetic code. This, however, is not the case. Although selection of correct tRNAs on the ribosome depends on base-pairing between mRNA codon and tRNA anticodon, this base-pairing need not always be strictly of the Watson–Crick type for all three base pairs. A slightly relaxed form of base-pairing in one of the base-pairs reduces the minimum number of tRNAs needed to read the genetic code to 32. The rules for base-pairing between the codon in mRNA and the anticodon in tRNA were formulated by Crick as the so-called **wobble hypothesis**. According to this hypothesis, the first two base-pairs, between the first and second nucleotides of the codon and the third and second nucleotides of the anticodon (**a** to **a'** and **b** to **b'** in Fig. 6.22) are of the Watson–Crick type, whereas the third base-pair (**c** to **c'**) can be more flexible (Section 2.1.2). The kinds of base-pairs allowed are listed (Table 6.4). G in a tRNA can pair with either U or C, a modified derivative of U in tRNA can pair with either A or G, and I in tRNA can pair with either U, C, or A in an mRNA. The wobble pairs that are proposed to form between tRNA and mRNA are G:U, I:U, I:C, and I:A (Figs 6.23 and 2.9). Thus, given the organization of the

Fig. 6.22 Base-pairing between the nucleotides of an mRNA codon and a tRNA anticodon. Base-pairing of **a** to **a'** and of **b** to **b'** is strictly of the Watson–Crick type whereas that between **c** and **c'** can be somewhat more flexible.

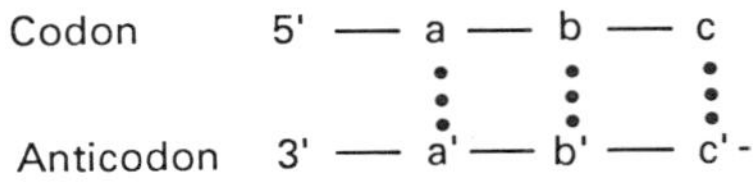

```
Codon       5' — a — b — c
                 :   :   :
Anticodon   3' — a'— b'— c'-
```

Table 6.4. Possible base-pairs between the third nucleotide of the mRNA codon and the first nucleotide of the tRNA anticodon

First nucleotide (5'-end) of anticodon	Third nucleotide (3'-end) of codon
C	G
G	U or C
U*	A or G
I	U, C or A

U*: a derivative of U

genetic code in which the four codons of a codon box specify either the same amino acid or two different amino acids, two tRNAs can read all four codons of a codon box. For example, a histidine tRNA with the anticodon sequence GUG can read both CAU and CAC codons for histidine. Similarly, a glutamine tRNA with U*UG can read both CAU and CAG codons for glutamine. Also in cases where a codon box specifies a single amino acid, two tRNAs can read all four codons. The two tRNAs could either have G to read U and C with U* to read A and G, or I to read U, C and A with C to read G. This choice depends on the organism.

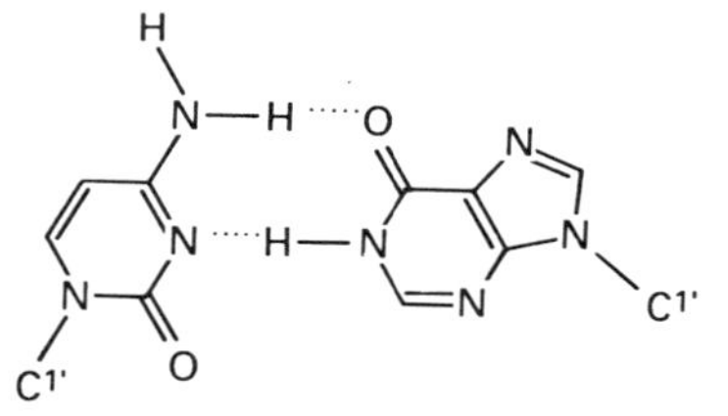

Fig. 6.23 Base-pairing between cytosine and inosine. Other wobble pairs are shown in Fig. 2.9.

In mitochondria, a simpler mechanism to read the genetic code further reduces the number of tRNAs from 32 to 24. A single tRNA with U in the first anticodon position can pair with either U, C, A, or G and read all four codons of a codon box. Since there are eight four-codon boxes each of which specifies a single amino acid (Fig. 6.20), this reduces the number of tRNAs needed by eight.

Mitochondrial protein synthesis is also unusual in providing an exception to the universality of the genetic code. First, there are some differences between the genetic code in mitochondria and the 'universal' genetic code and secondly there are variations in the genetic code among mitochondria from different organisms.

Summary

In protein synthesis, a linear sequence of nucleotides in mRNA is translated by the ribosome into a specific sequence of amino acids in protein. Amino acids are carried to the ribosome as aminoacyl-tRNA. Each tRNA is aminoacylated by a specific amino acid in a reaction catalysed by a specific aminoacyl-tRNA synthetase. One of the two methionyl-tRNAs is formylated and is involved in initiation of protein synthesis.

Ribosomes consist of two subunits. The mRNA and fMet-tRNA are bound by the 30S subunit. The aminoacyl-tRNA binding domain and the catalytic domain are mostly in the 50S subunit. Most bacterial mRNAs are polycistronic whereas most eukaryotic mRNAs are monocistronic. Each translational start site in polycistronic mRNA may be translated independently and may have a different accessibility to ribosomes. mRNAs for ribosomal proteins are usually translated sequentially. Sequences upstream of the AUG start site can subtantially affect the efficiency of translation of both prokaryotic and eukaryotic mRNAs.

The three distinct stages of protein synthesis are initiation, elongation, and termination. Each stage involves a series of reactions that require a number of protein factors. Initiation results in fMet-tRNA becoming bound to the P site on the ribosome. In elongation, an aminoacyl-tRNA becomes bound to the A site on the ribosome, the fMet (or peptide chain) at the P site is transferred to the amino acid of the aminoacyl-tRNA at the A site, and the elongated peptidyl-tRNA is translocated to the P site. After many cycles of elongation, a termination codon is reached which signals the release of the polypeptide and the last tRNA from the ribosome. A single mRNA is translated simultaneously by many ribosomes as a complex called a polyribosome. Many proteins are post-translationally modified before use.

The genetic code is the key that relates the sequence of nucleotide triplets (codons) on the mRNA to the sequence of amino acids in the corresponding protein. Because of flexibility in base-pairing between codons in the mRNA and anticodons in tRNA (the wobble hypothesis), only 32 tRNAs are required to read the 61 codons that specify the 20 amino acids found in proteins. In mitochondria, the number of tRNAs required is reduced to 24.

Further reading

6.1

Poulson, R. (1977). In Stewart, P.R., and Letham, D.S. (ed.) *The ribonucleic acids.* Springer, Weinheim, pp. 333–67.

6.2

RajBhandary, U. L. (1980). Recent developments in methods for RNA sequencing using *in vitro* ^{32}P-labelling. *Fed.Proc.*, **39**, 2815–21.

Stanley, J. and Vassilenko, S. (1978). A different approach to RNA sequencing. *Nature*, **274**, 87–9.

6.3

Ehresman, C., Baudin, F., Mougel, M., Romby, P., Ebel, J-P., and Ehresman, B. (1987). Probing the structure of RNAs in solution. *Nucleic Acids Res.*, **15**, 9109–28.

James, B. D., Olsen, G. J., Liu, J., and Pace, N. R. (1988). The secondary structure of RNase P RNA, the catalytic element of a ribonucleoprotein enzyme. *Cell*, **52**, 19–26.

Moazed, D., Stern, S., and Noller, H. F. (1986). Rapid chemical probing of conformation in 16S rRNA and 30S ribosomal subunits using primer extension. *J. Mol. Biol.*, **187**, 399–416.

Rich, A. and Kim, S-H. (1978). The three-dimensional structure of transfer RNA. *Sci. Amer.*, **238**, 52–62.

Woese, C. R., Gutell, R. R., Gupta, R., and Noller, H. F. (1986). Detailed analysis of higher order structure of 16S-like ribosomal RNAs. *Microbiol. Rev.*, **47**, 621–69.

6.4

McClain, W. H. (1977). Seven terminal steps in a biosynthetic pathway leading from DNA to transfer RNA. *Acc. Chem. Res.*, **10**, 418–25.

Sharp, P. A. (1987). Splicing of mRNA precursors. *Science*, **235**, 766–71.

Stryer, L. (1988). *Biochemistry*, 3rd ed. Freeman, San Francisco, pp. 703–6.

6.5

Cech, T. R. and Bass, B. L. (1986). Biological catalysis by RNA. *Ann. Rev. Biochem.*, **55**, 599–629.

Guerrier-Takada, C. and Altman, S. (1984). Catalytic activity of an RNA molecule prepared by transcription *in vitro*. *Science*, **223**, 285–6.

Symons, R. H. (1989). Self-cleavage of RNA in the replication of small pathogens of plants and animals. *Trends Biochem. Sci.*, **14**, 445–50.

Uhlenbeck, O. C. (1987). A small catalytic oligoribonucleotide. *Nature*, **328**, 596–600.

6.6

Crick, F. H. C. (1966). The genetic code. *Sci.Amer.*, **215**, 55–62.

Hou, Y-M., Francklyn, C., and Schimmel, P. (1989). Molecular dissection of a transfer RNA and basis for its identity. *Trends Biochem. Sci.*, **14**, 233–7.

Khorana, H. G. (1973). Nucleic acid synthesis in the study of the genetic code. *Nobel Lectures: Physiology or Medicine 1963–1970*, American Elsevier, pp. 341–69.

Nomura, M. (1984). The control of ribosome synthesis. *Sci.Amer.*, **250**, 102–14.

Stryer, L. (1988) *Biochemistry*, 3rd ed. Freeman, San Francisco, pp. 91–116.

Watson, J. D., Hopkins, N. H., Roberts, J. W., Steitz, J. A., and Weiner, A. M. (1988). *Molecular Biology of the Gene*, pp. 360–462. Benjamin-Cummings.

Zubay, G. L. (1988). *Biochemistry*, 2nd ed. Macmillan, New York, pp. 928–73.

Covalent interactions of nucleic acids with small molecules

7

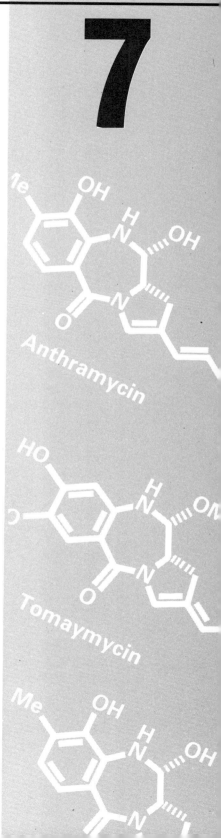

The simple purpose of this chapter is to provide an outline of the more important examples of covalent interactions of small molecules with nucleic acids. Topics relevant to modifications of intact nucleic acids have been chosen, especially as they relate to mutagenic and carcinogenic effects. While much of the early information has come from studies on nucleosides, in many cases the net effect of a reagent on an intact nucleic acid may be quite different from either the sum or the average of its interactions with separate components.

Above all, we have to recognize that studies on the subtle effects of DNA and RNA secondary and tertiary structures on covalent interactions are only in their infancy.

7.1 Hydrolysis of nucleosides, nucleotides, and nucleic acids

Nucleic acids are readily denatured in aqueous solution at extremes of pH or on heating. The phosphate ester bonds are only slowly hydrolysed (Section 3.2.2), but the N-glycosidic bonds are more labile. Purine nucleosides are cleaved faster than pyrimidines while ribonucleosides are more stable than deoxyribonucleosides. Thus, dA and dG are hydrolysed in boiling 0.1 M hydrochloric acid in 30 min, rA and rG require 1 hour with 1 M hydrochloric acid at 100°C, while rC and rU have to be heated similarly with 12 M perchloric acid (Fig. 7.1).

Formic acid has been used to prepare apurinic acid, which has regions of poly-pentose phosphate diesters linking pyrimidine oligonucleotides. Such phosphate diesters are relatively labile since the pentose undergoes a β-elimination in the presence of secondary amines such as diphenylamine. This gives tracts of pyrimidine oligomers with phosphate monoesters at both 3'- and 5'-ends. Total acidic hydrolysis with minimum degradation of the four bases is best achieved with formic acid at 170°C.

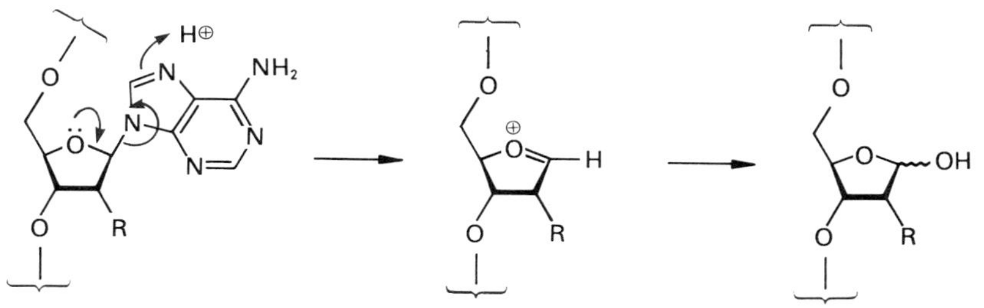

Fig. 7.1   Mild acidic hydrolysis of purine glycosides in DNA (R = H) and RNA (R = OH).

DNA is resistant to alkaline hydrolysis but RNA is easily cleaved because of the involvement of its 2'-hydroxyl groups (Section 3.2.2).

7.2 Reduction of nucleosides

Both the purine and pyrimidine bases are sufficiently aromatic to resist reduction under the mild conditions used, for example, in the hydrogenolysis of benzyl or phenyl phosphate esters. However, hydrogenation with a rhodium catalyst converts uridine or thymidine into 5,6-dihydropyrimidines. Alternatively, sodium borohydride in conjunction with ultraviolet irradiation gives the same products which can lead on by further reduction in the dark and cleavage of the heterocyclic ring.

Reduction of ribonucleosides directly to 2′-deoxynucleosides can be accomplished by one of the several Barton procedures involving tributyltin hydride. A nice example is the synthesis of a mixture of 2′- and 3′-deoxyadenosines which are easily separable (Fig. 7.2). This type of reduction has been widely employed to transform various extensively modified ribonucleosides and their nucleotide analogues into the corresponding deoxyribonucleosides.

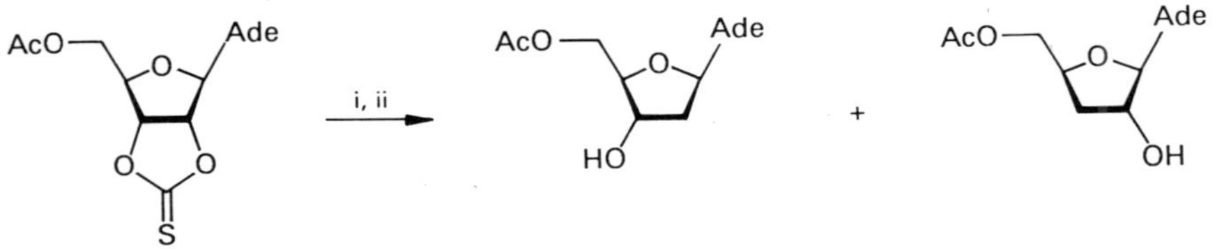

Fig. 7.2 Barton reduction of adenosine to 2′(3′)-deoxyadenosine.
Reagents: (i) Bu$_3$SnH, DMA, AIBN; (ii) NH$_3$, MeOH.

2′,3′-Dideoxynucleosides are valuable for use in DNA sequence analysis and also have some promise for AIDS therapy, both features being related to their chain-terminator activity (Section 5.2). One synthesis involves hydrogenation of 2′,3′-unsaturated nucleosides or an appropriate precursor (Fig. 7.3).

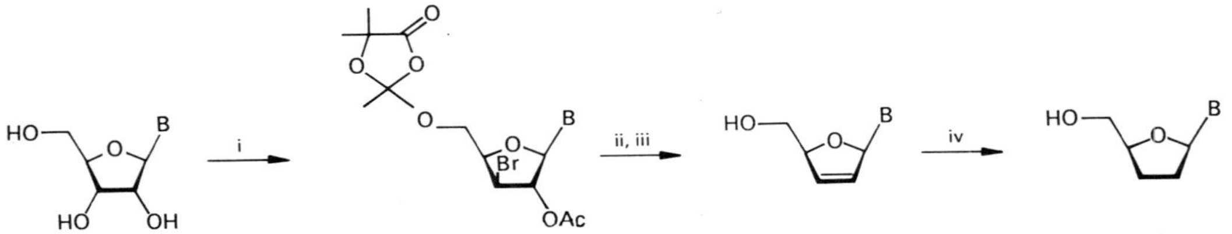

Fig. 7.3 Synthesis of 2′,3′-dideoxynucleosides by reduction.
Reagents: (i) Me$_2$C(OAc)COBr; (ii) Cr^{2+}, (CH$_2$NH$_2$)$_2$, −75°; (iii) KOH; (iv) H$_2$/PdC.

7.3 Oxidation of nucleosides, nucleotides, and nucleic acids

In general, nucleoside bases are destroyed by strong oxidizing agents such as potassium permanganate. Hydrogen peroxide and organic peracids can be used to convert adenosine into its N-1-oxide and cytidine into its N-3-oxide while the 5,6-double bond of thymidine is a target for oxidation by osmium tetroxide, forming a cyclic osmate ester of the *cis*-5,6-dihydro-5,6-glycol. This reaction is sensitive to steric hindrance and has been employed to study the details of cruciform structure in DNA (Section 2.3.3).

Recent studies on the oxidation of DNA with hypochlorate and similar oxidants has identified the formation of 8-hydroxyguanine residues as an important mutagenic event.

The pentoses are sensitive to free radicals produced by the interaction of hydrogen peroxide with Fe(III) or by photochemical means, which causes strand scission (Section 7.9.1). Dervan has made this process sequence specific *in vitro* by linking radical-generating catalysts to a groove-binding agent (Section 8.4.3) and he has also employed it as a 'footprinting' device by linking an Fe–EDTA complex to an intercalating agent such as methidium. Other useful oxidative reactions of the pentose moieties are typical of the chemical reactions of primary alcohols and *cis*-glycols. In particular, periodate cleavage of the ribose 2',3'-diol gives dialdehydes that have been either stabilized by reduction to give a ring-opened diol or condensed with an amine or with nitromethane to give ring-expanded products (Fig. 7.4). Such procedures have been adopted frequently to make the 3'-terminus of an oligonucleotide inert to 3'-exonuclease degradation.

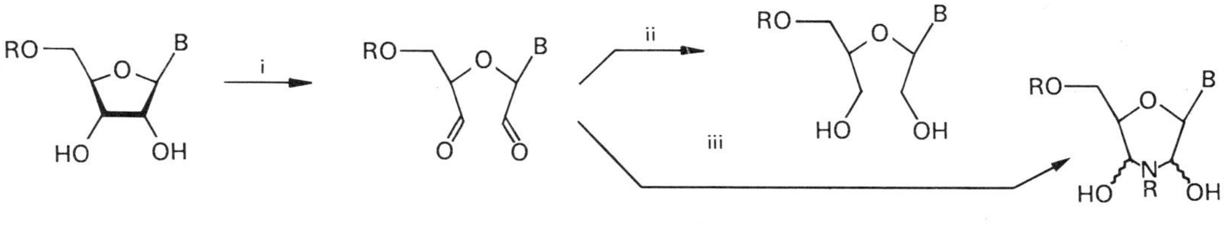

Fig. 7.4 Periodate cleavage of 3'-terminal nucleotide and subsequent modification.
Reagents: (i) IO_4^-, pH 4.5; (ii) $NaBH_4$; (iii) RNH_2.

Summary

The glycosidic bond in the nucleosides is hydrolysed by acids and shows increasing stability: dA, dG < rA, rG < dC, dT < rC, rU. RNA is readily hydrolysed in alkali *via* rapid formation of nucleoside 2′,3′-cyclic phosphates.

The 5,6-double bond of thymine is the most sensitive site for reduction of the bases. The pentoses are rather more difficult targets for reduction though some selective procedures provide access to unnatural deoxy-pentose nucleosides.

With appropriate oxidizing agents, nucleic acids can be oxidized at the thymine 5,6-double bond, at C-8 of guanine, or in the pentose. The latter is especially useful for effecting strand scission.

7.4 Reactions with nucleophiles

In general, nucleophiles can attack the pyrimidine residues of nucleic acids at C-6 or C-4 while reactions at C-6 of adenine or C-2 of guanine are more difficult. α-Effect nucleophiles, such as hydrazine, hydroxylamine, and bisulphite, are especially effective reagents for nucleophilic attack on pyrimidines.

Hydrazine adds to uracil and cytosine bases first at C-6 and then reacts at C-4. The bases are converted into pyrazol-2-one and 3-aminopyrazole respectively leaving an *N*-ribosylurea that can react further to form a sugar hydrazone These reactions are used in the Maxam–Gilbert method of DNA sequence determination (Section 5.2) where subsequent treatment of the modified ribose residues with piperidine causes β-elimination of both 3′- and 5′-phosphates at the site of depyrimidination (Fig. 7.5).

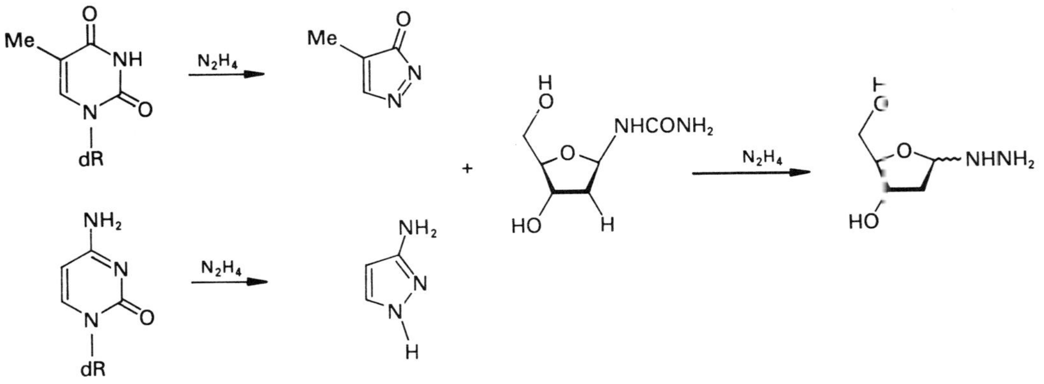

Fig. 7.5 Hydrazinolysis of pyrimidine nucleosides.

Under mild, neutral conditions, cytosine and its nucleosides react with hydroxyl-amine, semicarbazide, and methoxylamine to give N^4-substituted products. The mechanism of this process involves reaction with the cytosine cation, as illustrated for hydroxylamine (Fig. 7.6). The formation of N^4-hydroxydeoxycytidine is an important mutagenic event in DNA since this base exists to a significant extent in the unusual imino-tautomeric form (Section 2.1.2) which is capable of mis-pairing with adenine.

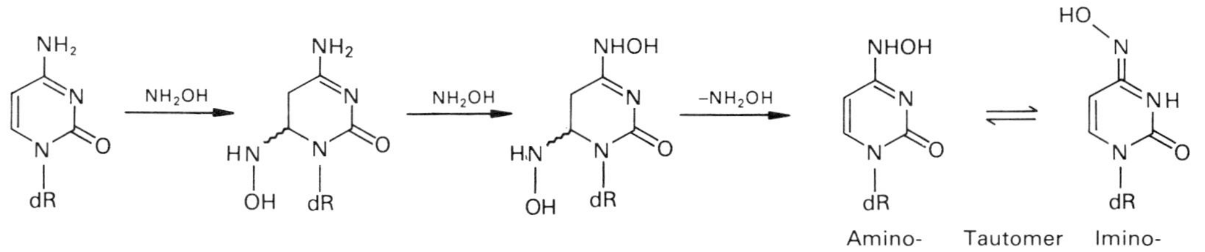

Fig. 7.6 Reaction of hydroxylamine with deoxycytidine leading to tautomerization of N^4-hydroxy-2′-deoxycytidine.

A third addition reaction at C-6 of cytosine and uracil residues involves bisul-phite anion, which adds reversibly. The resulting non-aromatic heterocycles can undergo a variety of chemical changes of which the most important are: trans-amination of cytosine at C-4 by various primary or secondary amines, hydrogen isotope exchange at C-5, and deamination of cytosine to uracil.

This third process explains the basis for the mutagenicity and cytotoxicity of bisulphite (which is equivalent to aqueous sulphur dioxide). Such mutations are best carried out at pH 5–6 to bring about deamination and then at pH 8–9 to elimi-nate bisulphite (Fig. 7.7). Other substitution reactions which require the addition of a thiol to C-6 are illustrated by the cysteine-catalysed *in vitro* incorporation of deuterium at C-5 into cytosine and the *in vivo* methylation of uracil at C-5 by thymi-dylate synthetase.

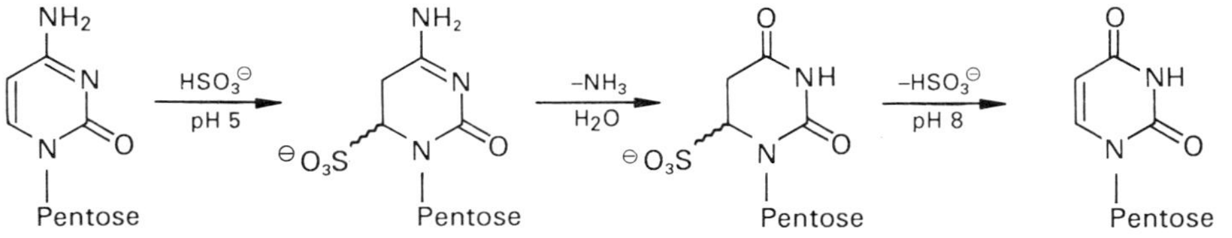

Fig. 7.7 Bisulphite-catalysed deamination of cytidine and deoxycytidine.

Summary

Cytosine is the most sensitive of the bases to nucleophilic attack, especially by α-effect nitrogen nucleophiles such as hydrazine, hydroxylamine, etc. Under mild conditions, transamination is easily accomplished. Under more vigorous conditions, transformation of the pyrimidine ring for C, T, and U gives a pyrazole with extrusion of a pentosyl-urea moiety.

Bisulphite addition to C-6 of cytosine is reversible, but the adduct can easily be deaminated which provides a mutagenic route from C to U.

7.5 Reactions with electrophiles

7.5.1 Halogenation of nucleic acid residues

Chlorine and bromine react directly with uracil, adenine, and guanine and so offer easy routes to 5-chloro- (or bromo-) uridines and 8-chloro- (or bromo-) purines (the latter are readily converted into 8-azidopurine nucleosides for use as photo-affinity labels). It is much more difficult to control the use of elemental fluorine, though fluorine gas can be used in anhydrous acetic acid **with care** to prepare 5- fluorouracil and 5-fluorouridine. The 5-iodouridines are best made by the action of iodine on 5-mercuri- or 5-palladium-uridines.

7.5.2 Reactions with nitrogen electrophiles

The normal reaction of nitrous acid with primary amino groups converts deoxy-adenosine into deoxyinosine, deoxycytidine into deoxyuridine, and deoxyguanosine into deoxyxanthosine. In each case, the reaction provides an alternative tautomeric form of the base which can lead to altered base-pairing. The transitions dA:dT→dI:dC and dC:dG→dU:dA are characteristic of the mutagenic action of nitrous acid (Fig. 7.8).

Aromatic nitrogen cations are a second important class of nitrogen electrophiles. These species are derived either from aromatic nitro-compounds by metabolic reduction or from aromatic amines by metabolic oxidation. In both cases, an intermediate hydroxylamine species interacts with purine residues in DNA or RNA either on C-8 or at N-7 (Section 7.6.1).

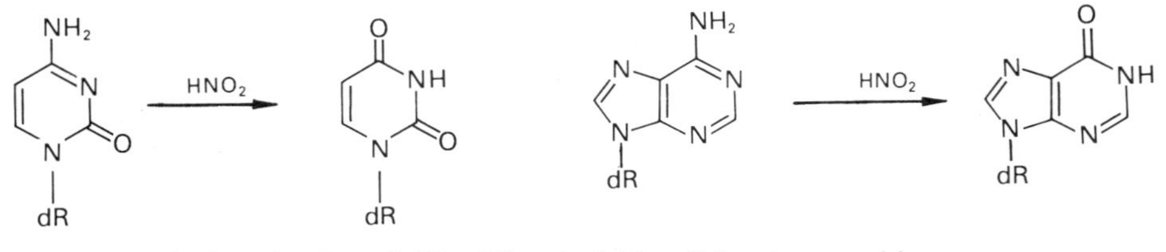

Fig. 7.8 Mutagenic deamination of dC→dU and of dA→dI by nitrous acid.

7.5.3 Reactions with carbon electrophiles

There is a huge variety of reagents which form bonds from carbon to nucleic acids. The simplest are species like formaldehyde and dimethyl sulphate. The most complex are carcinogens such as benzo[a]pyrene which require transformation by three, consecutive metabolic processes before they can become bound to purine bases in DNA or RNA. Not surprisingly, there is a wide range in selectivity for the sites of attack of these reactive species, some of which has been rationalized in terms of Pearson's **HSAB Theory** (HSAB = hard and soft acids and bases). Frontier orbital analysis can provide a more rigorous picture of the problem, but requires a deeper insight into theoretical chemistry. Other relevant factors may relate to the degree of steric access of the electrophile to exposed bases or to intercalation of reagents prior to bonding to nucleotide residues.

Formaldehyde

Covalent interactions of formaldehyde with RNA and its constituent nucleosides take place in a specific reaction of the amino bases. Formaldehyde first adds to the N^6-amino group of adenylate residues to give a 6-(hydroxymethylamino)purine and with guanylate residues to give a 2-(hydroxymethylamino)-6-hydroxypurine. These labile intermediates can react slowly with a second amino group to give cross-linked products which have a stable methylene bridge joining their two amino groups. All three possible species, pAdo-CH$_2$-pAdo, pAdo-CH$_2$-pGuo, and pGuo-CH$_2$-pGuo, have been isolated from RNA which has been treated with formaldehyde and then hydrolysed with alkali (Fig. 7.9). The detailed mechanism of formaldehyde mutagenicity is not yet clear.

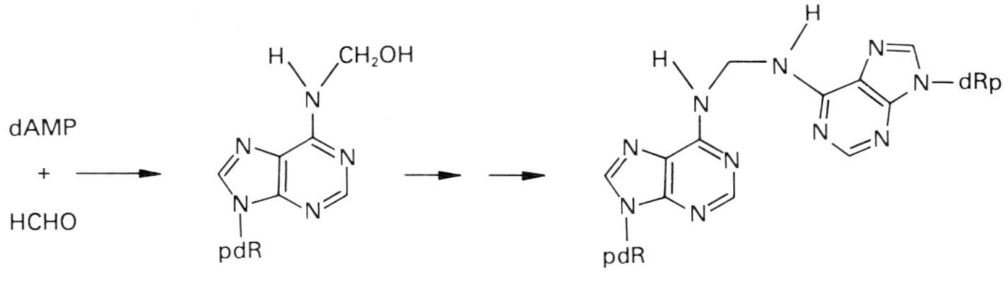

Fig. 7.9 Formaldehyde mutagenesis of adenine residues.

Alkylating agents

All of the nitrogen and oxygen residues of the nucleic acids bases, with the exception of purine-N^9 and pyrimidine-N^1, can be alkylated in aqueous solution at neutral pH. 'Soft' electrophiles, such as dimethyl sulphate (DMS), methyl methanesulphonate (MMS), and alkyl halides (such as MeI) react in an S$_N$2-like fashion. Alkylation takes place mainly at nitrogen sites with a general selectivity G-N^7 > A-N^1 > C-N^3 > T-N^3. A key parameter for 'softness' is the very low ratio of

G-O^6:G-N^7 methylation of 0.004:1. In double-stranded DNA, the major alkylation site for DMS with adenines is at *N*-3 with lesser substitution at *N*-7. 'Hard' electrophiles, such as *N*-methyl-*N*-nitrosourea (MNU) and its ethyl homologue ENU, are S$_N$1-like alkylating agents. MNU methylation of phosphate esters in nucleic acids can account for up to 50 per cent of total alkylation and also gives high ratios for G-O^6:G-N^7 products, ranging from 0.08 in liver to 0.15 in brain DNA. Other sites for *O*-alkylation include T-O^2, T-O^4, and C-O^2.

In contrast to the *C*-methylation of nucleic acids by various enzymes, such as thymidylate synthetase (Section 4.4), products arising from *C*-alkylation using chemical agents have not been detected.

Many alkylating agents are known to be primary carcinogens (agents that act directly on nucleic acids and that do not require metabolic activation). An extensive list includes DMS, MMS, and their ethyl homologues, β-propiolactone, 2-methylaziridine, 1,3-propanesultone, and ethylene oxide. The list of bifunctional carcinogenic agents includes bischloromethyl ether, bischloroethyl sulphide, and epichlorohydrin along with such 'first generation' anti-cancer agents as myleran, chlorambucil, and cyclophosphamide (Section 4.7.1). In general, 'hard' alkylating agents have been found to be a greater carcinogenic hazard than 'soft' ones.

Bis-(2-chloroethyl)sulphide

This is the mustard gas of World War I as well as of more recent conflicts. It is a typical bifunctional alkylating agent in addition to being a proven carcinogen of the respiratory tract. In the early 1960s, Brookes and Lawley showed that it can cross-link two bases either in the same or in opposite strands of DNA. The typical products isolated (Fig. 7.10) have a 5-atom bridge between N-7 of a guanine joined either to a second guanine or to adenine-N^1 or to cytosine-N^3. Similar products are formed on alkylation of DNA with 2-methylaziridine. These reagents show little sequence selectivity though nitrogen mustard, MeN(CH$_2$CH$_2$Cl)$_2$, shows some preference for alkylation of internal residues in a run of guanosines.

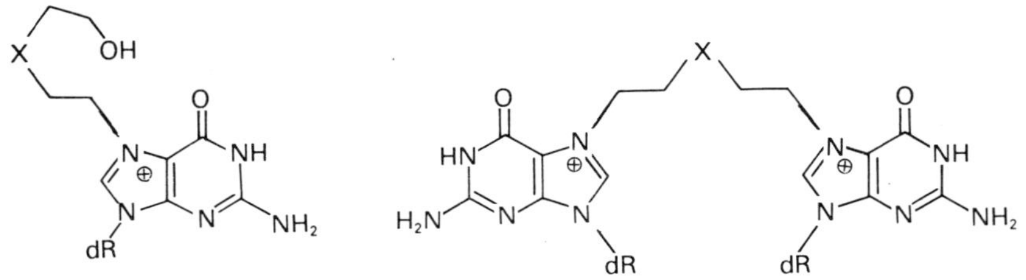

Fig. 7.10 Mono- and bi-functional products of dG with sulphur (X = S) and nitrogen (X = NH, NMe) mustard reagents.

Chloroacetaldehyde

This reagent combines the reactivity of formaldehyde and the alkyl halides. It reacts with adenine and cytosine residues, converting them into etheno-derivatives which have an additional five-membered ring fused on to the pyrimidine ring (Fig. 7.11). These modified bases are strongly fluorescent and have been used to probe the biochemical and physiological modes of action of a range of adenine and cytosine species.

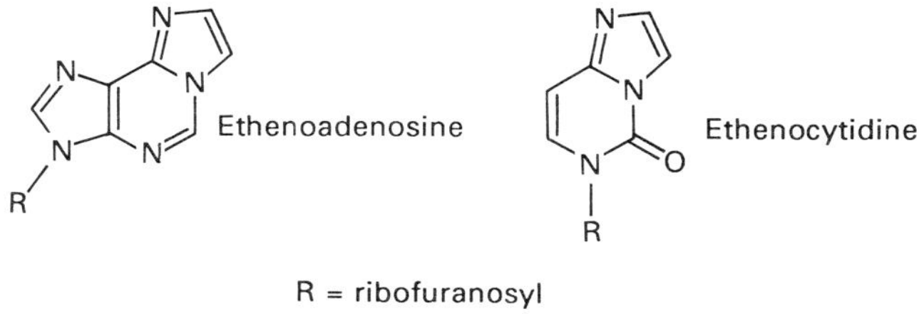

Ethenoadenosine   Ethenocytidine

R = ribofuranosyl

Fig. 7.11 Fluorescent etheno-derivatives of adenosine and cytidine from chloroacetaldehyde.

7.5.4 Metallation reactions

Mercury(II) acetate and chloride readily substitute C-5 of uridine and cytidine nucleosides. The products can be converted into organo-palladium species that are useful intermediates in the synthesis of a range of 5-substituted pyrimidine nucleosides. These include C-5 allyl-, vinyl-, halovinyl-, and ethynyluridines, all of which have been much studied for possible antiviral activity.

One of the most important recent applications of such chemistry is for the synthesis of fluorescent chain-terminating dideoxynucleotides, used in a rapid DNA sequencing method developed by Du Pont chemists (Fig. 7.12). 5-Iodo-2′,3′-dideoxyuridine is coupled to *N*-trifluoroacetylpropargylamine using palladium(0) catalysis and the resulting amine is then condensed with a protected succinylfluorescein dye. Deprotection provides the dideoxythymidine terminator species T-526 (which has a fluorescent emission maximum at 526nm). Related fluorescent derivatives of dideoxycytidine, C-519, dideoxyadenosine, A-512, and dideoxyguanosine, G-505, provide a family of chain-terminators which are incorporated with efficiencies comparable to those of unsubstituted ddNTPs and can be used for rapid, single-lane DNA sequencing (Section 5.2).

Rosenberg's discovery of the cytotoxicity of *cis*-diamminedichloroplatinum(II) has been carefully developed to make *cis*platin the reagent of choice for the successful chemotherapy of testicular (and other forms of) cancer. The reagent bonds to N-7 in one guanine residue and then joins it to a second purine. It binds selectively to d(p**G**p**G**) and d(p**A**p**G**) sequences, but not to d(p**G**p**A**) sites, and so forms intra-strand cross-links. *Cis*platin is also capable of binding to two guanines separated by another base, as in d(p**G**pNp**G**p**G**). X-Ray structures have been solved for its complexes with d(p**G**p**G**) and d(Cp**G**p**G**p) (Fig. 7.13) and show some agreement with solution structures determined by NMR. The platinum is *cis*-linked to N-7 in both guanines which lie in planes almost at right angles and this breaks up the base-stacking and the base-pairing patterns of the DNA helix. It is noteworthy that *trans*-diaminedichloroplatinum binds to DNA as well as the *cis*-isomer and while

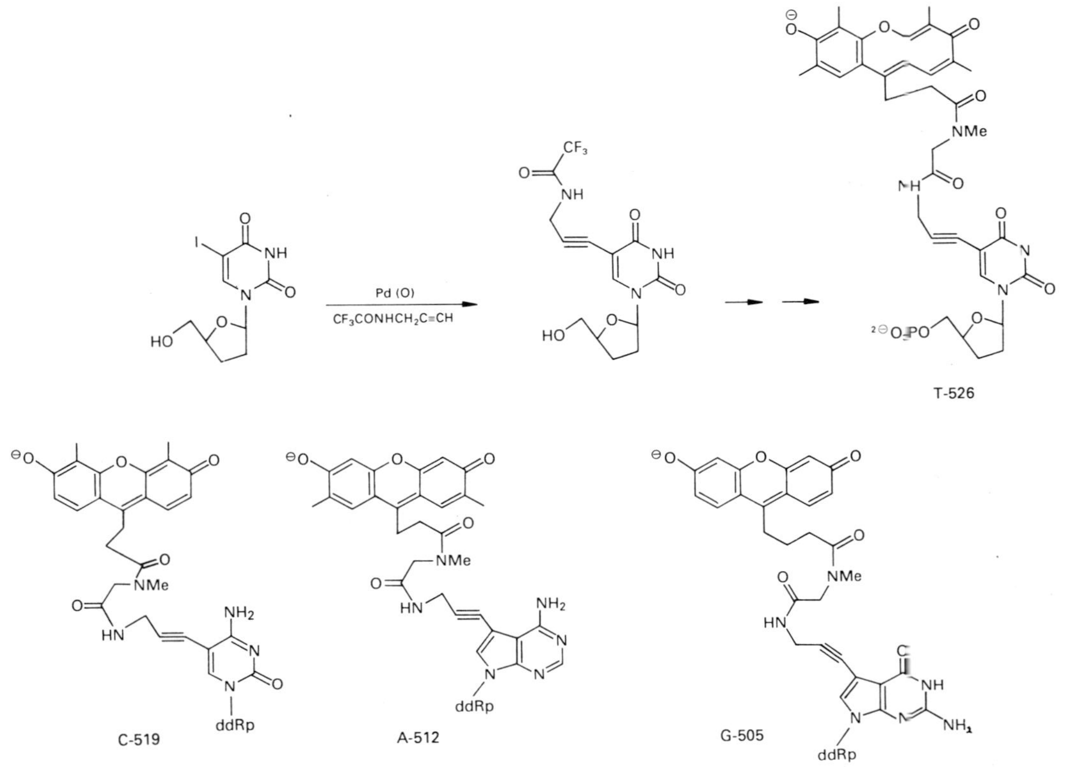

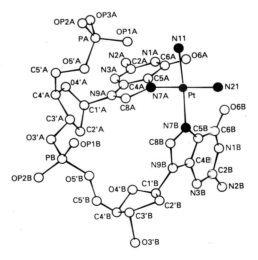

Fig. 7.12 Synthesis of fluorescent base T–526 using Pd(0) catalysis; structures of fluorescent dideoxy-nucleotides related to A, C, and G for use in rapid, single-lane DNA sequence analysis.

Fig. 7.13 Molecular structure of *cis*-[Pt(NH$_3$)$_2$d(pGpG)] showing the two guanines in perpendicular planes (adapted from Sherman, S. E., Gibson, D., Wang, A. H.-J., and Lippard, S. J. (1985). *Science*, **230**, 412–7. Copyright (1985) AAAS).

the resulting lesions may be more readily excised by repair enzymes than in the case of *cis*-platin, the *trans*-isomer has useful biological activity.

Summary

The pyrimidine bases are less aromatic than the purines and their chemistry is dominated by addition reactions to the 5,6-double bond, some of which have a mutational effect.

Electrophiles are capable of attacking most nitrogens and oxygens of the bases: the position of attack of soft electrophiles is N^7-G > N^1-A > N^3-C > N^3-T. For hard electrophiles there is an increasing proportion of alkylation on oxygen, especially at O^6-G. Aromatic nitrogen electrophiles bond to C-8 of guanine.

Many carcinogens directly act on DNA as alkylating agents, resulting in base-modification. Formaldehyde adds to the adenine N^6-amino group while bifunctional alkylating agents, such as *cis*platin and nitrogen mustards, mainly link two guanines together through N-7.

7.6 Reactions with metabolically activated carcinogens

Many synthetic chemicals and natural products are known to be carcinogens or mutagens. While these do not react directly with nucleic acids *in vitro*, they are transformed *in vivo* by metabolic processes to give electrophiles that bond covalently to DNA, RNA, and also to proteins. Most of these transformations are carried out by the mixed-function cytochrome P-450 oxidases, enzymes whose general function seems directed towards the detoxification of alien compounds! The following four classes of metabolically activated compounds are representative of an intensive study of a problem of very grave significance.

7.6.1 Aromatic nitrogen compounds

Investigations of the binding of *N*-aryl carcinogens to nucleic acids began in the 1890s with an investigation of the epidemiology of bladder cancer among workers in a Basel dye-factory. The list of chemicals banned today is extensive, but by no means complete and some examples of proscribed aromatic amines, nitro-compounds, and azo-dyes are illustrated (Fig.7.14).

Aromatic amines of this sort are substrates for oxidation by cytochrome P-450 isozymes which give either phenols, that are inactive and safely excreted, or hydroxylamines. Conjugation of the latter by sulphotransferase or acetate transferase enzymes converts these proximate carcinogens into ultimate carcinogens that bind covalently to nucleic acid bases, especially guanine.

The competition between such alternative 'safe' and 'hazardous' metabolic processes is illustrated for 2-acetylaminofluorene (Fig. 7.15). Two guanine nucleoside

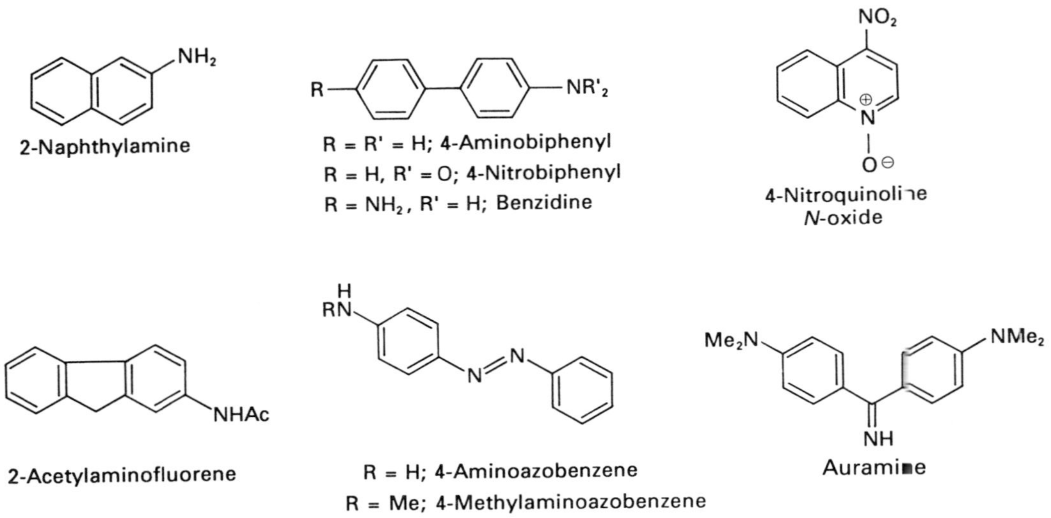

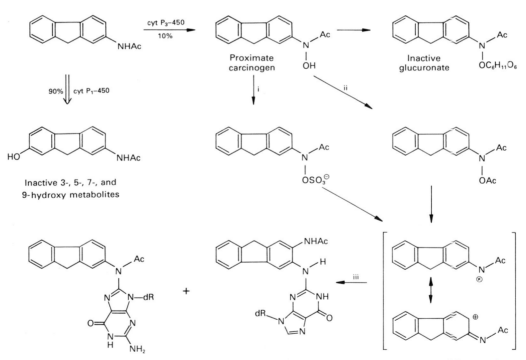

Fig. 7.14 Examples of *N*-aryl carcinogens.

Fig. 7.15 Metabolic activation of 2-acetylaminofluorene, AAF, and its binding to dG *via* a hypothetical nitrenium intermediate.

Processes: (i) sulphotransferase; (ii) acetyl transferase; (iii) DNA binding *in vitro* or *in vivo*.

adducts have been isolated and identified, and are formally derived from a hypo-thetical nitrenium ion. This, as an ambient cation, bonds either from nitrogen to guanine-C^8 or from carbon to guanine-N^2. Similar adducts have been identified for many other amines. Azo dyes are first cleaved *in vivo* by an azoreductase to aroma-tic amines and then activated as described above.

One recent example of considerable significance is the mutagenicity of two types of heterocyclic amine which are found in cooked meats, where they are formed by

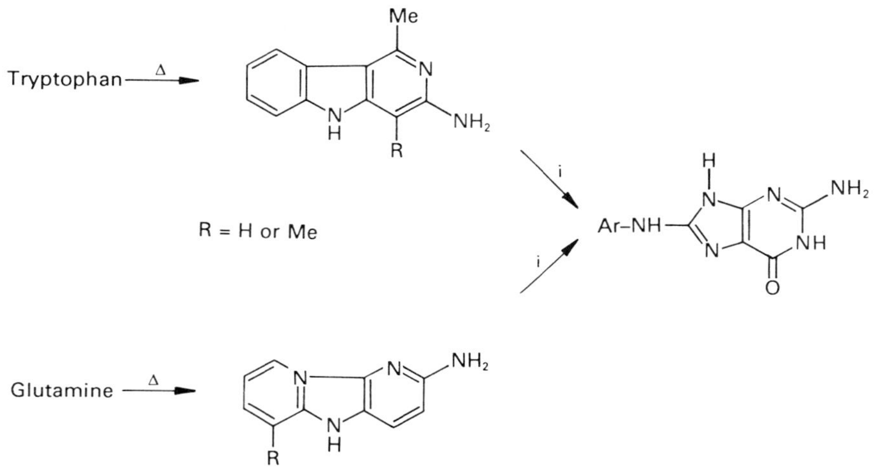

Fig. 7.16 Metabolic activation of heterocyclic amines from cooked food and their binding to DNA. **Processes:** (i) cytochrome P–450, DNA binding, hydrolysis.

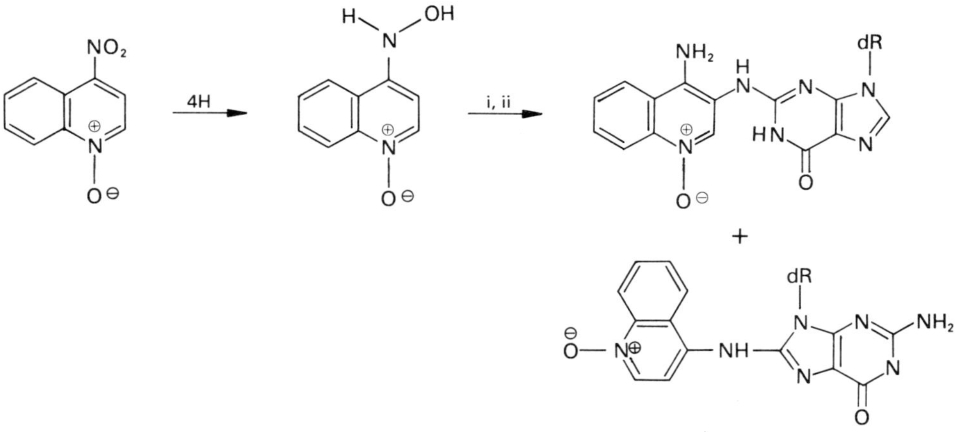

Fig. 7.17 Reductive activation of 4–nitroquinoline *N*-oxide and products resulting from its binding to DNA *in vivo*.
Reagents: (i) DNA *in vivo*; (ii) hydrolysis.

the pyrolysis of tryptophan and glutamine. Sugimura has identified guanine adducts which are generated by metabolic activation and binding to nucleic acids (Fig. 7.16). Grilled beef has been estimated to contain nearly 1 ppm of Trp-P-1 while up to 80 ng of this carcinogen has been isolated from the smoke of a single cigarette!

Aromatic nitro compounds are present in diesel engine emission, urban air particles, and photocopier black toners, and some have been identified as mutagens in the Ames Test. They can be reduced to aryl-hydroxylamines by anaerobic bacteria in the gut, by xanthine oxidase, or by cytochrome P-450 reductase to give substrates for the processes described above. The best-studied example is 4-nitroquinoline N-oxide, which is first reduced to a hydroxylamine and then binds to DNA *in vivo*, forming characteristic guanine adducts (Fig. 7.17).

7.6.2 *N*-Nitroso compounds

Nitrosoureas, nitrosoguanidines, and nitrosourethanes hydrolyse to give methyl-diazonium hydroxide, Me–N=N–OH, which is a 'hard' methylating agent. (In the case of methyl *N*-nitrosoguanidine (MNNG), thiol groups may catalyse the *in vivo* methylation of DNA by this carcinogen.) The same methylating species is produced as a result of the cytochrome P-450 oxidation of a wide range of *N*-methyl-nitrosamines, of which dimethylnitrosamine was the first to be identified as a carcinogen in 1956. The common metabolic pattern is hydroxylation of one alkyl residue to form a carbinolamine which breaks down to give methyldiazohydroxide (Fig. 7.18). Very many *N*-nitroso compounds of this type have proved to be carcinogenic in animals and lead to methylation, ethylation, or propylation of DNA, as described above (Section 7.5.3).

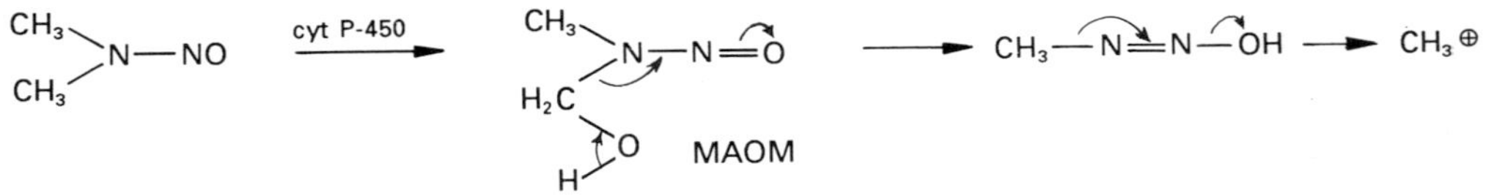

Fig. 7.18 Cytochrome P–450 oxidation of dimethylnitrosamine and its conversion into methyl azo-oxymethanol (MAOM) *en route* to DNA methylation.

7.6.3 Polycyclic aromatic hydrocarbons

The polycyclic aromatic hydrocarbons (PAHs) provided the first example of an industrial carcinogen, benzo[a]pyrene (BaP). Its identification marked the first stage in the molecular analysis of hydrocarbon carcinogenesis, which had begun with Percival Pott's study of chimney sweeps and scrotal cancer in 1775. BaP becomes bound to DNA *in vivo* following a series of three metabolic changes. In the first, cytochrome P_1-450 adds oxygen to BaP to give the two enantiomers of BaP-7,8-epoxide. These are used as substrates for an epoxide hydrase that converts them into the two enantiomeric *trans*-dihydrodiols. Finally, both diols are substrates for cytochrome P_1-450 and are converted into three of the four possible stereoisomers of the 9,10-epoxide, BPDE (Fig. 7.19).

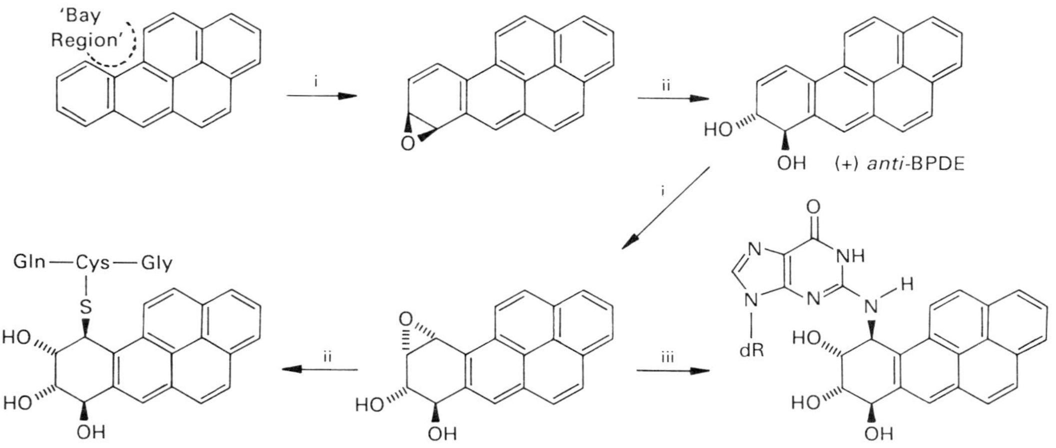

Fig. 7.19 Metabolic activation of B[a]P to BPDE (major stereoisomer illustrated) and its binding to DNA *in vivo* to give guanine adducts (major product shown).
Processes: (i) Cyt P_1–450; (ii) epoxide hydrolase; (iii) DNA binding *in vivo*; (iv) glutathione-S-transferase.

The carcinogenicity and DNA-binding capability of such dihydrodiol epoxides is closely linked to 'Bay Region' architecture. Among the products characterized are adducts with guanine-N^2 (the major product), guanine-N^7, guanine-O^6, and adenine-N^6 (minor products). These are formed as a result of a rapid intercalation of the BPDE into $d(AT)_n$-rich parts of the DNA helix, which is manifest as a red-shift in UV absorption of the hydrocarbon and a negative CD spectrum for the complex. A rate-determining protonation of the C-10 hydroxyl group then leads to the formation of a carbocation that reacts predominantly (90 per cent) with water to give the 7,8,9,10-tetra-ol or less frequently (10 per cent) binds to a proximate base, most often a dG residue.

The resulting covalent adducts appear to be of two distinct types. The minor 'site I' adducts have the hydrocarbon still intercalated in an intact DNA helix. The major 'site II' adducts appear to have the hydrocarbon lying at an angle to the helix axis, either in the minor groove of a DNA helix or forming a wedge-shaped intercalation complex. Similar results have been found for chrysene, while the Bay Region dihydrodiol epoxide of 3-methylcholanthrene appears to be too bulky to intercalate into DNA.

This type of epoxidation is not restricted to synthetic chemicals. One of the most potent groups of carcinogens are the aflatoxins, which are fungal products from *Aspergillus flavus*. A dose of less than 1 ppm of Aflatoxin B_1 can cause lung, kidney, and colon tumours in rats and is directly attributable to its oxidation to an epoxide that binds covalently to guanine residues in DNA (Fig. 7.20).

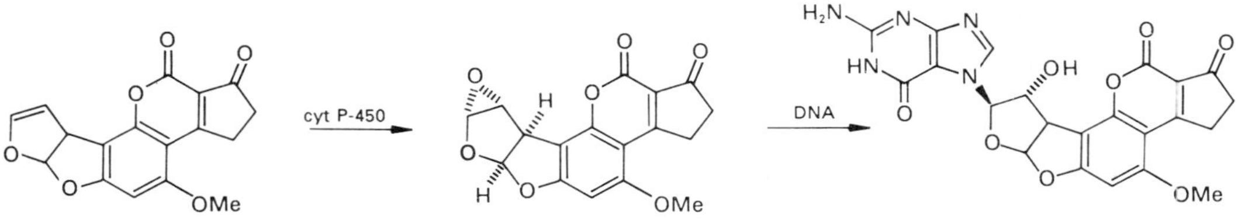

Fig. 7.20 Metabolic oxidation of Aflatoxin B$_1$ and epoxide binding to DNA.

Summary

Many carcinogenic compounds are converted by metabolic processes into alkylating agents which form covalent bonds to nucleic acid bases. For aromatic nitrogen compounds the hazardous species are aryl-hydroxylamine derivatives, which can be formed either from amines by biological oxidation or from nitro compounds by metabolic reduction. Activation of azo-dyes involves first reduction and then oxidation.

Dialkylnitrosamines are converted into alkyldiazonium hydroxides as a result of α-hydroxylation by cytochrome P-450, and these behave as 'hard' alkylating agents.

Many polycyclic aromatic hydrocarbons and some natural products, for instance the aflatoxins, are also metabolized by cytochrome P-450 oxygenation and lead directly or indirectly to epoxides as ultimate carcinogens.

7.7 Reactions with anti-cancer drugs

A large number of 'first generation' anti-cancer drugs were designed to combine a simple alkylating function such as a nitrogen mustard, an aziridine, or an alkanesulphonate ester with another function designed to direct the agent towards the target tissue. Most of these compounds turned out to be less tumour-selective than one might have hoped and, what is worse, many of them have proved to be carcinogens which have given rise to new tumours some time after the termination of chemotherapy for the original cancer. As a result, their general use is now viewed with some suspicion.

A group of 'second generation' compounds has emerged, many of them natural products, but now augmented by a growing number of rationally designed, synthetic drugs. Their common feature is that they appear to form an initial physical complex with DNA before covalently bonding to it. This heterogeneous group of compounds includes aziridines, such as mitomycin C, several pyrrolo[1,4]benzodiazepines, and spirocyclopropanes such as CC-1065. Their vital purpose is to kill bacteria by disrupting the synthesis of DNA and RNA, but many of them have also shown useful anti-tumour activity, which must arise from selective toxicity. This can be attributed to DNA-binding specificity or to preferential metabolic activation by tumour cells.

7.7.1 Aziridine antibiotics

An assortment of naturally occurring antibiotics, each having an aziridine ring, has been isolated from *Streptomyces caespitonis*. The most interesting of them in clinical terms is **mitomycin C**. This compound requires enzymatic reduction of its quinone function to initiate the processes that cause it to alkylate DNA. It seems likely that the second step is elimination of methanol which potentiates either monofunctional or bifunctional alkylation (Fig. 7.21). The antibiotic has been shown to interact with DNA at $O^6G > N^6A > N^2G$ and forms one cross-link for about every ten monocovalent links. A range of mechanisms has been discussed for the covalent interaction of the reduced form of mitomycin C with the nucleophilic centres of DNA while molecular mechanics simulations have given strong support to evidence for tight binding of the antibiotic in the major groove of DNA.

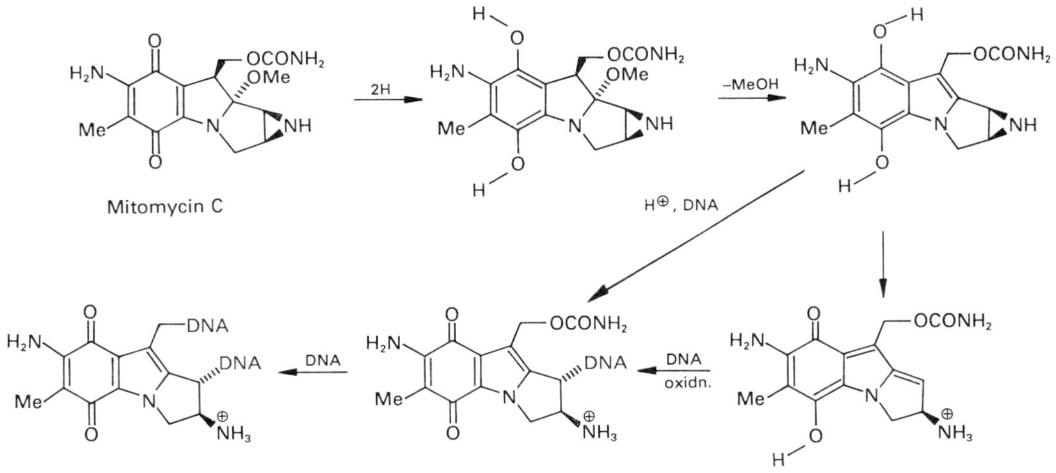

Fig. 7.21 Activation of mitomycin C by metabolic reduction and possible bifunctional alkylation of DNA.

Another example of an aziridine antibiotic is **carzinophilin A**, which is an anti-tumour metabolite produced by *Streptomyces sahachiroi*. It has two aziridine functions and produces cross-links in DNA, probably as a result of bis-intercalation (cf. Section 8.5) followed by acid-promoted opening of the aziridine rings to alkylate DNA.

Many drugs which act on DNA exhibit a requirement for reductive activation, including adriamycin, daunomycin, actinomycin, streptonigrin, saframycin, bleomycin (Section 8.7.1), and tallysomycin in addition to mitomycin C. While there is no common factor uniting the chemistry of DNA modification by these agents, the fact that tumour tissues seem to have a higher reducing potential than normal tissue has led to the concept of bioreductive drug activation.

7.7.2 Pyrrolo[1,4]benzodiazepines, P[1,4]Bs

Anthramycin and tomaymycin, along with sibiromycin and neothramycins A and B, are members of the potent P[1,4]B anti-tumour antibiotic group produced by various actinomycetes (Fig. 7.22). The first three of these compounds bind physically in the minor groove of DNA where they then form covalent bonds to N^2-G, showing a DNA sequence specificity for 5'-PuGPu sequences.

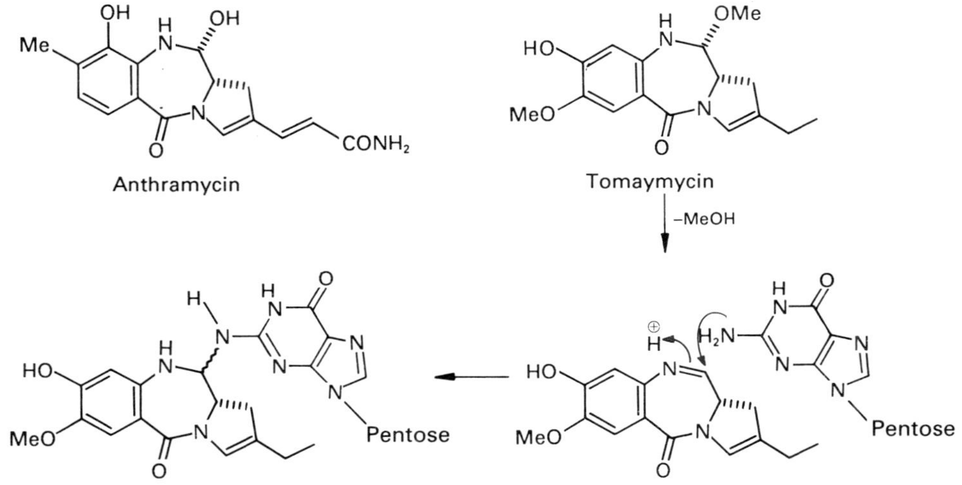

Fig. 7.22 Binding of P[1,4]B antibiotics to the N^2-amino group of guanine.

These P[1,4]Bs seem to interact with DNA in a biphasic process. Initially there is a rapid, non-covalent association which results from a close interaction of the antibiotic with the 'floor' of the minor groove of DNA (Section 8.3). Subsequent loss of water or methanol and covalent addition of N^2-G to C-11 then forms an aminal linkage that is well stabilized by favourable steric and electrostatic interactions. The structure of a condensation product between anthramycin and d(ATGCAT)$_2$ has been partially characterized by NMR analyses and much studied by molecular mechanics. Altogether, the picture that emerges is that the bonding from the guanine N-2 to C-11 is in the (S)-configuration and this makes the aromatic ring of the antibiotic lie in the DNA minor groove on the 3'-side of the modified guanine. In the case of tomaymycin, NMR studies have established the existence of two distinct tomaymycin-d(ATGCAT)$_2$ species in solution. These have the antibiotic oriented in opposite directions in the minor groove according to its (*R*)- or (*S*)-configuration at C-11. The resulting lesions appear neither to impede Watson–Crick base-pairing nor to distort the B-DNA helix structure (Fig. 7.23) so that they probably pose difficult recognition problems for DNA repair systems (Section 7.11).

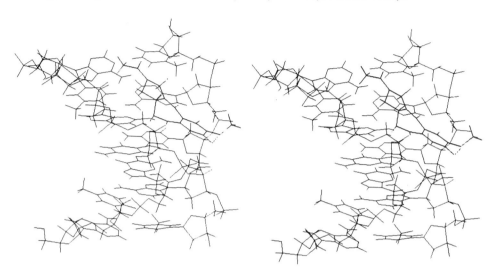

Fig. 7.23 Stereopair diagram of the 11–(*R*) condensation product of tomaymycin with d(ATGCAT)$_2$ (adapted from Cheatham, S., Kook, A., Hurley, L. H., Banklay, M. D. and Remers, W. (1988). *Journal of Medicinal Chemistry*, **31**, 583–90. Copyright (1988) American Chemical Society). (Stereopair for parallel viewing).

7.7.3 Spirocyclopropane antibiotics

The structurally novel antibiotic **CC-1065** is an extremely potent cytotoxin whose biological activity has been attributed to sequence-selective binding in the minor groove of DNA followed by covalent bonding. This involves attack by adenine N-3 on C-4 of the antibiotic to open the cyclopropane ring and aromatize the indole system (Fig. 7.24). CC-1065 is highly selective for AT rich sequences in DNA, with a strong affinity for 5'-PuNTTA and 5'-AAAAA while molecular modelling studies have suggested that the concave edge of the antibiotic interacts through close van der Waals contacts with the floor of the DNA minor groove. One consequence of alkylation of adenine at N-3 is that its glycosidic bond becomes very labile and thermal treatment of CC-1065:DNA adducts leads to single-stranded breaks on the 3'- side of modified adenine residues (Fig. 7.24).

While CC-1065 cannot be used as an anti-tumour agent because of its unacceptably toxic side effects, a synthetic analogue, **U-71184**, has enhanced anti-tumour activity and diminished side-effects. The desired biological activity is found only in the enantiomer that corresponds to the stereochemistry of CC-1065 (Fig. 7.24).

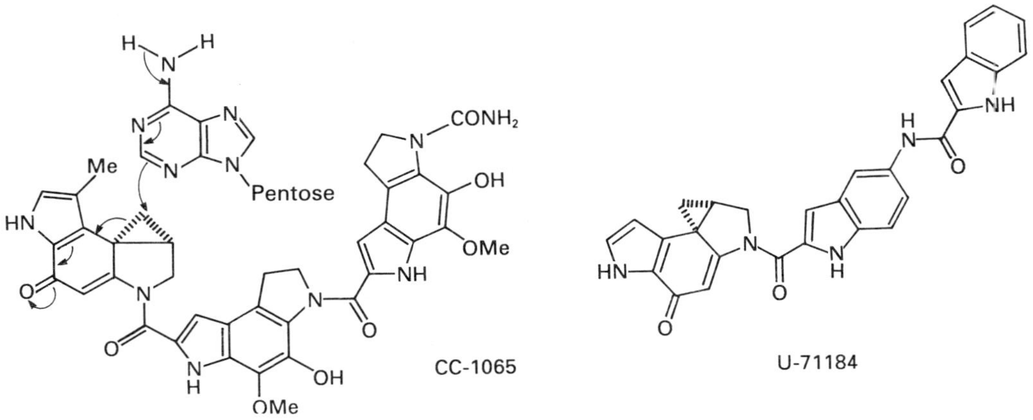

Fig. 7.24 Structures of antibiotics CC–1065 and U–71184 and mode of their covalent binding to adenine-N^3.

Summary

Considerable progress has been made towards the identification of the mode of action of a range of natural and synthetic anti-tumour antibiotics. Physical binding in either the major or minor groove of DNA gives target selectivity. Covalent bonding to DNA may be a consequence (a) of inherent, weak electrophilic character (CC-1065), (b) of elimination of water or methanol (anthramycin and tomaymycin), or (c) of metabolic reduction (mitomycin C). In some cases the covalently attached antibiotic resides in the DNA groove without distortion of its conformation (tomaymycin). In others, single-stranded breaks may result (CC-1065).

7.8 Photochemical modification of nucleic acids

Our very serious concern about the depletion of the global ozone barrier is directly related to the action of ultraviolet light on nucleic acids. Its effect is mutagenic at low doses, cytotoxic at high doses, and is linked to skin cancer where there is chronic, excessive exposure to sunlight among whites, albino blacks, or for people with repair gene deficiencies.

7.8.1 Pyrimidine photoproducts

Light of 240–280 nm excites the pyrimidine bases, C, T, and U, to give a higher single state ($^1P_1$) which has a lifetime of only a few picoseconds before it gives photohydrates (in which water has added to either face of the 5,6-double bond), decays, or passes into the triplet state. Uridine photohydrate (U*) dehydrates slowly to uridine in acidic or alkaline solution and is moderately stable at neutral pH ($t_{\frac{1}{2}}$ 9 h at 50°C). The cytidine photohydrate is some tenfold less stable ($t_{\frac{1}{2}}$ 6 h at 20°C) and either reverts to cytidine (90 per cent) or is deaminated to give U* (10 per cent) (Fig. 7.25). This process effects a net conversion of C into U (Section 7.11.2). The formation of photohydrates of thymine has a very low quantum yield and its biological consequences are not significant.

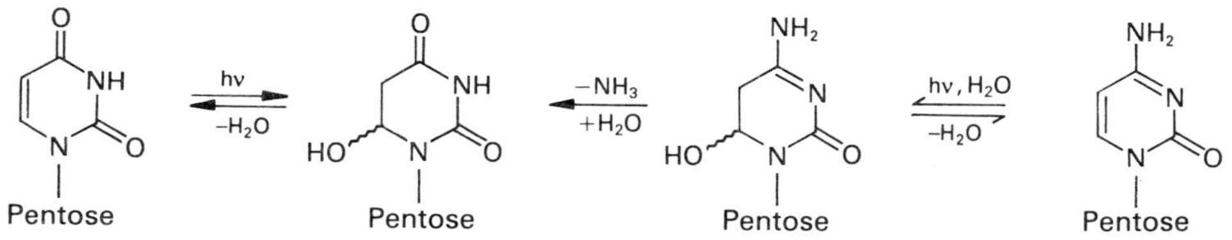

Fig. 7.25 Photohydration of uracil and cytosine nucleosides and deamination of cytosine photohydrate leading to uracil.

All of the major pyrimidines form cyclobutane photodimers on irradiation at 260–280 nm. The reaction is a [2+2] cycloaddition, mainly involving the triplet state. Of the four possible isomers for thymine dimer, T<>T, the *cis–syn* isomer is formed by irradiation of thymine in an ice-matrix and is known to be the major product (>95 per cent) formed by UV irradiation of native DNA. The *trans–syn* isomer, which is the major product produced by the photosensitized irradiation of thymidine in solution, accounts for some 2 per cent of the native DNA T<>T, although a larger proportion of this thymine dimer is formed in denatured DNA (Fig. 7.26).

The stereochemistry of these dimers has established that T<>T formation in native DNA is predominantly an intrastrand process with photoaddition involving adjacent thymines. Oligonucleotides containing a T<>T dimer have been analysed by molecular modelling and by NMR methods, which suggest that there is little distortion of the DNA helix apart from the local disruption of hydrogen-bonding to adenines. This unexpected result will no doubt be clarified by X-ray structural analysis. The *trans–syn* isomer may well result from regions of Z-DNA, and NMR analysis of an oligomer containing this lesion has indicated that in this structure it causes a larger distortion of helical structure.

These photodimers revert to monomers on irradiation at wavelengths shorter than 254 nm. Dimers containing cytosine residues are easily deaminated and subsequent photoreversion provides yet another source of C→U base transition as a result of the following reaction sequence:

$$C + U \xrightarrow{h\nu} C<>U \xrightarrow{H_2O} U<>U \xrightarrow{h\nu} U + U$$

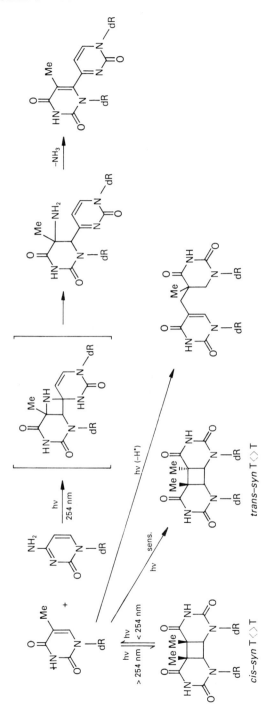

Fig. 7.26 Pyrimidine [2+2]photodimers and other photoproducts from DNA.

Many other types of pyrimidine photoproduct have been isolated from irradiated DNA. The most noteworthy are the bipyrimidine adducts and photoproducts formed from a radical generated by loss of hydrogen from the methyl group of thymine (Fig. 7.26).

7.8.2 Psoralen–DNA photoproducts

Psoralens are furocoumarins that have been widely used in the phototherapy of psoriasis and other skin disorders. Their photochemical cross-linking of DNA involves intercalation followed by two successive photochemical [2+2] cycloadditions which result in the formation of two cyclobutane dimers. Thymines are the preferred target so that cross linking occurs mainly at d(pTpA) sites. The products are predominantly the *cis–syn* stereoisomers which have an overall **S**-shape as a result of one thymine being above and the other below the plane of the psoralen (Fig. 7.27).

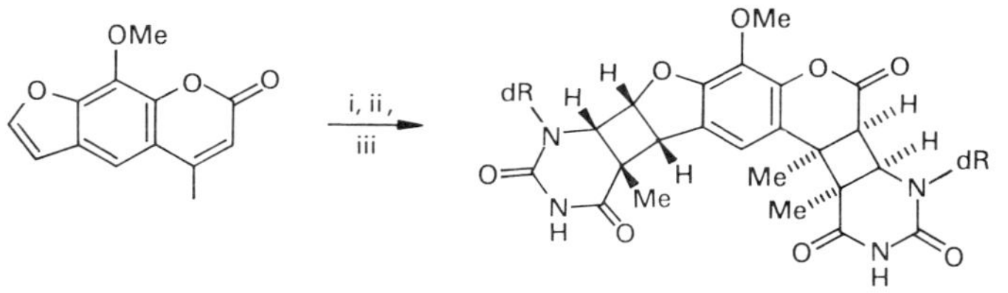

Fig. 7.27 Photochemical binding of 8-methoxypsoralen to DNA and isolation of dithymidine photoproducts.
Procedures: (i) DNA; (ii) hv 320–400 nm; (iii) H$^+$.

NMR analysis of the conformation of a psoralen adduct of the octanucleotide, d(GGGTACCC), has shown a degree of distortion of the B-DNA helix both for the psoralen-bonded thymine residues and the adenines. This distortion is also shared by their nearest neighbours. However, the outer G-1, G-2, C-7, and C-8 residues still form a regular B-helix stack. The overall effect of psoralen binding is to unwind the helix by 63° and impose a kink of 45° on its axis (Fig. 7.28).

7.8.3 Purine photoproducts

Adenine and guanine are intrinsically more photostable than the pyrimidines and tend to transfer photochemical excitation energy to neighbouring pyrimidines in DNA duplexes. General purine photoreactivity is most marked at C-8 as illustrated by three examples:

(a) Adenosine and guanosine form C-8 substitution products with secondary alcohols such as isopropanol by UV or γ-radiation induced processes;

(b) UV irradiation of 8-bromo-adenine or -guanine generates purinyl radicals that either couple to form dimers or add to an adjacent purine to form 8,8′-dipurinyl products;

(c) UV irradiation of coenzyme B$_{12}$ causes scission of the bond from cobalt to adenosine C-5′ and results in the formation of 8,5′-cyclo-5′-deoxyadenosine.

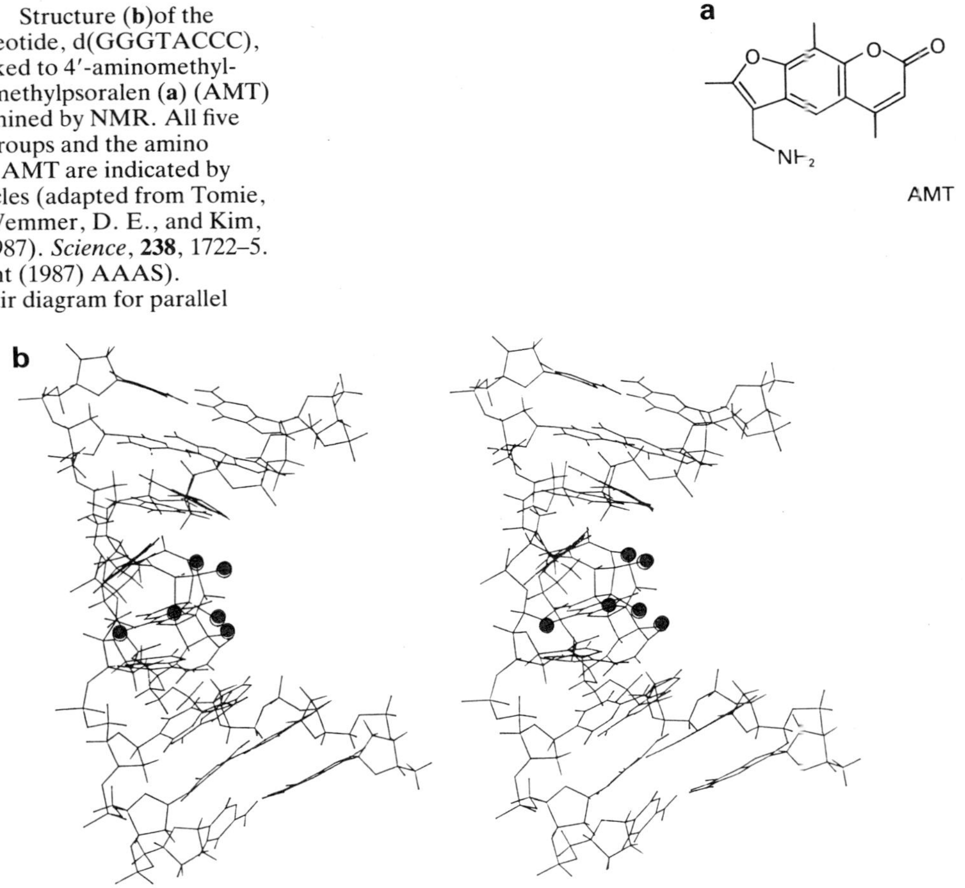

Fig. 7.28 Structure (**b**) of the octanucleotide, d(GGGTACCC), cross-linked to 4′-aminomethyl-4,5,8-trimethylpsoralen (**a**) (AMT) as determined by NMR. All five methyl groups and the amino group of AMT are indicated by filled circles (adapted from Tomie, M. T., Wemmer, D. E., and Kim, S.-H. (1987). *Science*, **238**, 1722–5. Copyright (1987) AAAS). Stereopair diagram for parallel viewing.

AMT

7.9 Effects of ionizing radiation on nucleic acids

X-rays, γ-radiation, and high-energy electrons all interact indirectly with nucleic acids in solution as a result of the formation of hydroxyl radicals, of solvated electrons, or of hydrogen atoms. In aerobic conditions, the most important processes result from HO˙ radicals which abstract a hydrogen atom to give a radical which then captures O_2. Measurements of the efficiency of these processes indicate that for every 1000 eV of energy absorbed, about 27 HO˙ radicals are formed of which 6 react with the pentoses and 21 react almost randomly with the four bases.

7.9.1 Deoxyribose products in aerobic solution

The hydroxyl radical can abstract a hydrogen atom from C-4′ or any of the other four carbons in the sugar. The resultant radicals capture oxygen to give a hydroperoxide radical at C-4′ or C-5′ which leads directly to cleavage of a phosphate ester. Similar reactions at C-1′ or C-2′ lead to alkali-labile phosphate esters that are

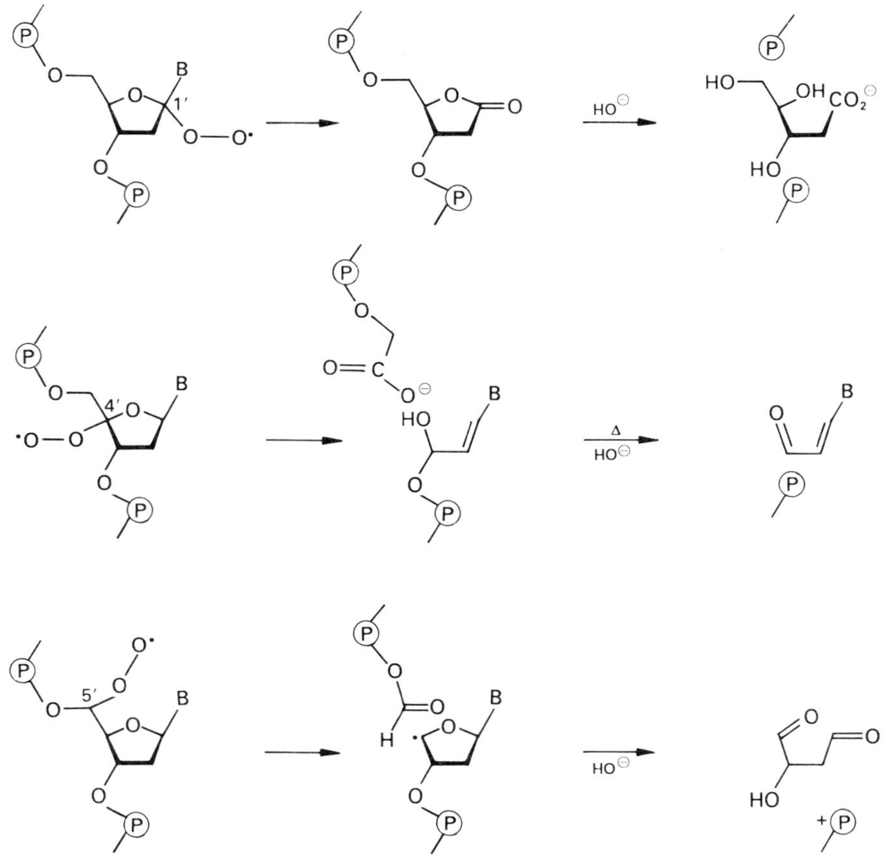

Fig. 7.29 Breaks in DNA resulting from hydrogen atom abstraction and peroxide radical formation at C–1' (upper), C–4' (centre), and C–5' (lower) in deoxynucleotides followed by mild alkaline treatment (0.1 M NaOH, 10 min. 20°C).

cleaved on incubation with 0.1 M sodium hydroxide at room temperature in 10 min. In both cases, the bases are released intact (Fig. 7.29).

7.9.2 Pyrimidine base products in solution

At least 24 different products have been isolated from irradiation of thymidine in dilute, aerated solution and many more are formed in anoxic conditions or in the solid state. Cytosine and the purines show a similar diversity. The situation is simplified by limiting the study to oxygenated solutions when the principle site for reaction with the pyrimidines is the 5,6-double bond. Under these conditions, hydroxyhydroperoxides are formed that are semistable for thymidine, but break down rapidly for deoxycytidine to give a range of products, of which the major ones are illustrated (Fig. 7.30). About 10 per cent of thymine modification occurs at the methyl group with the formation of 5-hydroxymethyldeoxyuridine.

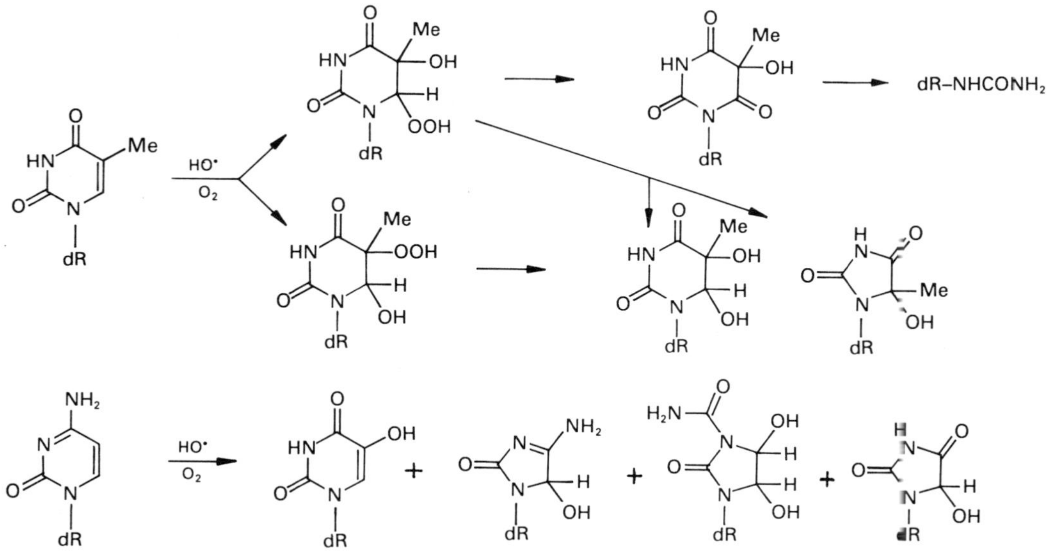

Fig. 7.30 Products resulting from radiolysis of deoxynucleosides in aerobic conditions

In anaerobic solution, γ-radiolysis gives 5,6-dihydrothymidine as the major product with a preference for formation of the 5(R)-stereoisomer.

7.9.3 Purine base products

The radiation chemistry of the purines is less well understood than that of the pyrimidines. Deoxyadenosine can add the HO˙ radical at C-8 either to give 8-hydroxydeoxyadenosine or, via cleavage of the imidazole ring, to give a 5-formamido-4-aminopyrimidine derivative of deoxyribose (Fig. 7.31). Guanine has been even less well studied but reports of the formation of 8-hydroxyguanine from anaerobic irradiation coupled to much interest in this modified base as a very mutagenic lesion may change this situation.

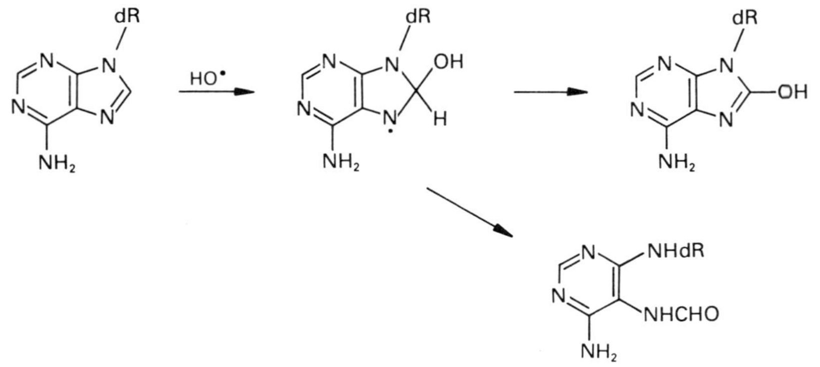

Fig. 7.31 Major γ-radiolysis products from deoxyadenosine.

Summary

The photochemistry of nucleic acids is dominated by the formation of thymine dimers, T<>T, of which the principle isomer obtained from DNA has the *cis–syn* cyclobutane structure. Uracil and cytosine can also form photodimers but these are less stable and dimers containing cytosine can be deaminated before monomerization. U and C form photohydrates, U* and C*, through addition of OH to position-6 and of H to position-5. Cytosine hydrate can be deaminated to give U* which through dehydration to U causes a C→U base transition.

Psoralens can act as photochemical cross-linking agents and form cyclobutane adducts involving two thymine residues in opposite strands of DNA.

While the purines are relatively photostable, ionizing radiation in aerated solution generates hydroxyl radicals which react with all four bases giving a wide range of products. The pentoses are also attacked by hydrogen atom abstraction, which leads to single-stranded breaks.

7.10 Biological consequences of DNA alkylation

7.10.1 *N*-Alkylated bases

The major site of DNA alkylation is at the N-7 position of guanine. However, methylation of this site appears not to change the base-pairing of G with C and so is an apparently harmless lesion. Moreover, the glycosidic bond of a 7-alkylguanine residue appears to be slowly and spontaneously hydrolysed, and so creates an apurinic site which is a target for repair (Section 7.11.2). By contrast, the cross-linking of two neighbouring guanines at N-7 by a nitrogen mustard or equivalent bifunctional alkylating agent is an important cell-killing event, both for interstrand and for intrastrand cross-links.

3-Alkyladenines are the major toxic lesion resulting from monofunctional alkylation of DNA. The alkyl group lies in the minor groove of the DNA double helix where it can block the progress of DNA polymerases (Section 7.11.3). 3-Alkylguanines have a similar physiological effect but are much less prevalent, being formed ten times less frequently than 3-alkyladenines (Fig. 7.32). Some 1-methyladenine is

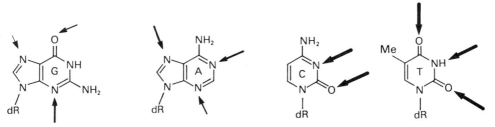

Fig. 7.32 Major sites of methylation of bases in double-stranded DNA by hard alkylating agents such as MNU and NMNG.

generated by the methylation of DNA, but while this lesion as a 'non-instructional base' must necessarily interfere with A:T pairing, it appears to be only slowly excised *in vivo*.

7.10.2 *O*-Alkylated lesions

O[6]-**Alkylguanines** are locked in the *enol*-tautomeric form while guanine in DNA is normally in the *keto*-form (Chapter 2.1.2). It appears that O^6-methylguanine does not base-pair either with C or with T. However, because it does not block DNA polymerase, it leaves the enzyme free to introduce either a C or a T in the daughter strand opposite an O^6-MeG residue. As the mismatch repair system fails to detect the O^6-MeG:T combination, the result is a G→A transition. This type of mutation is common in cells exposed to hard alkylating agents. In particular, it has been identified with a single base transition for activation of the Ha-*ras*-1 proto-oncogene in the process of initiation of mammary tumours in rats with methylnitrosourea (MNU), as a result of the specific conversion of G^{35} into O^6-MeG^{35}.

O[4]-**Alkylthymines** exist in the **enol**-tautomeric form and therefore they can, in principle, cause an A→G transition. However, alkyl-pyrimidines are very minor products of DNA alkylation and their biological effects appear to be of low significance.

Lastly, the *O*-alkylation of the phosphate diesters in DNA gives phosphate triesters but these are repairable (Section 7.11.1) and do not seem to be important either as cell-killing or as mutagenic lesions.

Summary

The alkylation of a single DNA base can have three consequences: either no change in its coding property, a change to pairing with a different base, **mispairing**, or a change to a non-coding base, a **non-instructional site**. These are illustrated by the formation of 7-methylguanine, 4-*O*-methylthymine, and 6-*O*-methylguanine respectively. Covalent bonding of a base with a larger molecule, such as acetylaminofluorene or BaP-diolepoxide, can have more complex effects.

The overall effect on DNA of chemical modification by other external agents is one of five types of damage: (a) introduction of strand breaks; (b) loss of a base leaving an unpaired partner; (c) covalent modification of a base, especially by the addition of bulky groups; (d) conversion of one base into another with changed Watson–Crick pairing; (e) covalent linking of bases in the same or opposite strands.

7.11 DNA repair

All living cells possess a range of DNA repair enzymes in order to correct damage resulting from radiation and external chemical agents. Humans appear to be more effective than rodents in repairing DNA and are also better able to resist mutagenic agents. A striking similarity has emerged between repair systems found in many species from bacteria to man, though much of our knowledge comes from studies on bacteria or yeasts. Three general types of repair process have been identified.

7.11.1 Direct reversal of damage

Photoreactivation is catalysed by an enzyme which uses tryptophan residues to sensitize the photochemical (300–400 nm) cleavage of *cis–syn* cyclobutane photo-products to their constituent pyrimidines. This photolyase does not cleave other types of pyrimidine photoproducts (Fig. 7.28) and it may be associated with a genetic deficiency in humans who suffer from the disease *Xeroderma pigmentosum*.

O-Demethylation is carried out by an O^6-methylguanine-DNA methyltransferase. This sacrificial enzyme uses the SH group of a Pro–Cys–His sequence near its C-terminus as a methyl acceptor (Fig. 7.33). The same enzyme can also dealkylate ethyl, 2-hydroxyethyl, and 2-chloroethyl O^6-derivatives of guanine. This repair reaction changes the enzyme into an S-alkylcysteine protein which is both inactive and also not reactivated. It follows that one molecule of the enzyme can only repair one O-methylated base. Human lymphoid cells which are resistant to alkyl-nitrosoureas have 10 000–25 000 copies of this enzyme per cell while repair-deficient cells have virtually none. It is easy to see that the threshold of tolerance for alkylating agents may vary greatly between different species of cells.

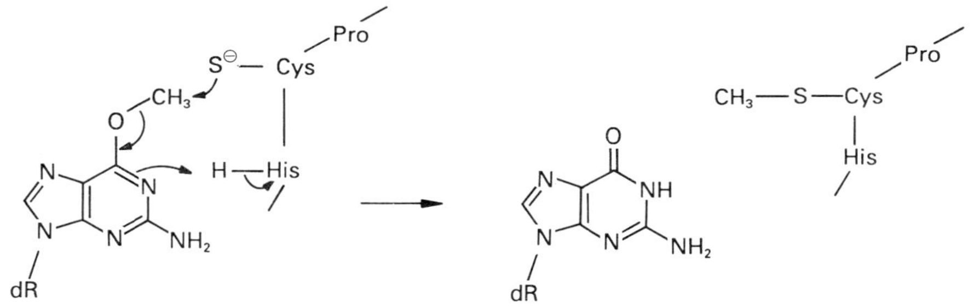

Fig. 7.33 O^6-Methylguanine demethylation by the Pro-Cys-His component of methyltransferase.

In *E. coli*, the methyltransferase from the ada^+ gene differs in three respects from the mammalian enzyme. Its C-terminal cysteine demethylates both O^6-MeG and O^4-MeT; there is a second active cysteine near the N-terminus which demethylates DNA phosphate triesters (but only of the (S)-configuration); and the system shows an adaptive response to alkylating agents which may be triggered by the demethylation of phosphate triesters.

7.11.2 Excision repair of altered residues

In mammalian cells, excision repair is the most important mechanism and involves enzyme recognition of the modified base. There are two distinct types of excision repair process and both appear to be error-free.

In the first, an endonuclease begins by making a single-stranded incis on close to the damaged nucleotide. The best understood repair system of this sort s that from *E. coli*, whose *uvrABC* genes have been cloned and their protein products purified. The *uvrA* protein (114 kDa) is an ATP-dependent DNA-binding protein that recognizes and binds to the DNA photolesion and to many other types of bulky base modification. The *uvrB* (84 kDa) and *uvrC* (70 kDa) proteins can now bind on and initiate the repair process by nicks in the damaged strand which are on either side of the lesion and some 12 bases apart. The excised nucleotide containing the lesion is now released while it is still bound to the protein complex. That leaves a single-stranded gap which is enlarged to 30–60 nucleotides long by exonuclease action. This enables DNA polymerase I to bind and fill the gap from the 3'-end until it approaches the 5'-terminus. DNA polymerase III then extends the patch up to the 5'-terminus and ligation completes this **short-patch repair** (Fig. 7.34).

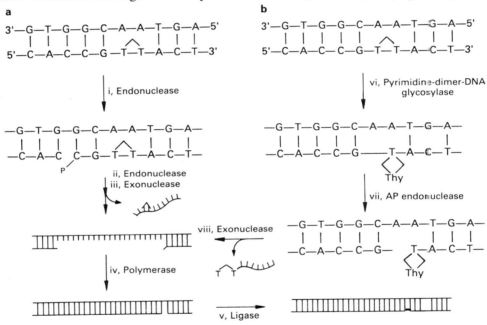

Fig. 7.34 Excision repair—illustrating two alternative modes of excising *cis–syn* thymine photo-dimers. (**a**) (left) (i) endonuclease makes an incision on 5'-side of dimer; (ii) incision of 3'-side and removal of a 10-base oligomer containing the lesion; (iii) gap enlarged by exonuclease; (iv) polymerase resynthesis; (v) ligation to complete patch (colour). (**b**) (right) (vi) a glycosylase cleaves the 5'-thymine from its deoxyribose; (vii) AP endonuclease hydrolyses the exposed phosphate ester; (viii) an exonuclease makes a gap of *ca*. 30 nucleotides, to allow (iv) and (v) as in (**a**).

The second type of excision-repair uses a DNA-glycosylase. These enzymes hydrolyse the N^9-$C^{1'}$ glycosidic bond of modified purine nucleosides. 3-Methyladenine is the principal target for a pair of 3-methyladenine-DNA glycosylases from *E. coli*, and one of them also acts on 3-MeG, O^2-MeC, and O^2-MeT residues.

Among several other glycosylases is a uracil-DNA glycosylase which efficiently cuts out deaminated cytosines in DNA. A pyrimidine dimer–DNA glycosylase works on thymine dimers, cleaving only the glycoside bond of the 5′-thymine base.

The next step is incision of the phosphate backbone by an AP endonuclease, which operates close to apurinic and apyrimidinic sites. Subsequent endonuclease action then excises nucleotides to create a gap at least 30 residues long. In some cases, such as endonuclease III from *E. coli*, the same enzyme acts both as a glycosylase towards reduced thymine residues and as an endonuclease.

In the repair step, this single-stranded region acts as a template for DNA polymerase to fill in the excised sequence, and repair is completed by DNA resynthesis and ligation as before (Fig. 7.34). This UVR repair system is also short-patch and, like the direct base repair systems, it is error-free. Such excision repair appears to be the dominant type of repair process in bacteria while, in humans, any genetic defect in this multi-gene repair system leads to abnormal sensitivity to sunlight and to the high incidence of skin cancer associated with *Xeroderma pigmentosum*.

7.11.3 Post-replication repair

The repair systems described in the two preceding sections maintain the integrity of the genetic message in the cell as long as repair is rapid and is accurately completed before DNA polymerase attempts to copy the damaged DNA. What happens if repair is slow or deficient? It appears that when replication by DNA polymerase is frustrated by a region of damage, a major response is the activation of 'error-prone repair'.

SOS repair is the name given to such enhanced repair (ER), which is induced by a wide variety of types of DNA damage. This is how it seems to work in bacteria. When the advancing DNA polymerase reaches a lesion which blocks its further progress, such as T<>T or Me^3A, a very sizeable gap is left opposite the lesion before replication starts again. This gap can either be made good by recombination processes, known as recombination-repair (Section 10.1), or by the induction of SOS repair.

The SOS process has been most thoroughly studied in *E. coli* and is one of the functions of the *recA* gene protein. Several *recA* proteins identify and bind to any long, single-stranded stretch of DNA which has been formed as a result of excision repair by the *recBC* proteins following the arrest of DNA polymerase, as described above. Through interaction with ATP, *recA* now becomes activated as a protease towards the *lexA* gene protein, which is known to be a multifunctional repressor in control of several DNA repair enzymes. Among the many consequences of hydrolysis of *lexA* are the derepression of *urvABC* and *recA* genes and a reduction in the 3′→5′ exonuclease proof-reading activity of DNA polymerase III. The net result is that pol III operates with decreased fidelity, not only in its ability to read through regions of damage such as T<>T but also elsewhere in the cell, and this situation persists until the activation of *recA* as a protease ends. At this point SOS repair is rapidly switched off.

The mechanism of repairing these long gaps, which can be over 1000 bases long, is highly variable. In particular, the specificity of elongation varies with several factors: the type of polymerase, the activity of the 3′–5′ editing nuclease, and the particular type of DNA lesion. For example, in cases where the base modification behaves as a 'non-instructional' site there is a strong tendency for the incorporation of an adenine residue. This explains why transversion mutations are seen when guanines are modified by aflatoxin or by acetylaminofluorene derivatives. Even then, further elongation in the gap is dependent on the base-sequence in the template strand on the 5′-side of the lesion. The overall result is that 'long-patch' repair seems likely to be idiosyncratic for each type of lesion in each mutable site and, above all, is error-prone.

Summary

Cells have a large repertoire of repair kits to deal with damage by various external agents and safeguard DNA against their consequences. The simplest act by direct reversion of damage, as in the photolyase cleavage of T<>T dimers and demethylation of O^6-MeG. Glycosylase enzymes specifically excise damaged bases from DNA and initiate error-free, short-patch repair processes.

When DNA replication gets going in advance of such repair, the progress of DNA polymerase is blocked by many of the base lesions we have described. This starts up an SOS repair system which activates the *recA* protein to switch on the *uvrABC* repair genes. At the same time there is a reduction in the fidelity of repair. These changes all enhance the chances of survival of the cell. Quite long patches are made in the daughter strand opposite base lesions and these are error-prone. The error-frequency remains generally high for all DNA synthesis and persists until levels of recA protein return to normal.

Genetic deficiencies in some repair systems are associated with the human disease *Xeroderma pigmentosum* that is associated with an exceptionally high sensitivity to sunlight and early mortality from skin cancer.

Further reading

7.1–7.5

Brown, D. M. (1974). Chemical reactions of polynucleotides and nucleic acids. In *Basic principles in nucleic acids chemistry* (ed. P.O.P. Ts'o), Vol. II. Academic Press, London, pp. 2–90.

Schwartz, A., Marrot, L., and Leng, M. (1989). Conformation of DNA modified at a (dGG) or a (dAG) site by the antitumour drug *cis*-diamminedichloroplatinum(II). *Biochemistry*, **28**, 7979–84.

7.6

Blackburn, G. M. and Kellard, B. (1986). Chemical carcinogens. *Chem. Ind. (Lond.)*, 607–13, 687–95, 770–9.

Harvey, R. G. and Geacintor, N. E. (1988). Intercalation and binding of carcinogenic hydrocarbon metabolites to nucleic acids. *Acc. Chem. Res.*, **21**, 66–73.

Singer, B. and Grunberger, D. (1983). *Nucleic acid alkylation and molecular biology of mutagens and carcinogens*. Plenum Press, London and New York.

Walker, G. C. (1984). Mutagenesis and inducible repair responses to DNA damage in *E. coli*. *Microbiol.Revs.*, **48**, 60–93.

7.7

Hurley, L. H. and Needham-Van Devanter, D. R. (1986). Covalent binding of anti-tumour antibiotics in the minor groove of DNA, mechanism of action of CC-1065 and the pyrrolo[1,4]benzodiazepines. *Acc. Chem. Res.*, **19**, 230–7.

Neidle, S. and Waring, M. J. (ed.) (1983). Molecular aspects of anti-cancer drug action, Vol. 3. Macmillan Press, London.

7.8

Tomic, M. T., Wemmer, D. E., and Kim, S.-H. (1987). Structures of psoralen cross-linked DNA in solution by NMR. *Science*, **238**, 1722–4.

Wang, S. Y. (ed.) (1976). *The photochemistry and photobiology of nucleic acids*, Vol. 1. Academic Press, New York.

7.9

Hutchinson, F. (1985). Chemical changes induced in DNA by ionising radiation. *Progr. Nucleic Acid Res.*, **32**, 115–54.

von Sonntag, C. (1986). *The chemical basis of radiation biology*. Taylor and Francis, London.

7.11

Bohr, V. A. and Wassermann, K. (1988). DNA repair at the level of the gene. *Trends Biochem. Sci.*, **13**, 629–33.

Collins, A., Downes, C. S., and Johnson, R. T. (eds.) (1984), DNA repair and its inhibition. *Nucleic Acids Symposium Series No. 13*, IRL Press, Oxford.

Friedberg, E. C. (1985). *DNA Repair*. W. H. Freeman, New York.

Kemball, R. F. (ed.) (1981). Mechanisms of DNA repair. *Progr. Nucleic Acid Res.*, **26**, 181–246.

Lindahl, T. (ed.) (1985). Carcinogens and DNA. *Cancer Surveys*, **4(3)**, 491–624.

Reversible interactions of nucleic acids with small molecules

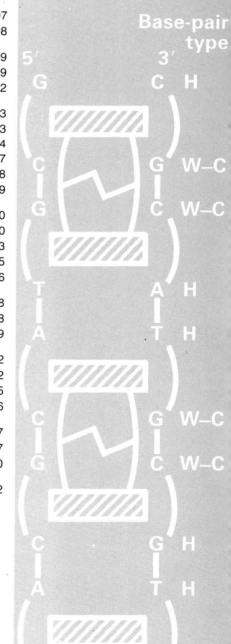

8.1 Introduction

Nucleic acids interact reversibly with a broad range of chemical species that include water, metal ions and their complexes, small organic molecules, and proteins. In this chapter we will focus on the non-covalent interactions of DNA with small molecules of molecular weight less than approximately one thousand. Molecules and ions in this group represent a wide variety of chemical types from simple to complex metal species, a variety of drugs, carcinogens, and complex antibiotics. Of the wide variety of examples which exist, only those will be chosen which give a fairly clear and detailed illustration of as many different features as possible.

At the outset, we should consider the importance of reversible interactions on nucleic acid structure and function. First, all of the intricate nucleic acid conformations which exist are stabilized by and are only possible because of reversible interactions with water, metal ions, and/or organic cations. Dramatic structural transitions in nucleic acids are brought about by changes in water activity, salt concentration (ionic strength), or by interaction with organic molecules. The B–Z transition, hairpin–cruciform formation (Section 2.5), and packaging of nucleic acids into virus particles and chromatin come particularly to mind as being quite sensitive to reversible interactions.

Second, one of the most important lines of drug development and of current chemotherapy against cancer, viral, and some parasitic diseases involves drugs which interact reversibly with nucleic acids. Natural antibiotics such as adriamycin and synthetic drugs such as amsacrine which interact with DNA are widely used in clinical treatment of a variety of neoplastic diseases. Much of the drive to understand nucleic acid interactions has come from the interest in understanding the mode of action of existing medicinal agents and from the desire to develop a new generation of superior drugs. Synthetic oligopeptides and oligonucleotides offer exciting new possibilities as nucleic acid recognizing drugs and as potential 'nucleases' of high sequence specificity.

Third, because of their relative simplicity, the interactions of small molecules with nucleic acids have provided much of our most accurate information about nucleic acid binding specificity, ligand-induced conformational transitions, the molecular basis of co-operativity in binding, the interaction of aromatic amino acid side chains with nucleic acid bases, and other similar critical features of nucleic acid interactions and chemistry.

It must be emphasized at this point that many more physical studies have been conducted on interactions of small molecules with DNA than with RNA. Part of this emphasis on DNA rather than RNA arises from the potential of DNA as a target for anti-cancer drugs. Due to the ease of synthesis of DNA relative to RNA oligomers and polymers, it is also possible to obtain better DNA than RNA model systems for high resolution physical studies. This chapter will, thus, necessarily emphasize studies with DNA.

8.1.1 Types of reversible interactions

Molecules and ions interact with duplex nucleic acids in three primary ways which are significantly different:

(a) binding along the exterior of the helix through interactions which are generally non-specific and are primarily **electrostatic** in origin;

(b) **groove-binding** interactions which involve direct interactions of the bound molecule with the edges of base-pairs in either of the (major or minor) grooves of nucleic acids; and

(c) **intercalation** of planar or approximately planar aromatic ring systems between base-pairs (Fig. 8.1).

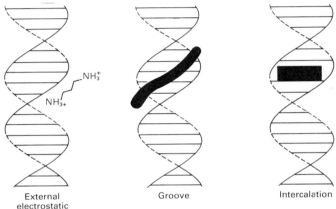

Fig. 8.1 The three primary binding modes are illustrated with a B-DNA double-helical section.

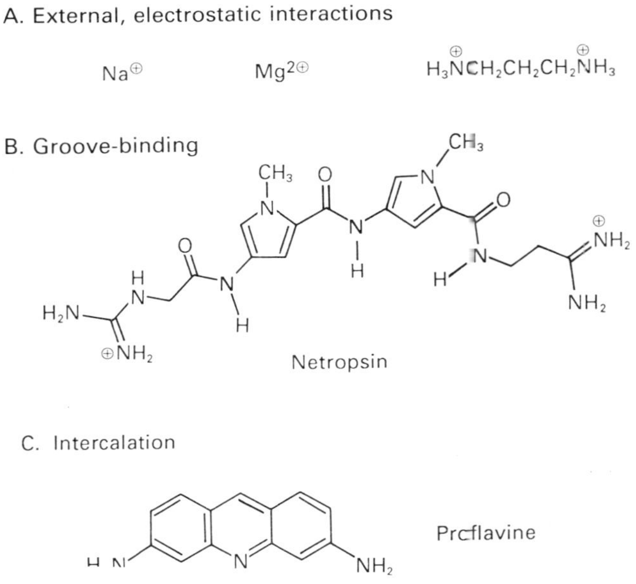

External electrostatic

Groove

Intercalation

A. External, electrostatic interactions

$Na^{\oplus}$ $Mg^{2\oplus}$ $H_3\overset{\oplus}{N}CH_2CH_2CH_2\overset{\oplus}{N}H_3$

B. Groove-binding

Fig. 8.2 Examples of cations which bind by the three primary modes.

Netropsin

C. Intercalation

Proflavine

The first two binding modes do not require a nucleic acid conformational change, but may induce structural transitions on complex formation. Intercalation requires changes in sugar–phosphate chain torsional angles for separation of adjacent base-pairs by a distance (typically near 3.4 Å) sufficient to allow insertion of the intercalating ring system. This can be accompanied by other changes in the helical parameters such as unwinding, bending, etc. Examples of the types of molecules and ions that bind to nucleic acids by the three different modes are shown (Fig. 8.2). These three types of interactions will be discussed separately.

8.2 External electrostatic interactions

8.2.1 Condensation type interactions

Nucleic acids are highly charged polyelectrolytes whose anionic phosphate groups strongly affect their structure and interactions. Manning and co-workers have shown that simple ions, such as those of the alkali metals, associate with nucleic acids largely as a function of the polymer charge density. In B-DNA, for example, if the molecule is modelled as a line of equally spaced charges there is one anionic charge per 1.7 Å distance (two charges per base-pair and approximately ten base-pairs per turn of the duplex). If the line charge spacing becomes less than approximately 7 Å, the molecule becomes unstable. Polymer conformations with charges that are more closely spaced must thus associate with counterions from solution to achieve stability. The association of ions with the polyelectrolyte is called **counterion condensation** and causes an unfavourable entropy term in the overall polymer conformational free energy summation. This unfavourable term is more than outweighed by the numerous favourable interactions in the folded polymer (such as the DNA double helix). Counterions 'condense' until the charge density is reduced to the stable level of approximately one charge per 7 Å. Additional counterions are associated with the remaining charges on the polyanion through Debye–Hückel type interactions. For a specific nucleic acid conformation the charge density and, thus, the amount of condensed counterion is constant. An important prediction of the condensation theory, which has received experimental support, is that the counterions condensed per phosphate charge remain constant as the solution salt concentration is varied over a fairly wide range.

The initial association of counterions is referred to as condensation, since the ions are associated with the general charge density of the polyelectrolyte and are not bound at specific sites. The ions retain their inner sphere water of hydration and move rapidly along the sugar-phosphate backbone of DNA. Secondary hydration layers of both the polyelectrolyte and counterion are affected by this interaction. The Poisson–Boltzmann equation, applied to nucleic acids as rod-like polyanions, predicts a similar number of strongly associated counterions. If the phosphate charges of DNA are 'deleted', for instance by esterification of the phosphate groups or by replacement with methylphosphonate linkages, the molecule still can form a

B-like double helix, but, as expected, it does not exhibit the salt-dependent properties of normal B-DNA. With the B-form duplex structure of DNA, the condensation theory predicts an average of 0.76 monovalent counterions (such as Na⁺) condensed per phosphate group and a total of 0.88 counterions associated per phosphate group in condensation plus Debye–Hückel type interactions. Predictions by other theoretical methods arrive at similar numbers of associated counterions.

The associated counterions reduce the effective charge on nucleic acids and strongly affect the solution properties and binding interactions of the polymer. A significant portion of the binding free energy of species ranging from small cations to large proteins can result from the neutralization of nucleic acid charges (ion pair formation) in the complex and the resulting favourable entropic effect of release of counterions. At a constant temperature, this entropic effect resulting from counterion release can lead to nucleic acid denaturation as well as to increased binding of cationic ligands as the bulk solution salt concentration is decreased. Thus, both the T_m of nucleic acids and the observed equilibrium constant for binding of cationic ligands depend strongly on salt concentration. Thomas Record and co-workers have developed a particularly useful formulation for the effects of salt concentration on nucleic acid equilibria. In particular they have shown that for a cationic ligand, L, binding to a nucleic acid site, D, the thermodynamic equilibrium in the presence of sodium counterions is:

$$D + L \rightleftharpoons C + m'\psi_c[Na^+], \tag{1}$$

where L makes m' ion pairs with DNA phosphate groups, C is the complex, and ψ_c is the average fraction of sodium ions condensed per phosphate group. The observed equilibrium is simply

$$D + L \rightleftharpoons C, \tag{2}$$

and is described by K_{obs}, an observed equilibrium constant. K_{obs} is experimentally evaluated at constant salt concentration and temperature by determining the concentrations of the species D, L, and C specified in equation (2). The variation of K_{obs} with sodium ion concentration is

$$\frac{\delta \log K_{obs}}{\delta \log [Na^+]} = -m'\psi, \tag{3}$$

where ψ represents the average fraction of a sodium ion associated with each phosphate group through condensation and Debye–Hückel type screening processes. ψ is related to the average linear phosphate group spacing, b, in nucleic acids through the dimensionless parameter ξ:

$$\xi = e^2/\varepsilon kTb, \tag{4}$$

where e is the electronic charge magnitude, ε is the bulk solution dielectric constant, k is Boltzmann's constant, and T is the temperature in K. The relations of ψ and ψ_c to ξ are:

$$\psi_c = (1 - \xi^{-1}) \tag{5}$$

$$\psi = (1 - (2\xi)^{-1}) \tag{6}$$

For B-DNA, b is 1.7Å and ψ and ψ_c are 0.88 and 0.76 respectively. As b increases on binding of specific cationic ligands or on DNA denaturation, ψ and ψ_c decrease.

The negative sign in equation (3) indicates that the observed equilibrium constant decreases with increasing sodium ion concentration. The magnitude of the decrease will depend on the charge on the ligand through m', the number of charged groups which can interact with phosphate groups on the nucleic acid. It is important to note that most experimental equilibrium constants, K_{obs}, have no meaning unless the solution conditions are carefully specified. It is also impossible to compare K_{obs} for different ligands unless they are measured under the same conditions. A better comparison method is to determine K_{obs} at several salt concentrations and plot $\log K_{obs}$ versus $\log[Na^+]$ according to equation (3). Comparison of several compounds can then be made under more defined standard state conditions, physiological salt concentration, etc. A plot of $\log K_{obs}$ versus $\log[Na^+]$ also allows a determination of m' from the slope.

As stated above, conformational changes of nucleic acids, such as denaturation, are also dependent on the salt concentration of the solution. Any ligand-binding process which causes a conformational change will thus have an additional salt dependence above that given by equation (3). For example, all intercalators lengthen the double helix of DNA and this causes release of counterions since the longer helix has a larger spacing between phosphate groups and a resulting lower intrinsic charge density. For a neutral intercalator, this will be the only source of counterion release since no ion pairs can form between the DNA phosphate groups and the uncharged ligand. Cationic intercalators release counterions from DNA both as a result of the conformational change induced in the double helix and because of phosphate neutralization through ion-pairing.

Multiply charged simple cations such as Mg^{2+} and cations of simple organic amines such as 1,3-diaminopropane interact with DNA more strongly than monovalent cations (Na^+, K^+, etc.) and displace the monocations from DNA. Much, and in some cases all, of the binding free energy of these simple multications is electrostatic in nature and their interaction with nucleic acids can be modelled as a condensation process. Cations of other metals, such as zinc, copper, mercury, etc., interact with DNA partly through a non-specific condensation type binding, but these more complex metal species can also bind directly to base-pairs to form a site-bound complex. Although condensation increases the T_m of DNA, the more direct base-binding interactions of metals frequently decrease the T_m of DNA since the bases in the denatured state can interact either more strongly or at additional sites with metals.

Any cation can associate with nucleic acids through a condensation type interaction. Since condensation creates a non-specific, mobile-type complex along the exterior of the double helix, the kinetics of association and dissociation of this complex are quite rapid. Kinetic studies indicate that complex cations, which at equilibrium bind to DNA through groove or intercalation complexes with very specific interactions, initially associate with the duplex through condensation. The association pathway may involve initial Debye–Hückel type interactions followed by condensation and fast diffusion along the duplex backbone to the specific binding site. This process is somewhat analogous to 'sliding' of protein molecules along the DNA helix as they search for their most favoured binding site. Dissociation could also involve condensation as an intermediate step between the site-bound ligand and release of the bound molecule into the bulk solution. The binding mechanism in specific cases may have additional steps.

Water is also bound along the exterior of nucleic acid structures (Section 2.2.2.). Its specific interactions with polar groups on bases and sugars as well as with the charged phosphate groups are essential for the stability of nucleic acid conformations. At a stage weaker than these specific interactions, the water of solvation begins more closely to approach the properties of bulk solvent. Release of strongly bound water by association of ligands at specific nucleic acids sites can provide both favourable (increase in entropy) and unfavourable (increase in enthalpy through loss of specific interactions) contributions to the free energy of binding. The relative magnitude of these contributions will depend on the number of water molecules released and the types of interactions which are broken during complex formation.

8.2.2 Non-specific outside stacking

Planar aromatic molecules can stack on each other to form dimers and higher aggregates. When the compounds are charged, as with proflavine (Fig. 8.2.), they repel each other electrostatically. If, however, the cations stack along the anionic DNA sugar-phosphate chain, the charge repulsion is decreased and this type of binding leads to nonspecific **outside stacking** of planar cations along the double helix. Such stacking has many charge interactions and releases a large fraction of condensed counterions. It is therefore very dependent on salt concentration and is generally quite weak at salt concentrations of 0.1 M and above. Because this binding mode is a type of extended self-association, it can be highly co-operative and will generally be more favourable at high ratios of the aromatic cation to DNA phosphate groups.

Summary

Nucleic acids are highly charged polymers which must 'condense' a significant number of cations from solution to exist in stable conformations. Partial release of these cations on denaturation or on binding of cationic ligands to specific nucleic acid sites accounts for the strong dependence of these processes on salt concentration.

8.3 Groove-binding molecules

8.3.1 Characteristics of groove-binding

The general concept of 'groove-binding' has been illustrated above (Fig. 8.1) and netropsin (Fig. 8.2) is a typical molecule which interacts with nucleic acids by this mechanism. The major and minor grooves differ significantly in electrostatic potential, hydrogen-bonding characteristics, steric effects, and hydration. Many protein and oligonucleotide molecules exhibit binding specificity primarily through major groove interactions while small, groove-binding molecules in general prefer the minor groove.

Typically, minor groove-binding molecules have several simple aromatic rings such as pyrrole, furan, or benzene connected by bonds with torsional freedom. This creates compounds which, with the appropriate twist, can fit into the helical curve of the minor groove with displacement of water from the groove. The minor groove is generally not as wide in A:T-rich relative to G:C-rich regions (Section 2.3.4) and may 'fit' aromatic molecules better at A:T than at G:C sequences. A molecule, given the correct twist of its linked aromatic rings, can fit snugly into the minor groove and form van der Waals contacts with the helical chains which define the 'walls' of the groove. Any specificity which arises in the binding comes from contacts between the bound molecule and the edges of the base pairs on the 'floor' of the groove. Hydrogen-bonds can be accepted by A:T base-pairs from the bound molecule to the C-2 carbonyl oxygen of T or the N-3 nitrogen of A. Although similar groups are present on G:C base-pairs, the amino group of G presents a steric block to hydrogen-bond formation at N-3 of G and at the C-2 carbonyl of C. The hydrogen-bond between the amino group of G and the carbonyl oxygen of C in G:C base-pairs lies in the minor groove and sterically inhibits penetration of molecules into this groove in G:C rich regions. Thus, the aromatic rings of many groove-binding molecules form close contacts with A-H^2 protons in the minor groove of DNA and there is no room for the added steric bulk of the G-NH$_2$ group in G:C base-pairs. Pullman and co-workers have shown that the negative electrostatic potential is greater in the A:T minor groove than in G:C-rich regions of DNA and this provides an additional important source for A:T-specific minor groove-binding of cations. It is possible to enhance G:C binding specificity by designing molecules which can accept hydrogen-bonds from the G-NH$_2$ group. Synthesis and analysis of such molecules is an active area of research.

A possibility with groove-binding molecules, which does not exist with intercalators, is that they can be extended to fit over many base-pairs along the groove and have very high sequence-specific recognition of nucleic acids. For example, an oligopyrimidine can be designed which forms a triple helix with a specific homopurine–homopyrimidine duplex sequence (Section 2.4.5). By linking DNA cleavage reagents to the oligopyrimidine, a highly specific 'nuclease' can be created (Section 8.7.2). Oligonucleotides are also being used as 'antisense' anti-viral, and anti-cancer drugs which specifically recognize single-stranded cellular nucleic acids.

8.3.2 Netropsin and distamycin

Dickerson and co-workers have obtained a crystal structure of netropsin (Fig. 8.2) bound to the DNA duplex d(CGCGAATTCGCG) (Fig. 8.3). This structure has provided considerable molecular detail about complex formation in the minor groove. Netropsin binds at the AATT centre of the duplex and displaces the spine of hydration seen in that region of the free oligomer (Fig. 8.3). The three amide NH groups point inward and form bifurcated hydrogen-bonds with N-3 of A and O-2 of T. The molecule is held in the centre of the groove by van der Waals contacts with the atoms of DNA which form the walls of the groove. These contacts hold the pyrrole rings approximately parallel to the walls of the groove and, as a consequence of the helical twist of the groove, the two pyrrole rings are twisted by approximately 33° with respect to one another (Fig. 8.3).

The two cationic ends of netropsin are also centred in the minor groove and are associated with N-3 on the outer A bases of the central four A:T base-pairs. Steric interactions between the pyrrole-CHs and the DNA bases prevent netropsin from moving more deeply into the groove. As a consequence, some of the hydrogen-bond lengths between netropsin and the A:T base-pairs are quite long (3.3–3.8 Å) compared to standard values (less than 3 Å). Binding of netropsin causes a slight widening of the minor groove in the AATT region and a bending of the helix axis away from the site of binding. No other characteristic helical parameters are significantly changed in the complex. As indicated above, the amino group of G prevents molecules of this type from sliding deeply into the minor groove and forming hydrogen-bonds with the bases. Netropsin binding to G:C-rich regions of DNA is thus weaker than to A:T sequences. Other factors such as water and counterion release contribute to the overall free energy of binding of netropsin, but probably have little effect on binding specificity.

Lown and Dickerson have devised an interesting new series of sequence-specific binding reagents, the **lexitropsins**, which are designed to recognize both A:T and G:C base-pairs. They noted that the pyrrole ring CH-groups of netropsin point into the minor groove and make close contact with A:T base-pairs. In the lexitropsins

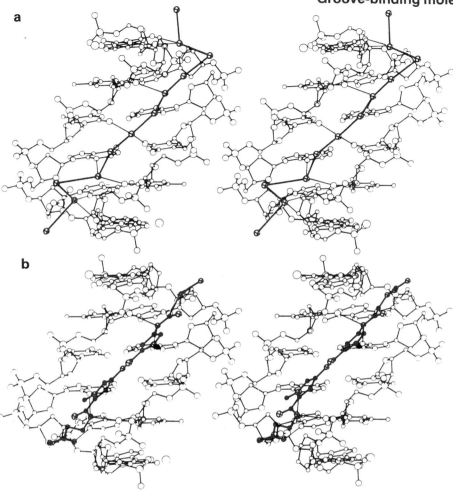

Fig. 8.3 Stereoviews of B-DNA minor groove binding of the spine of hydration (**a**) and the netropsin molecule (**b**). In each case, only the central 6 base-pairs are drawn for the dodecamer of sequence: CGCGAATT-BrC-GCG, as established by single-crystal X-ray structural analysis. Base-pair G4.C21 is at the top and C9.G16 is at the bottom (both chains numbered from the 5′-end). Open circles are DNA atoms and red circles are water or netropsin atoms. The guanidinium end of the netropsin molecule is at the top, and the amidinium end is at the bottom. Distances from adenine N-3 or thymine O-2 atoms to water oxygens, or to netropsin amide nitrogens, are drawn as thin lines only if they are 3.5 Å or less (adapted from Kopka, M. L., Yoon, C., Goodsell, D., Pjura, P., and Dickerson, R. E. (1985). *Proc. Nat. Acad. Sci. USA*, **82**, 1376–80). Stereopair figure for parallel viewing.

one or more of the pyrrole rings is replaced by hydrogen-bonding acceptor heterocycles such as imidazole. In particular, the exchange of a pyrrole-CH for an imidazole-N function alleviates the steric clash with the G–NH$_2$ group which prevents netropsin from binding in G:C regions. In its place, a specific hydrogen-bond can form between the G–NH$_2$ and an imidazole nitrogen which should confer G:C selectivity on appropriately substituted lexitropsins. Both NMR and footprinting experiments have confirmed increased G:C base-pair recognition specificity for synthetic lexitropsins although A:T base-pairs are still permitted at the imidazole substituted sites.

Distamycin (Fig. 8.4) has a structure and binding specificity similar to netropsin and its DNA complex has many characteristics quite similar to the netropsin complex. Rich and Wang have solved crystal structures for the oligomer d(CGCA$_3$T$_3$ GCG) and its distamycin complex. Distamycin has a crescent shape which closely matches the curvature of the minor groove which is the binding site of the drug. In the crystal structure, the molecule is twisted in a complementary manner to the natural helical twist of the minor groove. As with netropsin, the molecule makes close van der Waals contacts with the 'walls' of the minor groove and all solvent is displaced from the groove at the binding site.

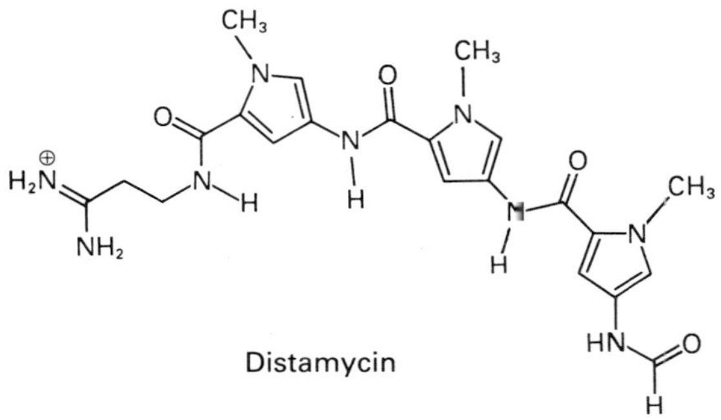

Fig. 8.4 Structure of distamycin. The crescent shape is similar to that observed in the crystal structure with the oligomer d(CGCA$_3$T$_3$GCG).

Distamycin

Five NH groups are on the inside of the 'crescent' shape of distamycin and these can form hydrogen-bonds with the N-3 of A and O-2 of T. At least three bifurcated hydrogen-bonds are formed (to N-3 and O-2) in the crystal structure. The geometric arrangement of NH groups on distamycin does not exactly match the helical displacement of hydrogen-bond accepting groups on A:T base-pairs in DNA, and consequently not all distamycin NH groups can be in optimum hydrogen-bonding positions simultaneously. The exact nature of the distamycin–DNA hydrogen-bonding will, no doubt, depend on the local sequence and helix geometry. The distamycin molecule has three pyrrole rings, compared to two in netropsin, and covers five A:T base-pairs in the complex rather than four in the case of netropsin. The amino group of G also blocks the hydrogen-bonding of distamycin in the minor groove and consequently weaker binding is observed in G:C base-pair regions. Because the minor groove is narrower in A:T-rich regions, van der Waals contacts of both distamycin and netropsin are better in A:T than in G:C regions of DNA.

The oligomer was found to have an unusual conformation in the central d(A$_3$ T$_3$) sequence of the crystalline complex. The base-pairs in this region have a high positive propeller twist which places the amino group of A between the O^4 group of its

Watson–Crick complementary T and the O^4 group of its 5′-neighbour T. Thus, this amino group of A can form bifurcated hydrogen-bonds with the two O^4 residues. A similar unusual conformation was found in crystals of this oligomer lacking distamycin and also by Klug and co-workers in the crystal structure of $d(CGCA_6GCG).d(CGCT_6GCG)$.

Both groove-binding molecules and intercalators exhibit unusual thermodynamic parameters, notably large positive enthalpies and entropies are found for binding to polydA.polydT. Enthalpies are generally negative for binding to other sequences. These unusual thermodynamic results can be explained if the polymer has bifurcated hydrogen-bonds for non-alternating A:T base-pair sequences, as observed for the oligomers. Any weakening or breaking of the precise geometry of these bifurcated hydrogen-bonds would result in a positive enthalpy contribution, but the enhanced flexibility of the structure could result in a positive entropy. The release of water molecules which are bound to the polydA.polydT helix and which are displaced on binding other molecules could also contribute to the unusual thermodynamic properties of this sequence.

8.3.3 Hoechst 33258

Hoechst 33258 (Fig. 8.5) is an antibiotic and chromosome stain. Both Dickerson and Wang and co-workers have solved crystal structures of Hoechst 33258 bound to the same DNA dodecamer as used with netropsin. The two structures for the complex are similar, but show some significant differences. In Dickerson's structure, the Hoechst molecule binds near the centre of the duplex sequence and displaces most of the spine of hydration, as does netropsin, but the actual binding site is proposed to be ATTC (rather than AATT observed in the case of netropsin). In this structure, the phenolic OH group of Hoechst 33258 is hydrogen-bonded to what is left of the original spine of hydration in the A:T region. The benzimidazole rings of Hoechst 33258 fit tightly into the minor groove in the A:T region similar to the pyrrole rings of netropsin and distamycin.

In Dickerson's structure there is a significant twist in the torsional bond connecting the piperazine and benzimidazole rings (the best mean planes for the two rings are almost perpendicular). Because of this molecular conformation, the piperazine

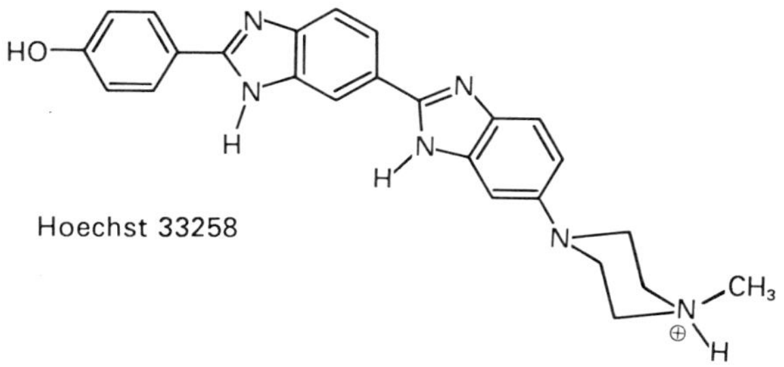

Hoechst 33258

Fig. 8.5 Structure of Hoechst 33258.

cannot fit into the narrow minor groove characteristic of A:T rich regions but, with some reduction in twist, can fit into the wider major groove in G:C regions. This leads to the proposed ATTC binding site for Hoechst 33258 where the piperazine ring is bound at the terminal C:G site of the ATTC sequence. If the bond connecting the two benzimidazole rings is rotated by 180°, the only apparent change in the complex is that the piperazine ring swings out of the groove. Both of those rotational conformers of Hoechst 33258 are thought to occur in approximately equal amounts in Dickerson's crystal structure.

In Wang's structure, the Hoechst 33258 molecule is located in the AATT centre of the oligomer as with netropsin. This can occur because in the Wang structure, the piperazine ring is almost parallel to the adjacent aromatic benzimidazole ring, rather than almost perpendicular to it as in the Dickerson structure. In the Wang structure, the binding of Hoechst 33258 to DNA involves hydrogen-bonds from the benzimidazole-NH groups to O^2 of T and N^3 of A and electrostatic interaction of the cationic dye with the anionic oligomer.

The Hoechst 33258 molecule is also highly curved in the Wang structure to match the curvature of the minor groove. The dye forms numerous favourable contacts with the walls of the minor groove and these interactions provide significant free energy of stabilization for the complex. The phenol ring of Hoechst 33258 makes an angle of only 8° with the benzimidazole ring to which it is attached, but the two benzimidazole ring planes are twisted 32° with respect to each other. The piperazine is only slightly puckered in the Wang structure and lies almost in the plane of the benzimidazole to which it is attached (dihedral angle 14°). The $O^{4'}$ atoms of deoxyribose in the minor groove are in a favourable position to interact with the π-electron system of the dye as it sits in van der Waals contact with the walls of the minor groove. This type of interaction may represent a general principle for binding of aromatic rings in the DNA minor groove. NMR studies may be required to determine the origin of the two different complex structures obtained for Hoechst 33258.

8.3.4 SN 6999

A range of biophysical studies with the drug SN 6999 (Fig. 8.6) indicate that it also binds in a minor groove complex with high A:T specificity. It is a very interesting compound with a fused bicyclic quinoline ring which, in the antimalarial drug chloroquine and other similar compounds, is capable of intercalation. The structures of the two quinoline compounds are shown (Fig. 8.6) for comparison.

2D-NMR methods can provide considerable structural detail about nucleic acids and their complexes. Leupin, Wüthrich, and co-workers have conducted a detailed investigation of the solution structure of the 1:1 complex of SN 6999 with the oligomer sequence d(GCATTAATGC) by such methods. Intermolecular nuclear Overhauser effect (nOe) results indicate that the compound binds in the minor groove and confirm that it does not intercalate despite the presence of the quinoline

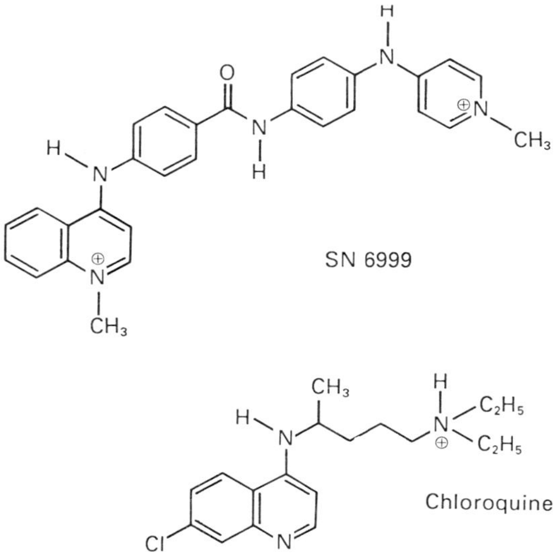

SN 6999

Fig. 8.6 Structure for the groove-binding compound SN 6999 and the intercalator chloroquine.

Chloroquine

ring. As with other groove-binding molecules, SN 6999 interacts with the central A:T base-pairs of the oligomer sequence. Bound and free SN 6999 are in slow exchange on the NMR timescale near 0°C, but move into fast exchange near 25°C. While the self-complementary oligomer is symmetrical when free, it becomes unsymmetrical in the complex with SN 6999 because of the lack of symmetry in the drug. The two exchange orientations of the drug in the oligomer A:T region and the intermolecular contact points between the base-pairs and the bound molecule have been deduced from nOe studies as shown (Fig. 8.7). The drug appears to be curved in a crescent shape to fit into the helical minor groove and to form hydrogen-bonds with A:T base-pairs as with other minor groove-binding molecules.

8.3.5 Groove-binding versus intercalation

The similarity in structures but difference in the binding modes of chloroquine and SN 6999 raise an important question: 'What determines the selection between intercalation and groove-binding interactions?' Based on model building studies, the quinoline ring of SN 6999 could form favourable contacts with the DNA base-pairs and perhaps electrostatic interactions with the DNA phosphate groups in an intercalation complex. Relative to groove-binding, it would lose significant free energy of hydrogen-bonding, of non-bonded contacts with the walls of the groove, and of solvent release from the groove. The same is true of netropsin and other similar groove-binding molecules. At least for structures of this type, binding in the minor groove is much more favourable than intercalation. Chloroquine, on the other hand, does not have an optimum structure for groove interactions and it binds to DNA in an intercalation complex. Even so, the intercalation binding constant of chloroquine is low and it is classified overall as a weak DNA-binding molecule.

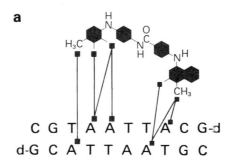

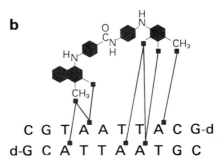

Fig. 8.7   Binding orientations and contacts for SN 6999 with the oligomer d(CGATTAATGC) as determined by nOe distance measurements (adapted from Leupin, W., Chazin, W. J., Hyberts, S., Denny, W. A., and Wüthrich, K. (1986). *Biochemistry*, **25**, 5902–10. Copyright (1986) American Chemical Society).

Summary

Outside-binding molecules typically are crescent shaped molecules with hydrogen-bonding N–H groups on the interior of the crescent. These NH groups form hydrogen-bonds with A:T base-pairs in the minor groove but are excluded from similar interactions with G:C base-pairs by the amino group of G.

Groove-binding molecules typically contain several small aromatic ring systems linked with torsional freedom to allow a twist complementary to that of the DNA minor groove. Electrostatic interaction of cationic groups with the negative electrostatic potential in the minor groove plus close van der Waals contacts with the 'walls' of the minor groove provide additional favourable components to the free energy of binding.

8.4 Intercalation

8.4.1 The classical model

In the early 1960s, Lerman described a number of physical studies on the interactions of DNA with planar aromatic cations. He concluded from these studies that planar aromatic molecules could bind to DNA by a process which he termed **inter-**

calation (Fig. 8.1). This mode of binding has now been established for a large number of polycyclic aromatic cations (example structures are shown in Figs 8.2 and 8.8). At the same time, the classical intercalation model has been extended in considerable molecular detail. Just as the classical B-DNA is now seen as only one of the many possible conformational states of double helical nucleic acids, the classical intercalation model is only one view of the way in which aromatic compounds can be accommodated between nucleic acid base-pairs.

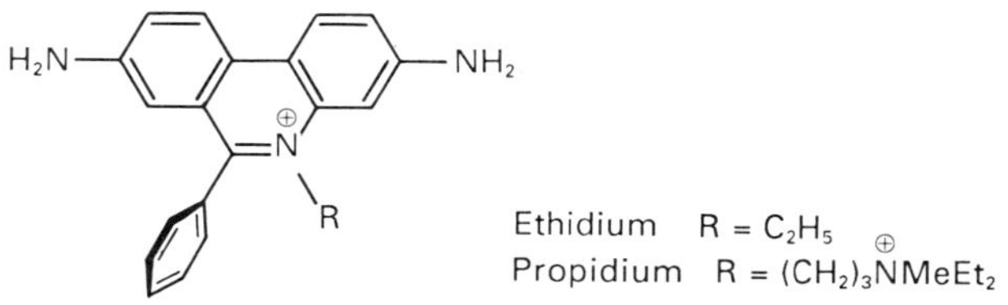

Ethidium R = C_2H_5
Propidium R = $(CH_2)_3\overset{\oplus}{N}MeEt_2$

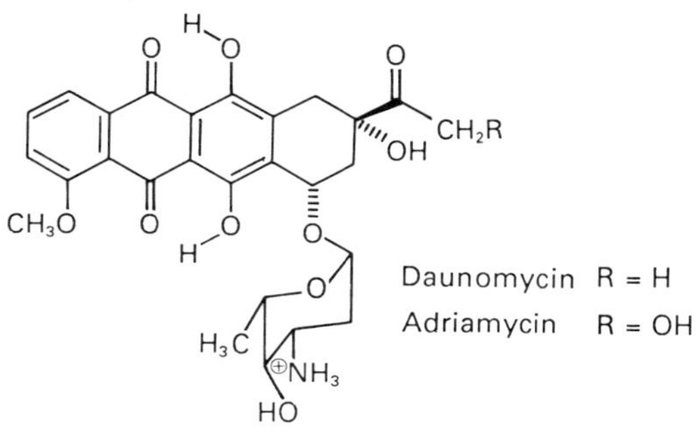

Daunomycin R = H
Adriamycin R = OH

Fig. 8.8 Structures of the phenanthridinium intercalators, ethidium and propidium, and the anthracycline intercalators, daunomycin and adriamycin.

As a result of rotation about torsional bonds in the DNA backbone, the creation of an intercalation site causes separation of base-pairs and a lengthening of the double helix. This increase in length can be detected by hydrodynamic methods, such as viscosity and sedimentation measurements, using short linear sections of DNA prepared by sonication or enzymatic methods. Compounds which bind in the DNA grooves such as netropsin, Hoechst 33258, and SN 6999 do not significantly increase the viscosity of sonicated DNA or unwind DNA base-pairs. In the classical model, the helix is lengthened by 3.4 Å, which is the thickness of typical aromatic ring systems. In practice, the observed length increase of the DNA complex is generally less than the 3.4 Å maximum. The helix is unwound at the site of an intercalation complex and the normal approximately 36° rotation of one base-pair with respect to the next is decreased as a result of intercalation. The amount of unwinding varies considerably with the intercalator structure and probably with the DNA sequence in a manner that is not well understood. Of the compounds illustrated

(Figs 8.2 and 8.8), ethidium and propidium give unwinding angles of 26°, proflavine and related acridines unwind DNA by 17°, and the anthracycline drugs daunomycin and adriamycin unwind DNA by 11° per bound molecule.

The amount of unwinding in DNA produced by intercalation is conveniently measured with closed circular supercoiled DNA (Section 2.3.4, Fig. 2.25) and the unwinding angle calculated from such experiments is averaged over all of the intercalation sites in the random sequence DNA. However, actual unwinding angles at each binding site may vary with the local sequences.

An additional test for intercalation is provided by dichroism methods. The planar ring systems of intercalators stack with the DNA base-pairs and thus have similar dichroism values to the base-pairs. The dichroism of groove-binding molecules is frequently the opposite of that of the base-pairs since the groove complexes are bound along the edges of the DNA base-pairs rather than stacked between base-pairs, as is the case with intercalators (Fig. 8.1).

Crothers and his group have measured the dichroism both of DNA bases and of bound intercalators in double helical nucleic acids when the complexes are oriented by means of a strong electric field (electric dichroism). Based on such electric dichroism results, they proposed a model in which the propeller-twist of base-pairs adjacent to the planar intercalator is flattened in order to stack better with the intercalated ring system. The intercalator and adjacent base-pairs are then tilted by 20–25° so as to optimize stacking interactions with propeller-twisted base-pairs adjacent to the complex. The overall DNA duplex seems not to be bent significantly by the bound intercalators. Therefore there must be some other compensating changes which occur to counteract the tilting which is induced at the intercalation site.

Crystallographic studies could potentially give further molecular detail concerning intercalator-induced tilt and long-range, compensating effects in the DNA duplex. However, all except one of the available mono-intercalator crystal structures involve dinucleotides which cannot provide long range information. With the dinucleotide crystals, only molecules with 5'-pyrimidine-purine-3' sequences form Watson–Crick base-paired duplex structures with intercalators. Such duplex dinucleotide structures provide support for and extend the detailed pictures of the classical intercalation model. In general, drugs such as proflavine and ethidium (Figs 8.2 and 8.8) are stacked with their long axes parallel to the long axes of the adjacent base-pairs in these crystal structures. The exocyclic amino groups of the intercalators point towards diester oxygens of the DNA phosphate groups at the intercalation site and provide additional electrostatic and hydrogen-bonding stabilization of the complex. The base-pair centres are separated by 6.9 Å in the complex (3.4 Å space for the intercalator) and stack well with the intercalated ring system. The base-pairs may be slightly kinked, especially for ethidium complexes where the out-of-plane phenyl group limits full intercalation of the cationic phenanthridinium ring. The phenyl and ethyl substituents of ethidium lie in the minor groove of the complex. Base-pair unwinding angles in these dinucleotide models have ranged from 0° for proflavine to 26° for ethidium. It is not necessary for them to relate closely to DNA-unwinding values because of the local sequence effects on unwinding angles in the high molecular weight nucleic acid and end-effects in the dinucleotides.

8.4.2 Daunomycin—d(CGTACG)

The first crystal structure with a monointercalator and oligonucleotide was obtained by Wang, Rich, and co-workers for a complex of the antibiotic daunomycin (Fig. 8.8) and the oligomer d(CGTACG) (Fig. 8.9). Unlike intercalators studied at the dinucleotide level, daunomycin binds to the duplex with its long axis almost perpendicular to the long axes of adjacent base-pairs at the intercalation site. The daunomycin amino-sugar, which is attached to ring A of the anthracycline ring system, lies in the minor groove while ring D, which bears a methoxy group, protrudes into the major groove. The daunomycin core, rings B and C, lies between base-pairs. The hexamer has two daunomycin molecules intercalated at each of the two C:G sites at either end of the duplex. These intercalation sites are presented (Fig. 8.9a) with the daunomycin molecules deleted so that the geometry of the intercalation site can be seen more easily. The central A:T base-pairs retain a general B-DNA-like geometry in the complex, but with some backbone distortions. The same view of the duplex is shown containing the intercalated drugs (Fig. 8.9b). Views of the complex into the major and minor grooves (Fig. 8.9c and 8.9d respectively) are also shown.

As expected, the C:G intercalation site is opened by 3.4 Å to create the intercalation space. The cationic amino-sugar substituent and ring A largely fill the minor groove and displace water molecules and ions from it. The hydroxyl group on ring A donates a hydrogen-bond to N-3 of G and is a hydrogen-bond acceptor from the $-NH_2$ group of the same G. This gives increased binding free energy and both base-pair and orientational specificity for the intercalated drug. The conformations of ring A and of the amino-sugar of daunomycin change relative to the unbound molecule and these changes facilitate the snug fit of the molecule into the right-handed minor groove.

The conformation of the duplex is also significantly changed relative to the B-DNA helical structure to accommodate the antibiotic. In addition to the increased separation of base-pairs at the intercalation site, the G:C base-pairs are also shifted laterally towards the major groove so that the helix axis changes position. There is no unwinding of base-pairs at the intercalation site, i.e. they maintain the usual 36° helical twist, but base-pairs at adjacent sites are unwound by 8°. While this agrees reasonably well with the net 11° unwinding angle determined from studies in solution with superhelical DNA, the fact that the unwinding occurs at the base-pairs adjacent to the intercalation complex is unexpected. Such long-range, induced

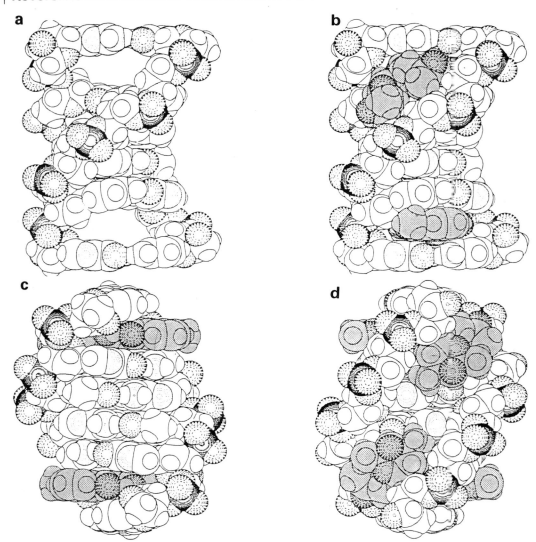

Fig. 8.9 Space-filled diagrams of the daunomycin-d(CGTACG) complex. (**a**) The DNA hexamer intercalation cavity with the daunomycin molecule deleted. This view is perpendicular to the molecular 2-fold axis, which lies horizontally in the plane of the paper. The minor groove of the distorted right-handed B-DNA is at the upper left of the figure, while the major groove is at the lower right. (**b**) The drug–DNA complex viewed from the same direction as in (**a**). The amino sugar of the daunomycin (colour) fills the minor groove of the double helix. Ring D of the intercalated aglycon chromophore skewers through the base-pairs and protrudes into the major groove. (**c**) A view looking down the 2-fold axis from the major groove. The degree of penetration of the aglycon ring through the base-pairs is evident. (**d**) A view looking down the 2-fold axis from the minor groove. The amino sugars of the daunomycins largely fill the minor groove of the double helix. Note that the disposition of the sugar relative to the aglycon keys the daunomycin for a right-handed helix (adapted from Wang, A. H., Ughetto, G., Quigley, G. J., and Rich, A. (1987). *Biochemistry*, **26**, 1152–63. Copyright (1987) American Chemical Society).

conformational changes illustrate the flexibility of DNA and the significant variations that can occur in duplex conformations to accommodate a large intercalator which has structurally complex groove-binding substituents.

The interaction of daunomycin at the C:G sites of d(CGTACG) should not be taken as an indication of high C:G specificity for this drug. Daunomycin requires a three base-pair binding site and, therefore, the only symmetrical way to bind two intercalators in the hexamer is to place the daunomycin molecules at the C:G sites with the amino sugars pointing into the centre of the oligomer. Unsymmetrical orientations probably occur in solution but would not be expected to crystallize as readily as the symmetrical complex. In fact, daunomycin is known to bind quite well to sites containing only A:T base-pairs.

8.4.3 Binding specificity

There are ten possible dinucleotide combinations which form different, right-handed, antiparallel intercalation sites for simple intercalators (Fig. 8.10). For intercalators with substituents which also contact non-adjacent base-pairs or which cause distortions in sequences neighbouring the binding site, the number of specific possibilities is even larger. Most intercalators display either no binding preference or a slight G:C base-pair preference and this contrasts with the general A:T preference of outside-binding compounds. It has been suggested that the general G:C preference of intercalators is due to the larger intrinsic dipole moment of G:C relative to A:T base-pairs and the resulting ability of G:C base-pairs to induce polarization in the ring system of intercalators. Nonetheless, intercalators with an A:T preference have been synthesized. This shows that overall binding specificity must depend not only on dipole interactions with polarizable intercalators but also on hydrogen-bonding interactions, on the size, hydration, and electrostatic potential of the grooves, and on other similar factors.

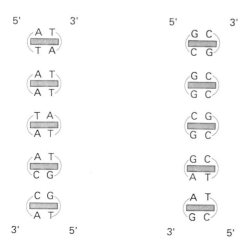

Fig. 8.10 The ten possible dinucleotide intercalation sites for an antiparallel double helix.

In comparing groove-binding molecules and intercalators, it is clear that groove-binders, as a class, display significantly greater binding selectivity than intercalators. Intercalation cavities created at A:T or at G:C base-pairs are quite similar in their potential for interaction with planar aromatic ring systems. Electrostatic, van der Waals, hydrophobic, etc. contributions to binding are similar for the two sites. On the other hand, groove-binding molecules can contact more base-pairs as they lie along the groove in a DNA complex and this gives them an inherently greater recognition potential. As discussed above, the grooves are quite distinct in A:T and G:C regions and this adds to the potential for specific interactions by groove-binding agents.

Hélène and co-workers have dramatically enhanced the binding specificity of intercalators by linking them covalently to oligonucleotides of specific sequence. These covalent adducts can locate complementary single-stranded regions on other nucleic acid molecules with high precision. The intercalator provides increased free energy of binding as well as additional possible specificity in complex formation. In a related approach, Dervan and co-workers have linked EDTA derivatives to intercalators, to groove-binding molecules, and to oligonucleotides to obtain selective recognition cleavage of DNA sequences the presence of Fe(II) and oxygen (Section 8.7.2). In the same manner, intercalators or other DNA-binding agents can be linked to an oligopyrimidine strand which is capable of sequence-specific recognition of polypurine–polypyrimidine duplex segments of DNA through triple helix formation. The third (pyrimidine) strand fits into the major groove of the original duplex and the new pyrimidines form Hoogsteen base-pairs with purine bases of the duplex (cf. Section 2.4.5).

8.4.4 Neighbour exclusion

Daunomycin molecules in the d(CGTAGG) crystal structure bind at both C:G sites with the amino-sugar groups pointing inwards and essentially filling the minor groove (Fig. 8.7). In agreement with this structure, solution studies indicate that each bound daunomycin molecule covers three DNA base-pairs. By contrast, it appears that simpler intercalators with reduced steric constraints such as proflavine and ethidium (Figs 8.2 and 8.8) could bind at every potential intercalation site between base-pairs. At saturation of the double helix with the intercalator this would yield a binding stoichiometry of one base-pair per bound intercalator.

While models of duplex DNA with reasonable backbone torsional angles can indeed be constructed with intercalators at *all* possible sites between base-pairs, solution studies indicate that even the simplest intercalators reach saturation at a maximum of one intercalator per two base-pairs. This empirical observation has led to the **neighbour exclusion principle** which states that intercalators can at most bind at alternate possible base-pair sites on DNA, giving a maximum of one intercalator between every second site. Initially, all spaces between base-pairs are potential binding sites for a non-specific intercalator. When an intercalator binds at one particular site, the exclusion principle states that it becomes impossible to bind another intercalator at either adjacent site.

Various ideas have been developed to explore the molecular basis of neighbour exclusion of binding. **Bisintercalators** are molecules with two intercalating rings covalently attached to a linking chain. Some bisintercalators synthesized to test this principle have been reported to violate neighbour exclusion. One suggestion for the molecular basis of neighbour is that intercalator binding induces conformational changes at adjacent sites in DNA and the new conformation makes it impossible to intercalate another monointercalator at the adjacent site.

An alternative, suggested by Friedman and Manning, is that since intercalators bind to DNA and neutralize some of its charge, the favourable release of condensed ions is reduced as intercalators bind to the helix and the observed equilibrium constant is also reduced. This leads to a curvature in binding isotherms which is similar to that predicted by neighbour exclusion. Since the local release of ions due to intercalation would be greater than ion release at a distance along the DNA molecule, the local release of counterions could also lead to exclusion of binding at neighbouring sites. It may well be that both conformational and electrostatic forces contribute to the observed neighbour exclusion.

It would seem that these various theories could be tested with simple uncharged intercalators which could not ion-pair with DNA phosphate groups. The difficulty, however, has been to design a simple, planar, uncharged aromatic molecule, e.g. similar to anthracene, which is soluble enough in water for its binding site size with DNA to be investigated quantitatively. Obviously, the validity and molecular origin of the neighbour exclusion principle will remain a topic of experimentation and discussion until its molecular basis is better understood.

Summary

Typical intercalators are planar, aromatic cations which bind by insertion of the aromatic ring system between DNA base-pairs. They may have non-planar substituents, either cationic or neutral, which protrude into one of the DNA grooves.

Generation of the intercalation site causes extension of the DNA duplex, local unwinding of the base-pairs, and other possible distortions in the DNA backbone (such as bends) which are characteristic of the intercalator species.

Specificity in intercalation binding generally favours G:C base-pairs, but its magnitude in most cases is much less than the specificity observed for groove-binding molecules.

8.5 Bisintercalators

8.5.1 Synthetic bisintercalators

Bisintercalators have two covalently linked intercalating ring systems with connecting chains of variable length (Fig. 8.11). It is also possible to link three or more ring systems together in the same way. The synthesis of these multiple ring compounds was partially stimulated by the idea that the medicinal activity of intercalating drugs could be enhanced by the significantly higher DNA-binding constants and slower dissociation rates from DNA expected for bisintercalators relative to monointercalators. Another feature anticipated for these molecules, as discussed above, is that with short linkers the two intercalating rings might be induced to bind at adjacent base-pair sites and violate the neighbour exclusion principle.

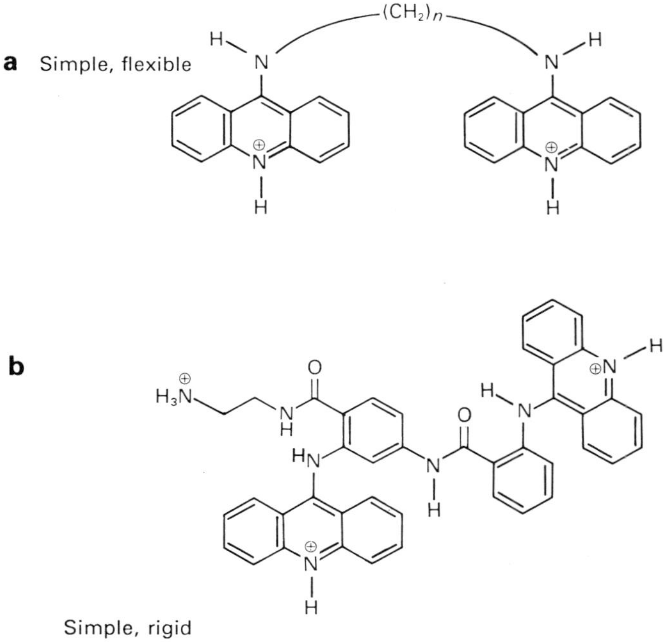

Fig. 8.11 (a) Simple, flexible acridine bisintercalators. (b) A more rigid acridine bisintercalator.

Such unusual complexes would facilitate an investigation of the molecular basis of neighbour exclusion. For example, with a long linker, both planar rings are capable of intercalation without violation of the neighbour exclusion principle (Fig. 8.12a). With a short linker, both the rings may intercalate at adjacent sites in violation of neighbour exclusion (Fig. 8.12b).

One obvious way to change the linker length in bisintercalators is to vary the number of methylene units as shown by varying n in the bisacridine in Fig. 8.11a. Compounds of this type have been synthesized and it has been found with viscometric methods that for $n < 4$ the molecule can bind only with one ring intercalated

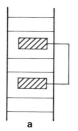 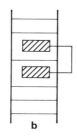 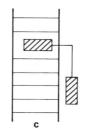

Fig. 8.12 The effect of linker length on the possible binding modes of bisintercalators.

(as in Fig. 8.12c). When $n = 6$, the molecule apparently binds by bisintercalation with violation of neighbour exclusion (as in Fig. 8.12b). When $n > 8$, the molecule binds by bisintercalation but without violation of neighbour exclusion (as in Fig. 8.12a).

The results for the $n = 6$ compound are not in accord with NMR studies on complexes with oligonucleotides. Analysis of imino proton chemical shifts on adding the bisacridines to the self-complementary oligomer d(AT)$_5$ indicated that while bisintercalation occurred for $7 \leqslant n \leqslant 10$, only monointercalation occurred for $4 \leqslant n \leqslant 6$.

It is possible that the highly flexible linker of the bisacridines (Fig. 8.11) could allow the $n = 6$ molecule, with some local distortion of DNA, to bind by bisintercalation without violation of neighbour exclusion. If such a distortion of the oligomer was not possible with the sequence used or under the conditions of the NMR experiments, this could explain the disagreement between viscometric and NMR methods.

To eliminate the problem of linker flexibility, Denny and co-workers used a more rigid compound (Fig. 8.11b) which is also a better model for natural antibiotics than are bisintercalators with rigid linkers (Fig. 8.13). Because of its rigid structure, it cannot bind by mono-intercalation of the type shown in Fig. 8.12c. Since the two rings of this species are insufficiently separated to allow bisinteracalation with an intervening site (as in Fig. 8.12a), the compound must either bind as in Fig. 8.12b or by a non-intercalative mode. Viscosity results indicate that the rigid compound does intercalate and, in view of the geometric considerations discussed above, suggest that there is in this case bisintercalation which violates the neighbour exclusion principle.

Crystallographic and high resolution NMR studies of complexes of these molecules with DNA oligomers are difficult because of the low solubility of the complexes. It will be quite interesting to see what perturbations in normal DNA structure are induced by bisintercalation of this type when such high resolution studies become available.

8.5.2 Natural bisintercalators

The quinoxaline antitumour antibiotics, echinomycin and triostin A (Fig. 8.13a), have been studied in considerable detail. X-Ray structural analysis indicates that the cross-linked cyclic peptides form a relatively rigid plate with the two quinoxaline rings perpendicular to the plate, parallel to each other on the same side and at opposite ends of the plate (Fig. 8.13b). The two aromatic rings are oriented in an optimum configuration for bisintercalation, separated by 10–11 Å, and can accommodate two base-pairs between the rings as required for binding by the neighbour exclusion principle.

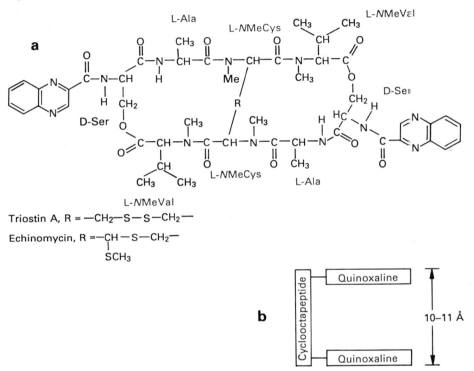

Fig. 8.13 (a) Structures for the natural bisintercalators triostin A and echinomycin. (b) Schematic diagram of the orientation and spacing between the quinoxaline rings.

In studies with supercoiled DNA, these molecules unwind the duplex by 40–50° and also produce a length increase approximately twice as large as for monointercalators. Both observations are in agreement with a bisintercalation binding mode. Binding and footprinting experiments suggest that echinomycin and triostin A intercalate preferentially around CpG sites in DNA and cover a six base-pair binding region of the type NNCGNN.

Crystal structures of echinomycin and triostin A complexes with c(CGTACG) and of triostin A bound to d(GCGTACGC) have been determined with two antibiotic molecules bound per oligomer in all three cases. In all of the structures the quinoxaline rings are bisintercalated with the CG sequence sandwiched between the intercalated rings (Fig. 8.14) in agreement with solution studies. The cyclic peptides lie in the minor groove and displace all water molecules and cations from the groove at the binding site. the alanine residues of the cyclic peptides are key components in the binding specificity because the NH groups of both L-alanine residues of the cyclic peptide point in to the DNA base-pairs where they form hydrogen-bonds with *N*-3 of the two G residues at the binding sites. The carbonyl group of the L-alanine on the outside edge of the complex also forms a hydrogen-bond to the amino group of G. The other amino acids of the linker make extensive van der Waals contacts with the minor groove of DNA, but the G:C specificity of the bisintercalators primarily results from the Ala-G hydrogen-bonds.

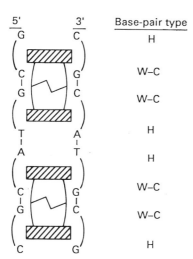

Fig. 8.14 Illustration of the binding sites and positions of Hoogsteen (H) and Watson–Crick (W-C) base-pairs in the crystal structure of triostin A (red) with the oligomer d(GCGTACGC).

The two A:T base-pairs in the centre of the oligomer have an unusual conformation in the crystal structure. The two adenines are in *syn*- rather than the standard *anti*- conformation and, as a result, they form Hoogsteen-type hydrogen-bonded base-pairs with their complementary thymines (Fig. 8.14). The width across the helix is less for a Hoogsten than for a Watson–Crick base-pair and as a result the chains are approximately 2 Å closer in the centre of the complex than they would be in an equivalent Watson–Crick helix. This reduction of helical width provides a better close packing geometry for the ends of the antibiotic molecule with the DNA minor groove. Because the quinoxaline aromatic system consists of only two fused intercalating rings, it also gives a better stacking arrangement with the reduced width Hoogsteen than with normal Watson–Crick base-pairs. (By contrast, fused tricyclic intercalators fill the space between Watson–Crick base-pairs better than bicyclic ring systems.) The distance between the base-pairs at the intercalation side in the X-ray structures is generally greater than the optimum 3.4 Å separation and suggests that much of the free energy of binding of complex formation with these antibiotics must come from peptide-base interactions in the minor groove rather than from base–intercalator interactions.

In the triostin A d(GCGTACGC) crystal structure, the central A:T and the terminal G:C base-pairs adopt the Hoogsteen conformation. This was particularly unexpected with the G:C base-pair because formation of the Hoogsteen base-pair requires protonation of C and, even then, only two hydrogen-bonds are formed instead of the three obtained with Watson–Crick G:C base-pairing.

These observations emphasize two important points. First, DNA is a highly flexible molecule and can assume a wide variety of conformations. Secondly, the binding free energy for complex formation with small molecules is sufficient for conversion of DNA into any of a variety of structures which, under normal conditions, are only slightly less stable than the normal B-family of conformations of DNA. It should be noted, however, that studies with specific chemical probes have not provided evidence for Hoogsteen base-pairs in longer DNA sequences. Such Hoogsteen base-pairs have been identified in NMR studies of echinomycin complexes with some short DNA sequences but not in others. The importance of Hoogsteen base-pairs in DNA–bisintercalator complexes may thus depend very strongly on the length and sequence of the DNA under study.

Summary

Bisintercalators have two potential intercalating ring systems connected with linkers which can vary in length and rigidity. The interaction of the ring systems with DNA base-pairs is controlled to a large extent by the characteristics of the linker. It has been proposed that some synthetic bisintercalators, with short linkers, can bind to DNA in violation of neighbour exclusion.

Quinoxaline antibiotics also bind to DNA by bisintercalation and exhibit specificity for C:G sequences. The antibiotics cause extensive structural distortions on binding to DNA.

8.6 Non-classical intercalators

8.6.1 Intercalators with multiple bulky substituents

Intercalators which have substituents on opposite sides of the ring system must thread one of the substituents between the base-pairs at the intercalation site in the binding mechanism. If the substituent is bulky and/or polar or charged, this could be an unfavourable step which might profoundly affect the kinetics of association and dissociation of the intercalator. Examples of intercalators of this type are naphthalene-bisimides and cationic porphyrins (Fig. 8.15).

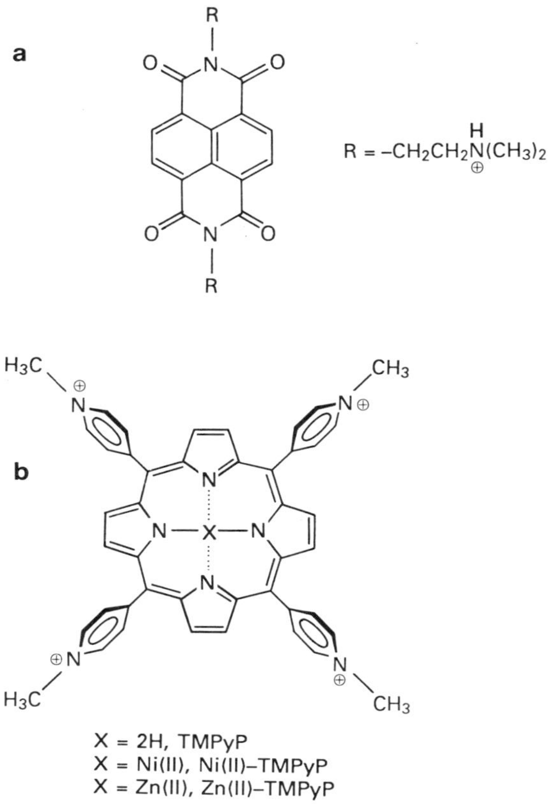

Fig. 8.15 Structures for (**a**) a naphthalene-bisimide (R can be an aliphatic amine or large cation) and (**b**) cationic porphyrins. The non-metallo or Ni(II) derivatives are intercalators and the Zn(II) derivative is a groove-binding compound.

The antibiotic nogalamycin, and perhaps the anti-cancer drugs mitoxantrone and bisantrene (Fig. 8.16), also fall into this category. It should be emphasized that the unfavourable step for binding of these compounds refers to the kinetics and not to the thermodynamics of complex formation.

The binding mechanism for *simple* intercalators such as proflavine (Fig. 8.2) involves two steps. First, the cationic intercalator interacts with DNA through an external electrostatic complex with the anionic backbone of the double helix. The intercalator can then diffuse in the anionic potential along the surface of the helix until it encounters gaps between base-pairs which have separated as a result of thermal motion to create a cavity for intercalation. The molecule can then bind in an intercalation complex. Molecules with large side chains require wider openings with significant distortions or breaking of base-pair hydrogen-bonds before they can form an intercalation complex. These larger, dynamic motions of the base-pairs are less likely and the kinetics of intercalation are thus much slower for molecules with bulky side chains.

These types of molecules (Figs 8.15 and 8.16) all have very high DNA-binding constants indicating that once the side chain slides between the base-pairs, the

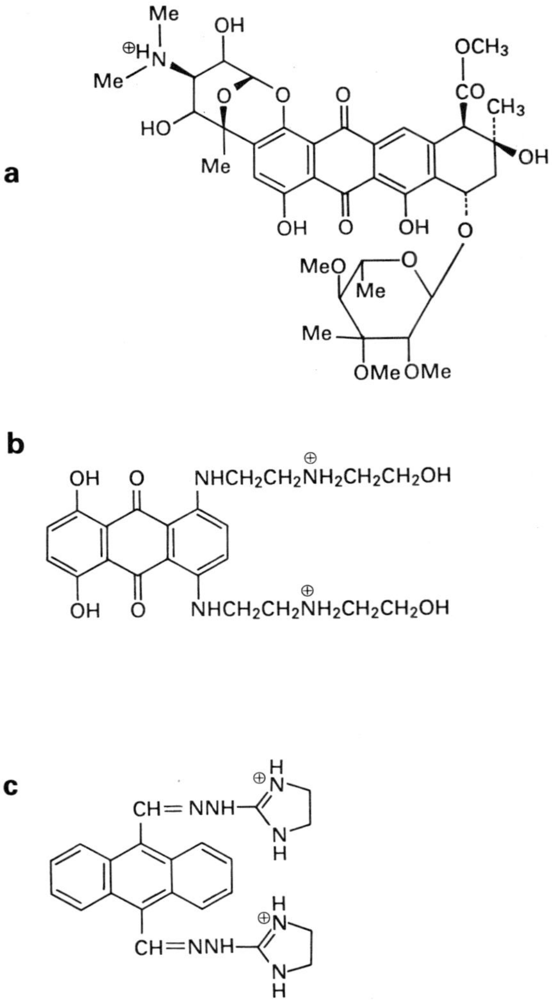

a

Fig. 8.16 Structures
of (**a**) nogalamycin,
(**b**) mitoxantrone, and
(**c**) bisantrene.

b

c

DNA molecule can assume a conformation that gives a very favourable free energy of complex formation. The side chains present a kinetic barrier to binding, but have favourable interactions in the final complex after they have passed through the double helix. The kinetic barrier to binding depends on the size of the side chain, its orientation, and its polarity.

Porphyrins (Fig. 8.15b) are particularly interesting molecules since they appear to intercalate in G:C rich regions of DNA, but bind in the minor groove in A:T-rich regions. Their binding properties vary substantially with the cationic substituent and type of metal bound centrally in the porphyrin. Oligonucleotide NMR studies by Marzilli, Wilson, and co-workers have suggested that the intercalating porphyrins have very high selectivity for binding at CpG sequences. This is unusual since the porphyrins do not have the hydrogen-bonding capability observed for other molecules which show high selectivity for binding at specific DNA sequences.

8.6.2 Propeller-twisted intercalators

Several examples of aromatic cations with unfused bicyclic (netropsin) and tricyclic (distamycin and Hoechst 33258) aromatic rings have already been described. These compounds which are all groove-binding molecules, have torsional freedom to optimize their twist to fit the minor groove of DNA. Strekowski, Wilson, and co-workers have designed, synthesized, and evaluated the DNA interactions of several new classes of unfused aromatic cations. The structure of one such derivative is shown (Fig. 8.17). Both molecular mechanics and crystal structure determination indicate that this molecule has a significant intrinsic twist of 20° to 25° between the

a

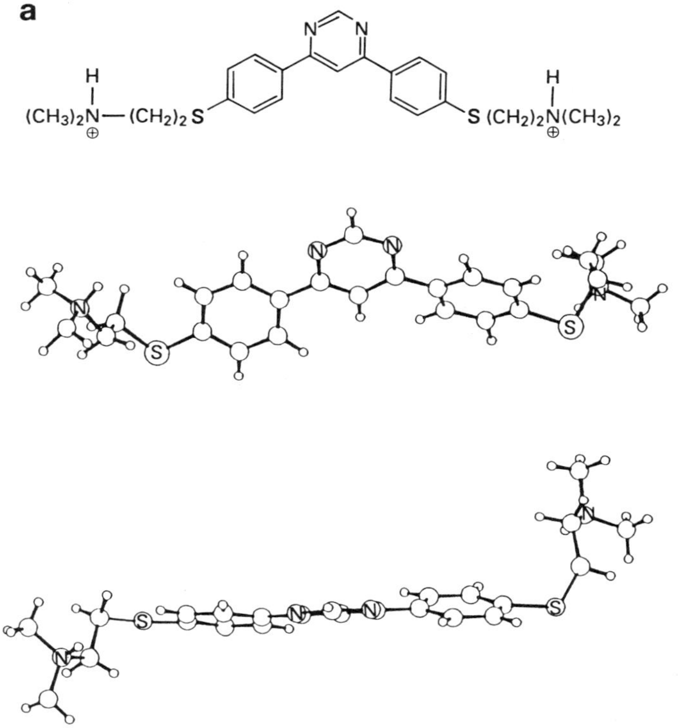

Fig. 8.17a   Crystal structure for a propeller-twisted intercalator. The compound was synthesized by Dr L. Strekowski and the structure was solved by Drs S. Neidle and G. Webster.

b

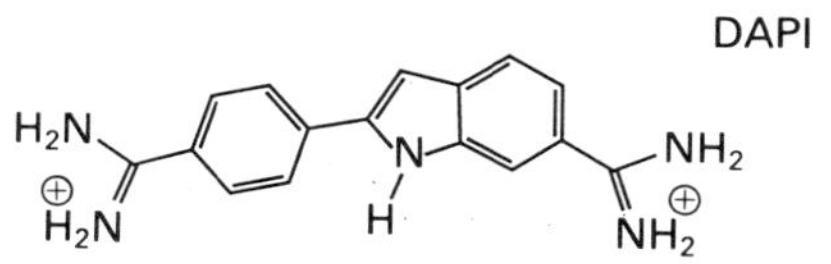

DAPI

Fig. 8.17b   The structure of DAPI.

benzene and pyrimidine planes. The twisted rings and terminal basic functions match the paradigm for a groove-binding molecule. Solution studies, however, indicate that the 4,6-diphenylpyrimidine (Fig. 8.17) is a strong intercalator. It unwinds supercoiled DNA, lengthens linear DNA, and has NMR and linear dichroism characteristics which unequivocally establish it as an intercalator. It is interesting that all of these twisted tricyclic intercalators increase the rate of the bleomycin-catalysed cleavage of DNA (Section 8.7).

8.6.3 Intercalation versus groove-binding

A comparison of the factors which control groove-binding versus intercalation was made earlier and it is worthwhile to consider these factors again after the above discussion of intercalators. It is now well known that the base-pairs of DNA can have significant intrinsic propeller twist. Retrospectively, it seems quite reasonable that intercalation of unfused aromatic ring compounds might complement or perhaps enhance the interactions of base-pairs responsible for propeller twist. What is not yet completely clear is why the 4,6-diphenylpyrimidine (Fig. 8.17a) is an intercalator while other related compounds such as Hoechst 33258 bind in the DNA minor groove. There are several possible reasons for this difference. First, the major groove is too wide to form many favourable contacts with all of these molecules and it is rejected as a significant binding site. Secondly, in the minor groove the amino group of G provides a steric block to prevent significant binding of many molecules. This leaves either A:T regions in the minor groove or intercalation as preferred binding sites. Thirdly, the choice between minor groove-binding at A:T sites and intercalation probably strongly depends on the availability of hydrogen-bond donating groups, such as NH, in the small molecule. *Appropriately placed* NH groups can form hydrogen-bonds with N-3 of A and/or O-2 of T to form a very strong minor groove complex. With the 4,6-diphenylpyrimidine molecule (Fig. 8.17), however, there are no good hydrogen-bond donor groups in or connecting the unfused rings and so there is no strong minor groove-binding mode. The 4,6-diphenylpyrimidine obviously can form a good stacked complex with DNA base-pairs, however, and it forms a strong intercalation complex with DNA.

Results with 4',6-diamidino-2-phenylindole (DAPI, Fig. 8.17b) have provided insight into the method by which an unfused aromatic cation selects an intercalation or a groove-binding mode. Numerous biophysical studies have shown that DAPI binds in the minor groove at AT sequences much as netropsin does. Footprinting experiments indicate that DAPI spans approximately three base-pairs in the minor groove which is reasonable based on the molecular length of DAPI. This molecule is structurally similar to many unfused intercalators discussed above and a variety of biophysical experiments have provided the surprising result that DAPI binds to G:C base-pairs by intercalation, not by groove interactions.

On review this finding agrees quite well with what we have learned about DNA structure and interactions. Both A:T and G:C base-pairs form good intercalation sites. A:T but not G:C sequences, however, also have a very favourable minor-groove-binding site for unfused aromatic cations like DAPI and netropsin. DAPI thus selects intercalation as the binding mode at G:C sites but forms a minor-groove complex in A:T regions. The binding constant for DAPI at G:C intercalation sites is similar to the binding constant for other intercalators such as proflavine (Fig. 8.2). In the A:T minor groove, however, DAPI covers more base-pairs with more specific contacts than at the intercalation site. DAPI thus binds particularly strongly at A:T sites. This type of site selection by DAPI illustrates that intercalation and groove-binding should be viewed as a continuum. The binding mode with the lowest energy will depend on the DNA sequence and structure as well as on the molecular features of the bound molecule.

Summary

Classical intercalators are planar aromatic cations with no bulky substituents. Non-classical intercalators fall into two general classes. First, there are molecules with bulky substituents, such as porphyrins, which perturb the kinetics of interaction with DNA, but enjoy good stacking of the intercalated ring system with DNA base-pairs. Secondly, some molecules with non-fused, twisted, aromatic ring systems and terminal cationic functions, which more closely resemble the paradigm for a groove-binding molecule, bind to DNA by intercalation. The distinction between intercalation and groove-binding for unfused-aromatic cations probably depends on (i) placement of hydrogen-bonding groups, (ii) stacking interactions with DNA base-pairs relative to interactions with the walls of the minor groove, and (iii) the degree of twist in the unfused ring system.

8.7 DNA cleavage reagents

8.7.1 Antibiotics

The most thoroughly studied of the DNA cleavage reagents is the glycopeptide antibiotic and anti-cancer drug bleomycin A2 (Fig. 8.18). A number of closely related antibiotics have been discovered with slight modifications in different regions of the bleomycin molecule.

The 2,4'-bithiazole rings, a common feature of bleomycins, are linked to a small side chain which usually has a cationic charge as in bleomycin A2. The extensive remaining portion of the molecule, linked at the opposite side of the bithiazoles, provides a metal-complexing site which is responsible for DNA strand cleavage. Removal of the bithiazole rings and cationic side chain leaves a metal-binding domain which neither significantly binds to nor degrades DNA at reasonable concentrations. Despite considerable study, the mechanisms by which the bithiazoles and side chain interact with DNA are not clear. NMR and hydrodynamic studies show that the bithiazole rings upon cleavage from the metal-binding domain can

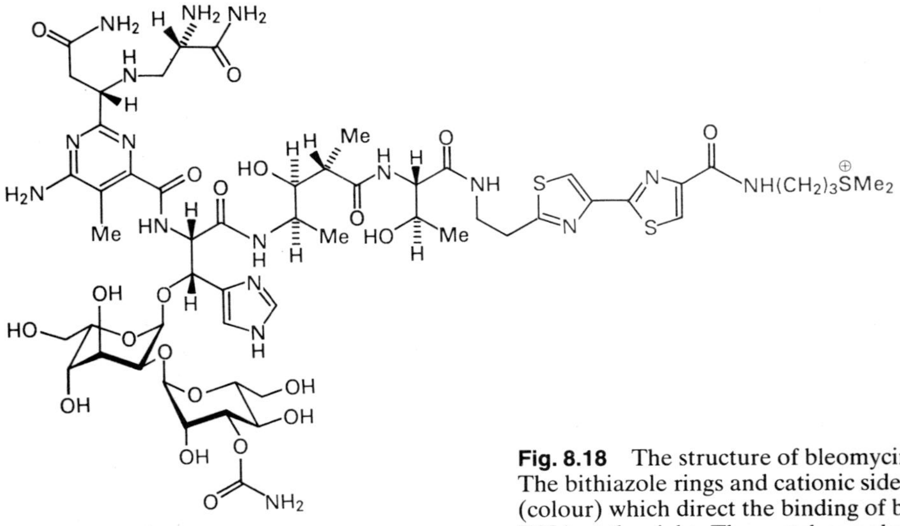

Fig. 8.18 The structure of bleomycin A2. The bithiazole rings and cationic side chain (colour) which direct the binding of bleomycin to DNA at the right. The metal-complexing portion of the molecule is to the left.

bind to DNA in an intercalation complex. Other such studies, however, strongly suggest that the bithiazoles bind to DNA through a groove-binding mechanism when they are part of the antibiotic. It may be, as with DAPI, that both binding modes are possible.

There is a selectivity for bleomycin cleavage of DNA at purine–pyrimidine sequences with a higher than expected occurrence of double-strand nicks (e.g. cleavage of both DNA strands at sites in close proximity). Bleomycin can complex with several different metal ions to cleave DNA, but most work has been done with the Fe(II)–bleomycin species. The initial step in degradation of DNA is very selective abstraction of a deoxyribose 4'-hydrogen (cf. Figure 7.29). Since H-4' lies in the minor groove of DNA, this site-selectivity suggests that bleomycin binds in the minor groove.

The species directly responsible for the removal of the 4'-hydrogen is an 'activated' bleomycin–Fe(II)–O$_2$ ternary complex whose activation requires a one-electron reduction. The electron can come from another bleomycin–Fe(II)–O$_2$ complex (producing the activated species and an inactive bleomycin–Fe(III) complex) or from organic reducing agents such as ascorbic acid or thiols (Fig. 8.19). Hydrogen peroxide can produce the activated complex directly from bleomycin–Fe(III). In the absence of DNA, the activated species slowly decomposes with destruction of the bleomycin molecule. A particularly interesting feature of the bleomycin-induced degradation of DNA is the fact that the rate of strand cleavage by bleomycin can be significantly enhanced by some compounds such as the unfused intercalator shown above (Fig. 8.17a). The unfused intercalators have been shown to bind with their side chains in the DNA major groove so that they do not block bleomycin entry to the minor groove. Formation of the intercalation complex expands the grooves and facilitates bleomycin cleavage of DNA in the minor groove. Since clinical use of many anti-cancer drugs is limited by toxic side-effects not necessarily related to DNA interactions, the ability to design amplifiers which increase the effects of the drugs at DNA but do not increase the toxic side-effects offers the potential of significantly increasing the therapeutic value of the drugs.

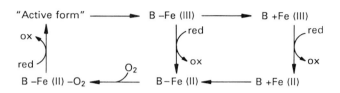

Fig. 8.19 A catalytic cycle for the bleomycin (B)-induced oxidative degradation of DNA. This scheme requires the presence of organic reducing reagents (red). Depending on the reducing agent and conditions, B-Fe(III) may be directly reduced or may first dissociate to free bleomycin and Fe(III).

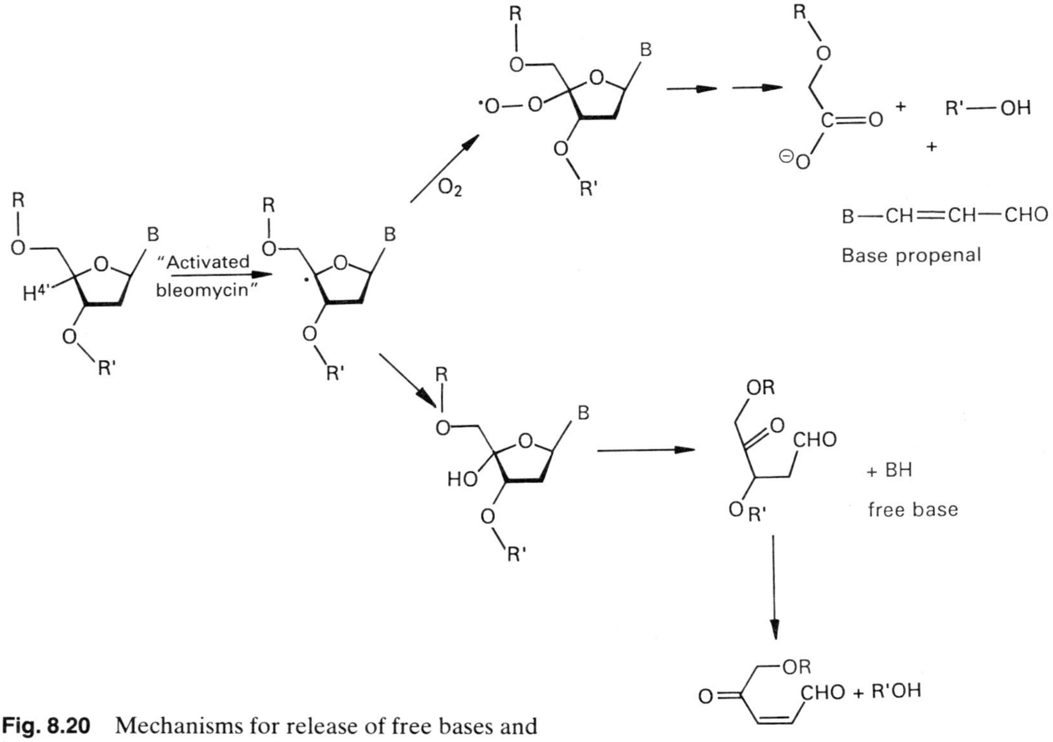

Fig. 8.20 Mechanisms for release of free bases and base-propenals by attack of 'activated' bleomycin on the C-4′ position of DNA.

In the basic bleomycin–DNA reaction scheme (Fig. 8.19) catalytic cleavage of DNA in the presence of organic reducing agents is a two-electron process involving two one-electron reductions of bleomycin–iron species. The selective reaction at the C4'–H bond of deoxyribose in DNA suggests that a specific orientation of the bleomycin complex in the minor groove of DNA is responsible for the chemistry and that no mobile intermediates, such as hydroxyl radicals (cf. Section 7.9), are involved in the reaction. The oxidative degradation of DNA produces two types of base products: free bases and base-propenals. Formation of base-propenals requires additional oxygen, but formation of free bases does not. Mechanisms for formation of both products, through a common reaction intermediate at C-4', are shown (Fig. 8.20). Both processes result in single chain breaks by the destruction of the deoxyribose sugar.

Neocarzinostatin (Fig. 8.21b), a peptide antibiotic quite different from bleomycin, also causes strand cleavage of DNA. The active part of the antibiotic is a non-protein chromophore containing a substituted naphthalenecarboxylic acid capable of intercalation with DNA. The strand cleavage of DNA by the neocarzinostatin chromophore is activated by O_2 and thiol reagents. Cleavage is proposed to occur through removal of a 5'-hydrogen from deoxyribose groups in DNA.

8.7.2 Synthetic cleavage reagents

Synthetic metal-complexing reagents such as porphyrin derivatives, a copper complex with two molecules of 1,10-phenanthroline, and an analogue of ethidium with an Fe(II)–EDTA containing system bonded to the *para*-position of the phenyl substituent (Fig. 8.21a) have also been found to bind to DNA and cause strand cleavage.

This MPE derivative (Fig. 8.21b), designed by Dervan and co-workers, is now in widespread use as a DNA-footprinting reagent because of its relatively non-selective cleavage of DNA in the presence of Fe(II), O_2, and a reducing agent. It is proposed to act through an initial non-selective intercalation complex of the phenanthridinium ring system with DNA in which the EDTA–Fe(II) side chain lies in the DNA minor groove. It is this group which is entirely responsible for cleavage of DNA.

Unlike the highly specific proton abstraction observed with the natural antibiotics, MPE–Fe(II)–O_2 produces a diffusible species, possibly a hydroxyl radical, which can react at several DNA sites (Section 7.9.1). The species is short-lived and attacks the deoxyribose groups of DNA in close proximity to the intercalated ring of MPE. As expected, a larger variety of products are produced than with bleomycin cleavage. Free bases are observed in the product mixture in amounts related to their percentage in the DNA sample. The Fe(III) produced in formation of hydroxyl radical can be reduced again to MPE–Fe(II), so that DNA cleavage can occur repeatedly if a reducing agent and oxygen are available. No DNA cleavage occurs if Fe(II) in MPE is replaced by Ni(II) or Zn(II).

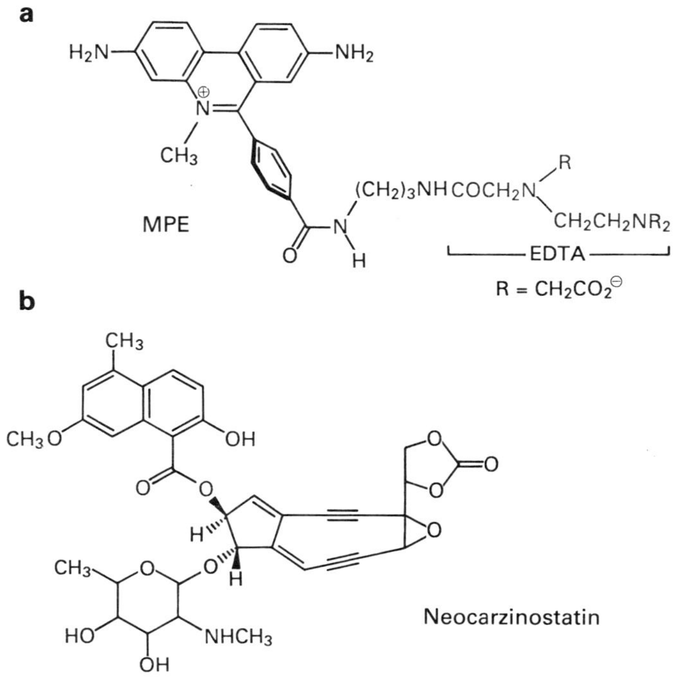

Fig. 8.21 Structures for (a) the ethidium analogue, methidiumpropyl-EDTA (MPE) (top) and (b) the neocarzinostatin chromophore (bottom).

The EDTA group has also been attached to distamycin and related A:T specific groove-binding molecules and, in the presence of Fe(II) and O_2, these reagents cause DNA cleavage with high specificity in A:T rich regions. More recently the EDTA group has been covalently linked to homopyrimidine oligonucleotides. These reagents can cause DNA cleavage through selective binding to sequences of double-helical DNA which can form a triple-helix with the homopyrimidine–EDTA strand. The chemistry of DNA cleavage is similar for all of these reagents and depends on the formation of diffusible radicals at the EDTA–Fe(II)–O_2 complex.

Summary

A number of natural and synthetic reagents has been found which have the ability to bind to DNA and cause strand cleavage. A metal ion associated with the bound molecule frequently initiates the chemistry of the cleavage reaction and O_2 is involved in some activated form. Highly specific cleavage of DNA can be obtained by linking a metal-complexing reagent to a molecule which binds to DNA with high intrinsic specificity.

8.8 Recognition of nucleic acids and drug design

The discussion in this chapter illustrates how reseach in reversible interactions of small molecules with nucleic acids can be divided into three primary areas: (i) analysis of model systems at high resolution for the development of a better understanding of the molecular basis for nucleic acid interactions in general; (ii) development and analysis of drugs which interact with nucleic acids; and (iii) design of new highly specific 'nucleases' whose sequence-recognition properties can be selected at will by molecular biologists.

Improvements in DNA and RNA synthetic techniques as well as design and synthesis of additional DNA interactive drugs and model compounds have made available molecules which are providing new information on the structure and energetics of nucleic acid complexes and are dramatically advancing the field of reversible nucleic acid interactions. In addition, it is clear that both DNA and RNA are prime targets as receptors in drug development. There is particularly exciting progress in the design of oligopeptides and oligonucleotides for use as drugs and as reagents for the recognition of specific nucleic acid sequences. Oligonucleotides can recognize single-stranded DNA and RNA as well as duplex nucleic acids through triple-helix formation. This gives oligonucleotides the potential to act as highly specific drugs, as drug carriers, or, when linked to a cleavage reagent such as Fe(II)-EDTA, as site-specific nucleases.

The base-pair specificity and nucleic acid binding constants for the types of molecules discussed in this chapter can vary quite widely. Simple monointercalators, such as an anthracene ring system with a cationic side chain, have DNA binding constants of $\sim 10^4$ M^{-1} at neutral pH and 0.1 M NaCl, and there is very little specificity in their DNA interactions. Bisintercalators or monointercalators with more complicated side chains, which make specific contacts in the DNA grooves, can have significantly higher binding constants and recognition specificity. Groove-binding molecules, such as the oligopeptides and oligonucleotides discussed above, can be designed to interact with the hydrogen-bond accepting and donating groups of base-pairs which face into the DNA grooves and can thus have very high DNA binding constants and recognition specificity. The ability to extend such groove-binding agents over many base-pairs gives them the ability to recognize a single sequence in the DNA of any organism.

From the standpoint of drug design, however, it is not clear that high binding constants and recognition specificity are a necessary requirement for biological activity. Fairly simple, synthetic monointercalators, for example, have shown excellent anti-cancer activity and are some of the most promising candidates in clinical trials. It is clear that, alongside highly specific groove-binding agents, simple intercalating molecules will continue to be of interest as nucleic acid interactive drugs.

Further reading

8.1

Dervan, P. B. (1986), Design of sequence-specific DNA-binding molecules. *Science*, **232**, 464–71.

Hurley, L. H. and Boyd, F. L. (1987). Approaches toward the design of sequence-specific drugs for DNA. *Ann. Rep. Med. Chem.* **22**, 259–68.

Kopka, M. L., Pjura, P. E., Goodsell, D. S., and Dickerson, R. E. (1987). Drugs and minor groove binding in G-DNA: netropsin and Hoechst 33258. In *Nucleic acids and molecular biology* (ed. F. Eckstein and D. M. J. Lilley), Vol. 1, Springer, Berlin, pp. 1–24.

Neidle, S. N., Pearl, L. H., and Skelly, J. V. (1987). DNA structure and perturbation by drug binding. *Biochem. J.*, **243**, 1–13.

Wang, A. H.-J. (1987). Interactions between antitumor drugs and DNA. In *Nucleic acids and molecular biology* (ed. F. Eckstein and D. M. J Lilley), Vol. 1. Springer, Berlin. pp. 53–69.

Waring, M. (1981). Inhibitors of nucleic acid synthesis. In *The molecular basis of antibiotic action* (ed. E. F. Gale *et al.*), second edn. Wiley, London, pp. 274–341.

Wilson, W. David (1987). Cooperative effects in drug–DNA interactions. *Progr. Drug Res.*, **31**, 193–221.

8.2

Bloomfield, V. A., Crothers, D. M., and Tinoco, I. (1974). *Physical chemistry of nucleic acids*. Harper and Row, New York.

Manning, G. S. (1978). The molecular theory of polyelectrolyte solutions with applications to the electrostatic properties of polynucleotides. *Q. Rev. Biophys.*, **11**, 179–246.

Record, M. T., Jr., Lohman, T. M., and deHaseth, P. (1976). Ion effects on ligand–nucleic acid interactions. *J. Mol. Biol.*, **107**, 145–58.

8.3

Coll, M., Frederick, C. A., Wang, A. H.-J., and Rich, A. (1987). A bifurcated hydrogen-bonded conformation in the d(A.T) base pairs of the DNA dodecamer d(CGCAAATTTGCG) and its complex with distamycin. *Proc. Nat. Acad. Sci. USA*, **84**, 8385–9.

Kopka, M. L., Yoon, C., Goodsell, D., Pjura, P., and Dickerson, R. E. (1985). The molecular origin of DNA-drug specificity in netropsin and distamycin. *Proc. Nat. Acad. Sci. USA*, **82**, 1376–80.

Lavery, R., Zakrzrewska, K., and Pullman, B. (1986). Binding of non-intercalating antibiotics to B-DNA: a theoretical study taking into account nucleic acid flexibility. *J. Biomol. Str. Dyn.*, **3**, 1155–1170.

Lee, M., Hartley, J. A., Pon, R. T., Krowicki, K., and Lown, J. W. (1988). Sequence specific molecular recognition by a monocationic lexitropsin of the decadeoxyribonucleotide d(CATGGCCATG): structural and dynamic aspects deduced from high field proton NMR studies. *Nucleic Acids Res.*, **16**, 665–84.

Leupin, W., Chazin, W. J., Hyberts, S., Denny, W. A., and Wuthrich, K. (1986). NMR studies of the complex between the decadeoxynucleotide d(GCATTAATGC) and a minor-groove-binding drug. *Biochemistry*, **25**, 5902–10.

Nelson, H. C. M., Finch, J. T., Luisi, B. F., and Klug, A. (1987). The structure of an oligo(dA).oligo (dT) tract and its biological implications. *Nature*, **330**, 221–6.

Neuhaus, D. and Williamson, M. P. (1989). *The NOE effect in structural and conformational analysis*, Chapter 12. Verlag Chemie, Weinheim.

Pjura, P. E., Grzeskowiak, K., and Dickerson, R. E. (1987). Binding of Hoechst 33258 to the minor groove of B-DNA. *J. Mol. Biol.*, **197**, 257–71.

Teng, M.-K., Usman, N., Fredrick, C. A., and Wang, A. H.-J. (1988). The molecular structure of the complex of Hoechst 33258 and the DNA dodecamer d(CGCGAATTCGCG). *Nucleic Acids Res.*, **16**, 2671–90.

Zimmer, C. and Wahnert, U. (1986). Nonintercalating DNA-binding ligands: specificity of the interaction and their use as tools in biophysical, biochemical and biological investigations of the genetic material, *Prog. Biophys. Molec. Biol.*, **47**, 31–112.

8.4

Assa-Munt, N., Denny, W. A., Leupin, W., and Kearns, D. R. (1985). Proton NMR study of the binding of bis(acridines) to d(AT).d(AT). 1. Mode of binding, *Biochemistry*, **24**, 1441–9.

Chaires, J. B., Fox, K. R., Herrera, J. E., Britt, M., and Waring, M. J. (1987). Site and sequence specificity of the daunomycin–DNA interaction. *Biochemistry*, **26**, 8227–36.

Hogan, M., Dattagupta, N., and Crothers, D. M. (1978). Transient electric dichroism studies of the structure of the DNA complex with intercalated drugs. *Biochemistry*, **17**, 280–8.

Rao, S. N. and Kollman, P. A. (1987). Molecular mechanical simulations on double intercalation of 9-aminoacridine into d(CGCGCGC).d(GCGCGCG): analysis of the physical basis for the neighbour exclusion principle. *Proc. Nat. Acad. Sci. USA* **84**, 5735–9.

Wang, A. H.-J., Ughetto, G., Quigley, G. J., and Rich, A. (1987). Interactions between an anthracycline antibiotic and DNA: molecular structure of daunomycin complexed to d(CGTACG) at 1.2-Å resolution. *Biochemistry*, **26**, 1152–63.

Wilson, W. D. and Jones, R. L. (1981). Intercalating drugs: DNA binding and molecular pharmacology. *Adv. Pharmac. Chemother.*, **18**, 177–222.

Wilson, W. D., Krishnamoorthy, C. R., Wang, Y.-H., and Smith J. C. (1985). Mechanism of intercalation: ion effects on the equilibrium and kinetic constants for the interaction of propidium and ethidium with DNA. *Biopolymers*, **24**, 1941–61.

8.5

Atwell, G. J., Stewart, G. M., Leupin, W., and Denny, W. A. (1985). A diacridine derivative that binds by bisintercalation at two contiguous sites on DNA. *J. Amer. Chem. Soc.*, **107**, 4335–7.

McLean, M. J. and Waring, M. J. (1988). Chemical probes reveal no evidence of Hoogsteen base-pairing in complexes formed between echinomycin and DNA in solution. *J. Mol. Recognition*, **1**, 128–51.

Wakelin, L. P. G. (1986). Polyfunctional DNA intercalating agents, *Med. Res. Rev.*, **6**, 275–340.

Wang, A. H.-J., Ughetto, G., Quigley, G. J., Hakoshima, T., van der Marel, G. A., van Boom, J. H., and Rich, A. (1984). The molecular structure of a DNA-triostin A complex. *Science*, **225**, 1115–21.

8.6

Marzilli, L. G., Banville, D. L., Zon, G., and Wilson, W. D. (1986). Pronounced proton and phosphorus NMR spectral changes on *meso*-tetrakis(*N*-methylpyridinium-4-yl) porphyrin binding to poly[d(G.C)].poly[d(G.C)] and to three tetradeca-oligodeoxyribonucleotides: evidence for symmetric selective binding to 5′CG3′ sequences. *J. Amer. Chem. Soc.*, **108**, 4188–92.

Strekowski, L., Wilson, W. D., Mokrosz, J. L., Strekowska, A., Koziol, A. E., and Palenik, G. J. (1988). A non-classical intercalation model for a bleomycin amplifier. *Anti-Cancer Drug Design*, **2**, 387–98.

Wilson, W. D., Tanious, F. A., Barton, H. J., Strekowski, L., Boykin, D. W., and Jones, R. L. (1989). Binding of DAPI to GC and mixed sequences in DNA: intercalation of a classical groove-binding molecule. *J. Am. Chem. Soc.*, **111**, 5008–10.

Yen, S.-F., Gabbay, E. J., and Wilson, W. D. (1982). Interaction of aromatic imides with deoxyribonucleic acid. Spectrophotometric and viscometric studies. *Biochemistry*, **21**, 2070–6.

8.7

Hecht, S. M. (1986). The chemistry of activated bleomycin. *Acc. Chem. Res.*, **19**, 383–91.

Hélène, C., Montenay-Garestier, T., Saison, T., Takasugi, M., Toulmé, J. J., Asseline, U., Lancelot, G., Maurizot, J. C., Toulmé, F., and Thuong, N. T. (1985). Oligonucleotides covalently linked to intercalating agents: a new class of gene regulatory substances. *Biochimie*, **67**, 777–83.

Lohman, T. M. (1985). Kinetics of protein–nucleic acid interactions: use of salt effects to probe mechanisms of interaction. *CRC Crit. Rev. Biochem.*, **19**, 191–245.

Moser, H. E. and Dervan, P. B. (1987). Sequence-specific cleavage of double helical DNA by triple helix formation. *Science*, **238**, 645-50.

Strekowski, L., Strekowska, A., Watson, R. A., Tanious, F. A., Nguyen, L. T., and Wilson, W. D. (1987). Amplification of bleomycin-mediated degradation of DNA. *J. Med. Chem.*, **30**, 1415–20.

Stubbe, J. and Kozarich, J. W. (1987). Mechanism of bleomycin-induced DNA degradation. *Chem. Revs.*, **87**, 1107–36.

8.8

Atwell, G. J., Baguley, B., and Denny, W. A. (1989). Potential antitumour agents. 57. 2- Phenylquinoline-8-carboxamides as 'Minimal' DNA-intercalating antitumor agents with *in vivo* solid tumor activity. *J. Med. Chem.*, **32**, 396–401.

Lee, M., Krowicki, K., Hartley, J., Pon, R. T., and Lown, J. W. (1988). Molecular recognition between oligopeptides and nucleic acids: influence of van der Waals contacts in determining the 3′-terminus of DNA sequences read by monocationic–lexitropsins. *J. Am. Chem. Soc.*, **110**, 3641–9.

Maher, L. J., Wold, B., and Dervan, P. B. (1989). Inhibition of DNA binding proteins by oligonucleotide-directed triple helix formation. *Science*, **245**, 725–30.

Praseuth, D., Doan, T. L., Chassignol, M., Decout, J.-L., Habhoub, N., Lhomme, J., Thuong, N. T., and Hélène, C. (1988). Sequence-targeted photosensitized reactions in nucleic acids by oligo-α-deoxynucleotides and oligo-β-deoxynucleotides covalently linked to proflavin. *Biochemistry*, **27**, 3031–8.

Zon, G. (1988). Oligonucleotide analogues as potential chemotherapeutic agents. *Pharm. Res.*, **5**, 539–49.

Interaction of nucleic acids with proteins

<big>**9**</big>

9.1 Perspective

9.1.1 Why study proteins?

The importance of nucleic acids in the cell has been pointed out many times, but the importance of proteins in the structure and function of nucleic acids is often ignored. Nucleic acids are rarely found alone: proteins package them and also mediate almost all the transformations that they undergo—polymerization, gene regulation, splicing, or degradation. Thus, although nucleic acids are the message that is in the cell, proteins are the medium through which that message is expressed. One means little without the other.

Rather than discuss the interaction of proteins and nucleic acids in general, we concentrate on the molecular basis of protein–nucleic acid interactions, with a strong bias towards systems where structures are known at high resolution. Consequently, although, for example, topoisomerases (Section 2.3.4) and DNA gyrase are biochemically fascinating, very little is know about them at high resolution and so these classes of proteins are not discussed. Similarly, kinases, ligases, and exonucleases are only briefly outlined in Section 9.4. The varied yet incomplete selection of systems we do discuss reflects how difficult it is to crystallize macromolecules and macromolecular complexes. Much more work needs to be done before X-ray crystallography can hope to offer a complete account of the structural basis of protein–nucleic acid interactions.

9.1.2 History of structure determination

To understand how a molecule works, one must know its structure. Probably the most famous example of this is the structure determination of DNA, which provided much more than a list of the positions of the atoms in the double-helix. It explained the stability of DNA and the Chargaff Rules, and provided a model of how DNA might work.

In comparison with DNA and RNA, proteins are much less regular, and so understanding them is more difficult. The first nucleic acid-binding proteins solved were nucleases, because they are stable and abundant. These structures provided much information about how proteins interact with single-stranded oligonucleotides. Work on more complex systems, such as repressors and polymerases, had to wait for molecular biologists to learn how to over-express normally scarce proteins and for chemists to learn how to synthesize large quantities of oligonucleotides. By the late 1970s it was reasonable to try to determine the structure of a nucleic acid, the protein it interacts with, and the complex between the two.

Some of the most exciting subjects of recent years have included the Klenow fragment of DNA polymerase I from *E. coli*, *Eco*RI complexed with DNA, bacteriophage 434 repressor complexed with DNA, and the increasing understanding of the nucleosome and the histone octamer.

9.1.3 The forces between proteins and nucleic acids

Intermolecular forces determine how proteins interact with ligands. Protein–nucleic acid interactions, however, are more complex than protein–ligand interactions because nucleic acids are large polymeric anions. One must consider not only the forces between the two molecules in isolation, but also the effects of the solvent and of the ions in solution (Section 8.2.1). Furthermore, both the size and arrangement of the attractive forces matter. They are arranged so that the protein and nucleic acid can interact in a unique way. How proteins and nucleic acids achieve this specificity is clearly related to the structures of the molecules involved.

The forces between proteins and nucleic acids can be classified into four types.

Electrostatic forces: salt bridges

Salt bridges are electrostatic interactions between groups of opposite charge. They typically stabilize the complex by about 40 kJ mol^{-1} per salt bridge. In protein–nucleic acid complexes they occur between the ionized phosphates of the nucleic acid and either the ε-amino group of lysine, or the guanidium group of arginine, or the protonated imidazole of histidine. Only histidine titrates under physiological conditions. Salt bridges are influenced by the concentration of salt in the solution: as it increases, the strength of the salt bridge decreases. A salt bridge is much

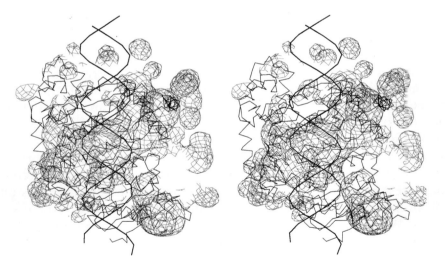

Fig. 9.1   The interaction of the Klenow fragment of DNA polymerase I with DNA—a view looking down on the active site from above. The positive electrostatic potential of the Klenow fragment resulting from the ionizable groups and the α-helix dipoles of the protein is shown in red. The α-carbon backbone of the protein and the phosphorus backbone of the model-built DNA is shown in black. The electrostatic potential is contoured at 2 kt/e: it follows the path of the major groove of the DNA in the DNA-binding cleft of the polymerase domain (adapted from Warwicker, J., Ollis, D. L., Richards, F. M., and Steitz, T. A. (1985). *J. Mol. Biol.*, **186**, 645–9).

stronger when there are no water molecules between the two ionized groups because water has a high dielectric constant.

Compared to the other forces between proteins and nucleic acids, salt bridges are relatively long-range. Their force is proportional to the inverse square of the separation of the two charges and it is insensitive to the relative orientation of the charges involved. They therefore do not confer a great deal of specificity on the interaction between protein and nucleic acid. Furthermore, changes in base composition do not substantially change the conformation of the phosphate groups in B-DNA. So salt bridges cannot simply distinguish one DNA sequence from another.

Although individual salt bridges are not very specific, the overall electrostatic field of the protein must orient the polyanionic nucleic acid on the protein correctly. Thus the surface of the protein touching the nucleic acid should have a positive electrostatic potential, whose three-dimensional shape can be used to help place the nucleic acid on the protein in a model structure when only the protein structure is known. This has been used in modelling the interaction of DNA with both catabolite activator protein and the Klenow fragment of DNA polymerase I (Fig. 9.1).

Dipolar forces: hydrogen-bonds

Hydrogen-bonds are due to dipole–dipole interactions:

$$\overset{-}{X}\!\!\!\longrightarrow\!\!\!\overset{+}{H} \;\text{- - - - - - - - -}\; \overset{-}{Y}\!\!\!\longrightarrow\!\!\!\overset{+}{R}$$

(donor)   hydrogen-bond   (acceptor)

The strength of a hydrogen-bond falls off with the inverse third power of the H–Y distance, and also decreases greatly if the bond is bent (i.e. if X, H, and Y are not in a straight line).

Hydrogen-bonds occur between the amino acid side-chains of the protein and the bases of the nucleic acid, both of which also form hydrogen-bonds to water. Since all linear hydrogen-bonds have similar free energies, they make little net contribution to the favourable free energy change when the protein and nucleic acid bind in solution. By contrast, forming poorly aligned hydrogen-bonds, or *not* forming some of them when the complex forms, carries a free energy penalty of about 4 kJ mol^{-1}. Thus hydrogen-bonds are one of the most important means of making sequence-specific protein–nucleic acid interactions.

Entropic forces: the hydrophobic effect

The hydrophobic effect is due to the behaviour of water at an interface. A non-polar molecule in water creates a sharply curved interface region where there is a layer of ordered water molecules. When non-polar molecules aggregate, the ordered water molecules which were at the interface are released and become part of the disordered bulk water. This stabilizes the aggregate by increasing the entropy of the system and hence decreasing its free energy.

Therefore, the surface of the protein will exactly complement the surface of the nucleic acid it binds to, so that there are no unnecessary bound water molecules left in the interface when the protein–nucleic acid complex forms.

Base-stacking: dispersion forces

Base-stacking is caused by two kinds of interaction: the hydrophobic effect mentioned above and dispersion forces. Molecules with no net dipole moment can attract each other by a transient dipole–induced dipole interaction first explained by London in the 1920s. Such dispersion forces decrease with the inverse sixth power of the distance separating the two dipoles, and so are very sensitive to the thermal motion of the molecules involved.

Despite their extreme distance dependence, dispersion forces are clearly important in the maintenance of the structure of double-stranded nucleic acids for they help to cause base-stacking (Section 2.5). Furthermore, base-stacking may well help single-stranded nucleic acids bind to proteins because aromatic protein side-chains can stack in between the bases of single-stranded nucleic acids in an intercalated fashion.

Summary

Electrostatic forces are long-range, not very structure-specific, and contribute substantially to the overall free energy of association. Hydrogen-bonds are dipolar short-range interactions; they do little to stabilize the complex, but do determine the specificity of a protein–nucleic acid interaction. Hydrophobic forces are due to changes in the solvation of the macromolecules when they interact: they are short range, sensitive to structure, and contribute to the free energy of association. Base-stacking, due to hydrophobic and dispersion forces, is very short-range and is most important in forming double-stranded nucleic acid complexes and complexes of protein with single-stranded nucleic acids. Clearly, a combination of forces is present when proteins and nucleic acids interact.

9.1.4 Geometric constraints imposed by the nucleic acid

Nucleic acids contain a number of different molecular groups: the phosphate backbone, the bases, and the sugars. The geometric arrangement of these groups, which depends on whether the nucleic acid is single-stranded or double-stranded, A-, B-, or Z-form (Section 2.2), influences how the protein and the nucleic acid interact.

Double-stranded DNA

Double-stranded right-handed B-DNA presents a highly repetitive, negatively charged polyphosphate surface to any molecule it interacts with. Thus a protein that binds B-DNA must have lysine and arginine residues positioned to neutralize the polyphosphate backbone by forming ionic bonds. Such bonds occur in complexes of protein with double-stranded DNA that are not sequence-specific, such as in the model for DNA polymerase interacting with DNA (Section 9.2.4), and in sequence-specific complexes, such as the *Eco* RI–DNA structure (Section 9.3.3).

However, for a protein to recognize a specific DNA sequence, it must also interact, via hydrogen-bonds, with bases either in the major or in the minor groove of the DNA. Long before the structures of any DNA-binding proteins were known, it was clear that an α-helix would fit well in the major groove of B-DNA (see Fig. 9.18) and that an anti-parallel β-ribbon would fit into the minor groove and form hydrogen-bonds to the phosphate backbone (Fig. 9.2) Both these and other modes of interaction with DNA have now been observed.

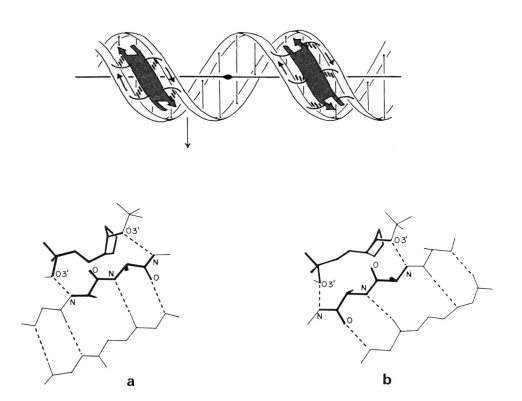

Fig. 9.2 Schematic representation of a β-sheet in the minor groove of DNA. The β-sheets are located in the minor grooves so that their dyad symmetries match that of B-DNA. In **a**, the polarities of the nucleic acid (5′-to-3′) and of the polypeptide (*N* to *C*) are parallel, whereas in **b**, they are antiparallel. In the more detailed drawings, hydrogen-bonds are shown as dashed lines and only one strand of the DNA is shown (from Church, G. M., Sussman, J. L., and Kim, S-H. (1977). *Proc. Nat. Acad. Sci. USA*, **74**, 1458–62).

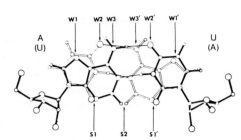
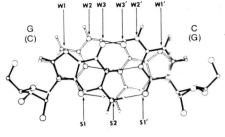
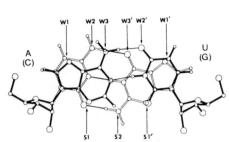
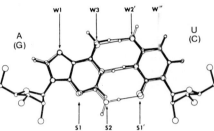

Fig. 9.3 A diagram showing which hydrogen-bond donors and acceptors are accessible in the grooves of B-DNA. All the four combinations of A:U and G:C base-pairs are shown, indicating where hydrogen-bonds can be used to discriminate between various base-pairs. The upper letters refer to the base-pair in solid lines, the lower letter (in parentheses) to the base-pair in hollow lines. W refers to potential hydrogen-bonds in the major groove; S refers to potential ones in the minor groove. W' and S' sites are related to W and S sites by the two-fold symmetry of DNA and RNA double-helices. The red sites have different hydrogen-bond donor and acceptor characteristics when the specific base change shown is made (from Seeman, N. C., Rosenberg, J. M., and Rich, A. (1976). *Proc. Nat. Acad. Sci. USA*, **73**, 803–8).

Model-building studies on double-stranded B-DNA showed that a larger number of potential hydrogen-bond donors and acceptors are accessible in the major than in the minor groove (Fig. 9.3). Furthermore, although the pattern of hydrogen-bond donors and acceptors accessible in the major groove for each of the four base-pairs is unique, no single site suffices to distinguish a base-pair uniquely. Thus a protein must form more than one hydrogen-bond to some of the base-pairs in the sequence it recognizes if it is to distinguish the correct sequence from all other possible sequences. One effective way of achieving this uses amino acids such as arginine or glutamine to form bridging hydrogen-bonds to two successive bases. Such bridging hydrogen-bonds are sensitive to the precise conformation of the DNA.

Single-stranded nucleic acids

Single-stranded nucleic acids are much more flexible than double-stranded DNA, and the predominantly hydrophobic bases are much more exposed. A single-stranded nucleic acid-binding protein must have enough positive side-chains to neutralize the charge of the phosphate backbone where it binds, and ideally it will have aromatic groups (tryptophans, tyrosines, or phenylalanines) which will intercalate with the nucleic acid bases.

9.1.5 The kinetics of forming protein–nucleic acid complexes

Two factors affect the rate of formation of protein–nucleic acid complexes that are not sequence-specific: random thermal diffusion and long-range, directional electrostatic attraction. They also affect the rate of formation of sequence-specific protein–nucleic acid complexes, but are not enough to account for the observed rates of formation. These are far faster than those predicted from the simplest model of random association of the protein on the nucleic acid, followed by dissociation and reassociation elsewhere if the sequence is not correct. This model is a three-dimensional random walk through the contents of the entire cell, and is clearly too slow.

An alternative model, which can account for the observed rate of formation of sequence-specific protein–DNA complexes, is a one-dimensional random walk. The protein first binds non-specifically to the DNA, and then diffuses or jumps along the DNA until it finds the appropriate sequence. How a protein which normally binds specifically can initially bind non-specifically is not clear. There must be relatively strong interactions which are *not* sequence-specific. The sequence-specific interactions to double-stranded DNA may be hydrogen-bonds between the protein and the DNA. While the protein is bound non-specifically, these hydrogen-bonds cannot form, thus decreasing the stability of the non-specific complex. Alternatively, as in *lac* repressor, more ionic interactions may occur in the specific complex than in the non-specific complex. Therefore such proteins may well exist in two conformations. The non-specific binding conformation would increase the rate of diffusion along the DNA, and the specific binding conformation, where hydrogen-bonds can be made to the bases, would form only when the correct DNA sequence is reached.

Summary

All nucleic acids have repeating polyanionic backbones, and so all proteins that bind to nucleic acids have strategically placed arginines and lysines that create an electrostatic field to neutralize the negative charge. To interact with B-DNA, the protein either (1) inserts an α-helix into the major groove, or (2) inserts a β-sheet into the minor groove, and forms hydrogen-bonds from the side-chains to specific bases. To interact with single-stranded nucleic acids, aromatic side-chains are used to stack against the nucleic acid bases.

Sequence-specific protein–nucleic acid complexes must be formed by the protein first binding loosely to the incorrect sequence and then diffusing along the DNA in a one-dimensional random walk until it finds the correct sequence. Thus all sequence-specific nucleic acid-binding proteins may exist in two conformations: one that allows tight, sequence-specific binding and one that allows looser, non-sequence-specific binding.

9.2 Non-specific interactions

9.2.1 The need for packaging

An *E. coli* bacterium contains about 3×10^6 base-pairs of DNA, which corresponds to about one millimetre of double-helical DNA (many times the circumference of the bacterium). The problem is even more acute in higher eukaryotes whose genomes, typically containing 6×10^9 base-pairs, would be two metres long! To avoid complete chaos in the nucleus, most organisms have some proteins whose sole purpose is to organize DNA into structures both more complex and more compact than the Watson–Crick double-helix (Section 2.6).

Packaging DNA: the eukaryotic nucleosome

In the eukaryotic nucleus, DNA is organized into a hierarchy of structures, some of which can be detected using a light microscope. However, study of the molecular architecture of the chromosome has only begun recently.

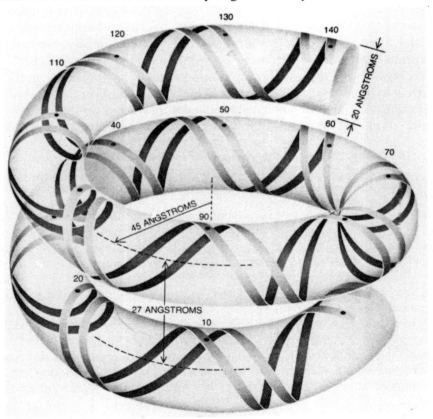

Fig. 9.4 Schematic view of the 20 Å-wide DNA double-helix in the nucleosome. It winds around the histone octamer to form 1.75 turns of left-handed supercoil (from Kornberg, R. and Klug, A. (1981). *Scientific American*, **244(2)**, 52–64).

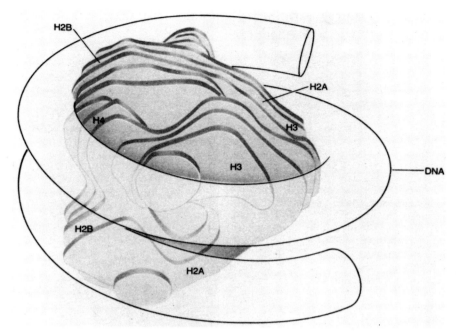

Fig. 9.5 Schematic view of the entire nucleosome DNA-histone complex. The individual proteins are shaded and their locations are shown (from Kornberg, R. and Klug, A. (1981). *Scientific American*, **244(2)**, 52–64).

When the metaphase chromosome is suspended in a medium of low salt concentration, it extrudes individual chromatin strands which, viewed under an electron microscope, look like beads on a string. The beads are nucleosomes: protein–DNA complexes which can be isolated by cleaving the intervening DNA (the string) with micrococcal nuclease (Section 2.6.1). The nucleosome is an octamer of about 146 base-pairs of DNA and two each of four, small, highly basic proteins (H2A, H2B, H3, and H4), called histones. The intact nucleosome has been crystallized and a low resolution (7 Å) electron density map obtained. In it the heart-shaped histone octamer occupies the centre of the nucleosome, with DNA wrapped around it to form 1.75 turns of left-handed (negatively supercoiled) superhelix (Figs 9.4 and 9.5). The DNA is right-handed B-DNA and the superhelix appears to be formed by a series of kinks rather than a smooth curve. Chemical cross-linking and other studies have shown that the core of the octamer is formed by two H3 and two H4 proteins (the tetramer) which by itself binds DNA. Thus the DNA forms a cylinder wrapped around the tetramer, with an H2A-H2B dimer sitting on each end.

Chromatin can also form higher order fibre-like structures (Fig. 9.6): the 10 nm filaments have a zigzag arrangement of nucleosomes and the 30 nm solenoid structure has a helical arrangement of nucleosomes. The H1 histone protein, released when chromatin is treated with micrococcal nuclease, may well be important in forming these structures (Section 2.6.2).

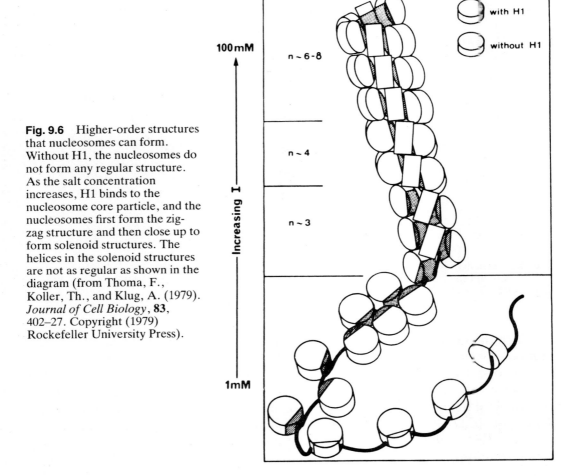

Fig. 9.6 Higher-order structures that nucleosomes can form. Without H1, the nucleosomes do not form any regular structure. As the salt concentration increases, H1 binds to the nucleosome core particle, and the nucleosomes first form the zig-zag structure and then close up to form solenoid structures. The helices in the solenoid structures are not as regular as shown in the diagram (from Thoma, F., Koller, Th., and Klug, A. (1979). *Journal of Cell Biology*, **83**, 402–27. Copyright (1979) Rockefeller University Press).

Packaging DNA in prokaryotes

There is no direct prokaryotic equivalent of the histone proteins, although there are a number of possible substitutes. These bacterial proteins do not form particles like the nucleosome, but they are basic, bind DNA, and seem to induce higher order structure in double-stranded DNA. The addition of DNA Binding Protein II (also referred to as HU or NS) can cause the formation of supercoils in closed circular DNA. The *E. coli* HU has been crystallized and its structure determined at high resolution. Compared to the wealth of sequence and solution data available for the histones, very little is known about HU. However, the structure does suggest how it might interact with DNA.

The *N*-terminal helices form the wedge-shaped core of the closely packed protein dimer (Fig. 9.7). Two strands of β-sheet from each monomer form arm-like structures that project from the central core. The crevice between the two arms is large enough to accommodate double-stranded B-DNA, and the twist of the β-sheets follows the model-built DNA. It is not yet clear, however, where the DNA and the arms are in contact.

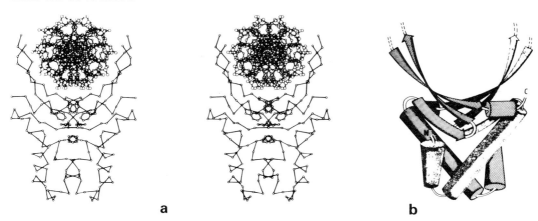

a b

Fig 9.7 Structure of DNA-binding protein II. (**a**) A stereopair drawing of the α-carbon backbone of the protein showing the proposed interaction with DNA. The view is down the helix axis of the model-built DNA. (**b**) A schematic representation of the secondary structure of DNA-binding protein II. Each monomer in the dimer is shaded differently and the disordered regions between the two strands are shown in dotted lines (from Tanaka, I., Appelt, K., Dijk, J., White, S. W., and Wilson, K. S. (1984). *Nature*, **310**, 376–81. Copyright (1984) Macmillan Magazines Limited).

The connection between the far ends of β-sheet of the arms (Fig. 9.7a) is not defined in the electron density map (probably because this region of the protein is flexible: it can assume more than one conformation). The flexibility of this region may help HU carry out its function. Once the protein has loosely bound the DNA, this peptide could close around the DNA and increase the energy of interaction between HU and DNA.

How then could HU package DNA? HU molecules might be placed as in Fig. 9.8 along the DNA so that the narrow end of their wedge-shaped cores point inward. This arrangement might help the DNA bend to form a circular or helical arrangement.

Other nucleic acid packages: viruses

RNA and DNA viruses exist with their nucleic acid in either single- or double-stranded forms. So all manner of protein–nucleic acid interactions exist. Furthermore, the structures of a number of virus coat proteins and intact viruses are

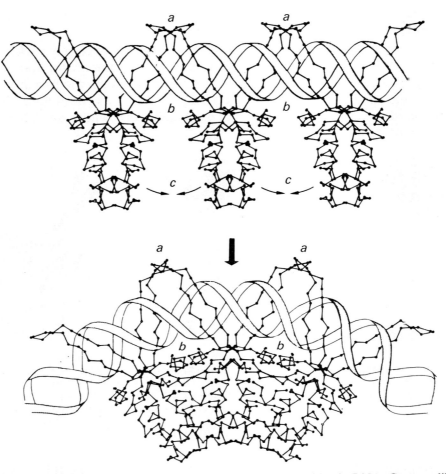

Fig. 9.8 Proposal for how DNA-binding protein II induces supercoiling in DNA. Contacts likely to be important in forming this structure are labelled: (**a**) between the ends of the arms of adjacent dimers (these arms are disordered in the crystal structure); (**b**) between the *C*-terminal helices; (**c**) between the wedge-shaped dimers (from Tanaka, I., Appelt, K., Dijk, J., White, S. W., and Wilson, K. S. (1984). *Nature*, **310**, 376–81. Copyright (1984) Macmillan Magazines Limited).

available, but we know very little about how proteins and nucleic acids interact in viruses.

The icosahedral viruses, such as polio virus and the common cold virus, have been crystallized, but their crystal structures give no indication of how the protein and nucleic acid interact, because the region of contact is not sufficiently ordered for X-ray analysis. Fibre-diffraction studies at 2.9 Å of the intact tobacco mosaic virus (TMV), which is a helical RNA virus, have shown that the phosphates of the RNA are neutralized by arginines on the coat protein. The coat protein subunits are arranged similarly in the fibre structure and in the crystal structure of the coat protein in the absence of RNA. In this crystal structure, however, the arrangement of the protein subunits is changed slightly so that they form rings rather than a helix.

Summary

The fundamental building block of chromatin in eukaryotes is the nucleosome, a protein–DNA complex. The complex is made from 146 base-pairs of DNA and four pairs of histone proteins. The DNA wraps around the histone octamer to form 1.75 turns of negatively supercoiled B-DNA.

The HU protein from *E. coli* is dimeric: its central crevice is formed by a pair of two-stranded β-sheets, one from each monomer. Double-stranded B-DNA can fit into the crevice.

Little is known about how viruses package DNA or RNA.

9.2.2 Single-stranded nucleic acid-binding proteins

Single-stranded DNA is formed during replication and most organisms produce proteins to bind this DNA. The high resolution structure of one single-stranded

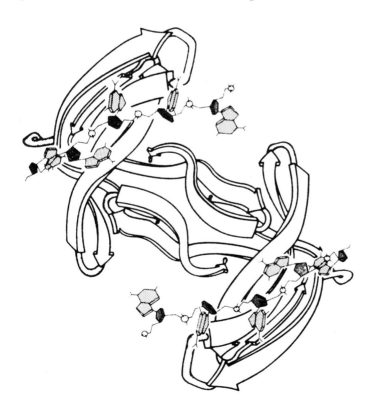

Fig. 9.9 Model of how gene-5 protein from bacteriophage fd interacts with single-stranded DNA. β-Strands are shown as ribbons and the paths of two strands of ssDNA are shown with the bases and sugars stippled differently (from Brayer, G. D. and McPherson, A. (1984). *Biochemistry*, **23**, 340–9. Copyright (1984) American Chemical Society).

DNA-binding protein is known, that of gene-5 protein from bacteriophage fd. During the replication of fd, gene-5 protein seems to bind to freshly synthesized viral DNA daughter strands, thus preventing other proteins, such as nucleases or polymerases, from gaining access to the viral DNA.

The protein is a closely associated dimer (Fig. 9.9), consisting almost entirely of β-strands which loop out from a hydrophobic core. One such loop forms the monomer–monomer interface, while the other two bind DNA. The DNA-binding surface of gene-5 protein is rather unusual because many aromatic side-chains are clearly exposed to solvent in the crystal structure. Any model of gene-5 interacting with DNA must explain this unusual feature.

This model of the complex proposes that the nucleic acid bases stack against protein aromatic side-chains, while nucleic acid phosphates are neutralized by basic protein side-chains (Fig. 9.10). The model is based not only on an X-ray crystal structure, but also on NMR, fluorescence quenching, and chemical modification studies. The crystal structure, however, is not completely consistent with the solution data. In particular, a small peptide in the protein must undergo a conformational change when DNA binds in order to account for some of the cross-linking data. The X-ray studies indicate that this peptide may be flexible. Again, conformational change seems important in protein–nucleic acid interactions.

Single-stranded DNA-binding proteins from other phages and from *E. coli* appear to have only a very low level of sequence similarity with gene-5 protein.

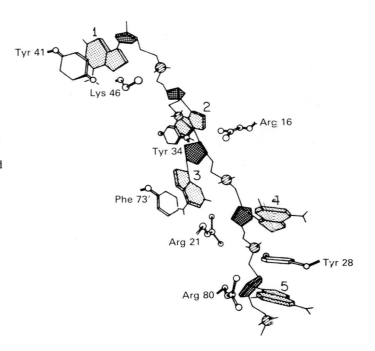

Fig. 9.10 Close-up of the interaction between gene-5 and one strand of DNA. The interactions shown were deduced from model-building and biochemical data. The amino acid side-chains are not stippled, bases are lightly stippled, and sugars heavily stippled (from Brayer, G. D. and McPherson, A. (1984). *Biochemistry*, **23**, 340–9. Copyright (1984) American Chemical Society).

However, the conserved residues appear to be those that interact with the DNA. This suggests that the type of single-stranded DNA-binding domain found in the gene-5 protein may be used to bind single-stranded DNA in other such proteins.

Summary

Single-stranded nucleic acid-binding proteins are unusual because they have a large number of aromatic side-chains on their surface. These side-chains stack against the exposed nucleic acid bases in a piece of single-stranded nucleic acid. This mode of protein–nucleic acid interaction is not seen in any other nucleic acid-binding protein.

9.2.3 Non-sequence-specific nucleases

All organisms must degrade nucleic acids during their life cycle. There is no single enzyme designed for this purpose, but rather a large number of enzymes with different specificities. These include exo- and endonucleases and enzymes specific for single- and double-stranded nucleic acids and for base sequences. The wide variety of specificities available makes these enzymes useful tools in biological research.

Ribonuclease A

Ribonuclease A (RNase A) cleaves single-stranded RNA. It is, perhaps, the best characterized of the nucleases with little sequence specificity. Its structure has been known for 20 years at high resolution and many substrate-binding studies have given a very detailed understanding of the active site and how it functions.

During the reaction, cleavage occurs on the 5′-side of the phosphate ester to yield a 3′-phosphate, and the scissile phosphate ester forms a pentaco-ordinate transition state (Fig. 9.11). Histidine-12, acting as a general base catalyst, forms a hydrogen-bond to the ribose 2′-hydroxyl, which attacks the phosphate to form the pentaco-ordinate species. A second histidine (His-119) and a lysine (Lys-41) form salt bridges to the phosphate ester oxygens.

DNA is a competitive inhibitor of RNase A, since it lacks the 2′-hydroxyl group. Thus RNase A once served as a primitive model of a single-stranded DNA-binding protein. Furthermore, DNA-binding studies and RNA digestion studies have shown that the RNA-binding site of RNase A is larger than the active site. About eleven nucleotides bind to the protein and inspection of the protein suggests where they might bind. Some distance from the active site a group of positively charged residues form an anion-binding site, and there is positive electrostatic potential from this site all the way to the active site. It was predicted that this pathway of positive electrostatic potential would bind single-stranded nucleic acids, and McPherson and Brayer have confirmed this by solving the structure of RNase A complexed with four d(pA)$_4$ oligonucleotides. Of the 16 nucleotides in the complex, only 12 bind directly to the protein. These form a continuous chain running

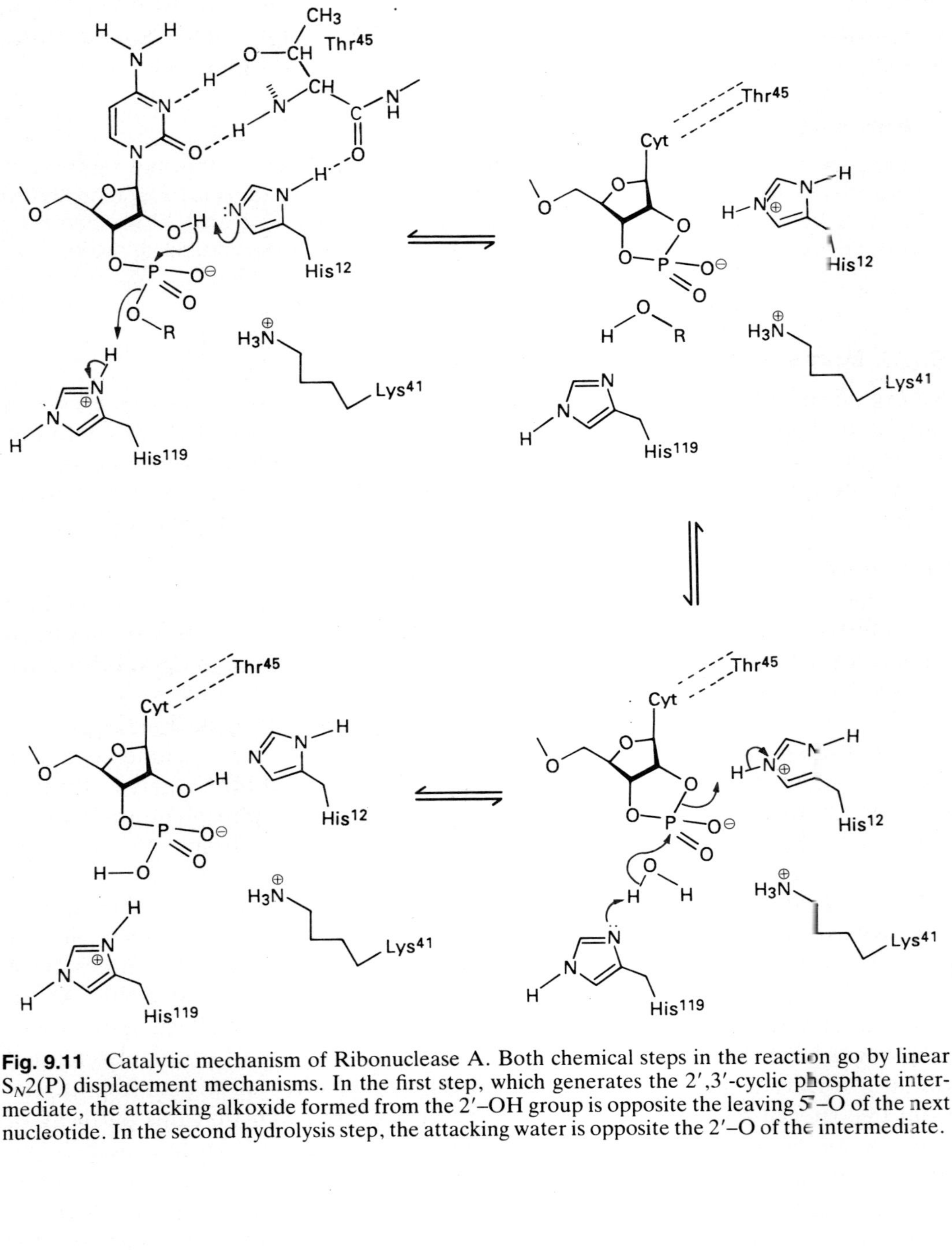

Fig. 9.11 Catalytic mechanism of Ribonuclease A. Both chemical steps in the reaction go by linear $S_N2(P)$ displacement mechanisms. In the first step, which generates the 2',3'-cyclic phosphate intermediate, the attacking alkoxide formed from the 2'–OH group is opposite the leaving 5'–O of the next nucleotide. In the second hydrolysis step, the attacking water is opposite the 2'–O of the intermediate.

5′-to-3′ across the surface of the protein (Fig. 9.12). The active site is at the 3′-end of the nucleotide chain while the anion-binding site is at the 5′-end. The main links between the protein and the DNA are electrostatic, resulting from ionic bonds

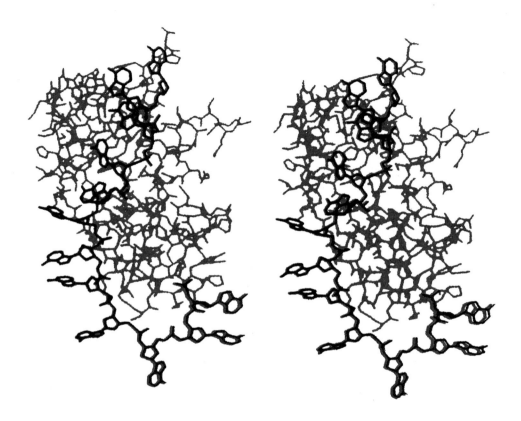

Fig. 9.12 Stereo-diagram of the single-stranded DNA–Ribonuclease A complex. In the stereo-diagram, the course of the 12 bases of DNA on the ribonuclease is shown. The oligonucleotide is black and the protein red. All the side-chains on the protein are shown (from McPherson, A., Brayer, G. D., Cascio, D., and Williams, R. (1986). *Science*, **232**, 765–8. Copyright (1986) AAAS). This stereo-pair diagram is for parallel viewing.

between positively charged protein side-chains and the polyphosphate DNA back-bone. The potential H-bond donors and acceptors on the DNA bases are mostly exposed to solvent, which explains why RNase A is not sequence-specific, although it is base-specific for the 3′-residue.

Some structures of other nucleases similar to RNase A have been solved. Ribonuclease T1 cleaves single-stranded RNA by the same mechanism as does RNase A but it is selective for a 3'-purine residue, while RNase A is 3'-pyrimidine selective (C > U). Staphylococcal nuclease can degrade both RNA and DNA and its structure has been solved for the protein complex with the inhibitor pdTp. These studies have shown that there are many different nuclease folds (unlike the nucleotide-binding fold) and that active sites vary quite dramatically from enzyme to enzyme.

Deoxyribonucleases

The structure of a double-stranded DNA nuclease is also known for DNase I complexed with pdTp at the active site. The current model of how it might interact with DNA (Fig. 9.13) suggests that the protein interacts with only one face of the DNA. There is a calcium ion at the active site which both holds the scissile phosphate in the correct orientation for nucleophilic attack by water and polarizes the scissile P–O bond. A loop on the surface of the protein may occupy the minor groove of DNA and form links to both strands of the DNA. It is unusual for the major contacts between the protein and DNA to be made in the minor groove of the DNA and this may explain the base-preference of the enzyme. In the proposed model, the tyrosine in the minor groove would experience considerable steric hindrance from the amino groups in the 2-position of a guanine in a G–C base-pair. Hence the enzyme would degrade A–T stretches far more readily than G–C stretches.

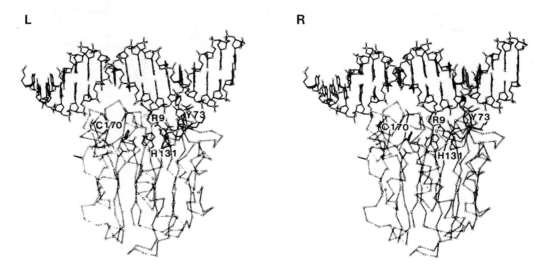

Fig. 9.13 Stereo-diagram of the model of DNAse I–DNA interaction. Only the α-carbon backbone of the DNAse is shown. The DNA and the regions of the protein thought to be in contact with the DNA are shown in solid lines and the rest of the protein is in hollow lines. The double circle by arginine-9 (R9) marks the position of the catalytically essential calcium ion (from Oefner, C. and Suck, D. (1986). *Journal of Molecular Biology*, **192**, 605–32). (For parallel viewing).

Summary

The mechanism of action of RNase and DNase is different because RNase uses the ribose 2'-hydroxyl group, not present in DNA, to displace the 5'-phosphate ester linkage. In RNase A, the RNA binding site is 12 bases long, but the interactions are entirely between the phosphate backbone and charged protein side-chains. Thus RNase A is not sequence-specific. Different RNases have different polypeptide folds. DNase I is a double-stranded DNase which inserts a polypeptide loop into the minor groove of B-DNA. Steric hindrance between a tyrosine in this loop and the 2-amino group on guanine may explain why DNase I degrades A–T tracts much better than G–C tracts.

9.2.4 DNA polymerase

DNA replication is a highly complex process which normally involves numerous proteins (Section 5.4). The key enzyme in DNA replication, the DNA polymerase, is frequently a multi-subunit enzyme, with different catalytic activities on different subunits.

One of the simplest polymerases, DNA polymerase I (Pol I) from *E. coli*, is a minor bacterial enzyme whose main function is to repair damaged DNA (Section 7.11.2). It has one polypeptide chain of molecular weight 108 kDa, which has three enzymatic activities: a DNA polymerase, a 3'-to-5' (proof-reading) exonuclease, and a 5'-to-3' exonuclease (to remove RNA primers). Pol I is small and inefficient compared to its larger, more complicated siblings, but the polymerase and 3'-to-5' proof-reading exonuclease reactions are similar to those of larger DNA polymerase molecules. Thus it is a good model system for studying the polymerases. Mild proteolysis cleaves Pol I into two fragments, the smaller of which contains the 5'-to-3' exonuclease activity. The larger fragment (the Klenow fragment), which has been studied by X-ray crystallography, has a molecular weight of 68 kDa and contains the DNA polymerase and the 3'-to-5' exonuclease activities, carried on two separate domains (Fig. 9.14): a small domain of about 200 residues and a large one of about 400 residues.

The small *N*-terminal domain has a typical 'Rossman' nucleotide-binding fold, with α-helices on the outside sandwiching six β-strands on the inside, most of which are parallel to each other. However, the precise arrangement of secondary structure seen here occurs only in the Klenow fragment. This domain is believed to contain the 3'-to-5' proof-reading exonuclease, firstly because it binds the exonuclease reaction product, nucleoside monophosphate, and secondly because site-directed mutants which produce single amino acid changes in this domain destroy the exonuclease activity completely.

The large domain of the Klenow fragment (Fig. 9.14) is rather unusual. Running through its centre is a cleft large enough to bind double-stranded DNA. However,

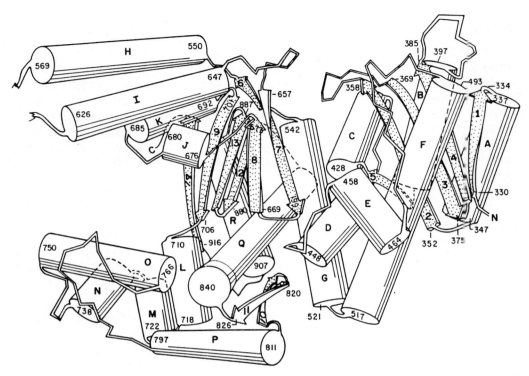

Fig. 9.14 Schematic representation of the Klenow fragment of DNA polymerase I. α-Helices are shown as cylinders, β-sheets as ribbons. The *C*-terminal domain (left) binds DNA and the *N*-terminal domain (right) has the 3′–5′ exonuclease activity. The cleft running through the *C*-terminal domain is believed to bind DNA and the two helices believed to protrude into the major groove are helices J and K (from Ollis, D. L., Brick, P., Hamlin, R., Xuong, N., and Steitz, T. A. (1985). *Nature*, **313**, 762–6. Copyright (1985) Macmillan Magazines Limited).

this only indicates that the DNA **might** bind here. If it is the DNA-binding cleft, then it should be highly conserved in other polymerases and, indeed, such conservation is observed when the sequences of Pol I and bacteriophage T7 DNA polymerase are compared. Although the general electrostatic field of the Klenow fragment is mostly negative and would tend to repel DNA, the DNA-binding cleft is mostly cationic and so could bind DNA (Fig. 9.1).

This structure also sheds light on some aspects of DNA replication. DNA polymerases are processive. The protein binds to the DNA and adds a number of bases to the primer strand before dissociating, which enables the enzyme to replicate long stretches of DNA efficiently. How might this occur? As shown (Fig. 9.14), there are two helices in the large domain, **J** and **K**, which protrude into the DNA-binding cleft. Although these helices are not like the helix-turn-helix motif found in repressors (Section 9.3.2) they do protrude, as in the repressors, into the major groove of the DNA in the model. In this position, the helices **J** and **K** prevent the protein from moving freely along the DNA and instead force it to follow the groove of the DNA helix. This helps Pol I to follow the DNA primer terminus. Secondly, the peptide chain linking helices **H** and **I** is disordered; it is probably flexible. This peptide is located so that it must swing out of the way when DNA binds into the active site groove of the Klenow fragment. Once the DNA has bound, however, the peptide could swing down and clamp the DNA into the active site groove, so that the protein and DNA could not dissociate easily.

Precisely what residues catalyse DNA polymerization is not yet know, because it has not yet been possible to bind DNA to the polymerase active site of the Klenow fragment in the crystal. However, photo-affinity labelling of dNTP analogues has been used to identify some of the residues in the active site. They are towards the N-terminus of the large domain and on the floor of the DNA-binding cleft (Fig. 9.14). Thus the presumed DNA polymerization active site is a long way from the proof-reading 3'-to-5' exonuclease site and it is not obvious how these two activities could interact to give the high fidelity found in DNA polymerases. A conformational change which brings the two domains together seems unlikely because it would involve substantial rearrangement of the protein. The current hypothesis suggests that when a base is incorrectly added to the primer strand, the polymerase stalls (presumably because the incorrect base-pair prevents formation of double-

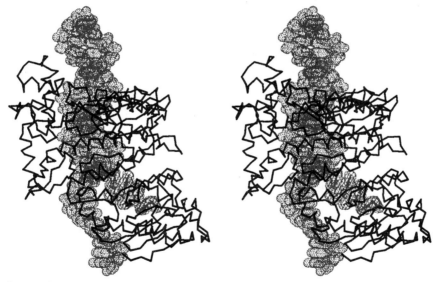

Fig. 9.15 Stereo-diagram of the model of the interaction of DNA with the Klenow fragment of DNA Polymerase I. The α-carbon backbone of the protein is shown in red whilst the DNA is in black. The van der Waals surface of the model-built DNA is stippled whilst the van der Waals surface of the four nucleotides seen in the co-crystal of polymerase and DNA is represented by a caged mesh. The nucleotides extend from the B-DNA which was model-built into the polymerase active site down to the exonuclease active site, showing how the DNA might slide between the two active sites (from Freemont, P. S., Friedman, J. M., Beese, L., Sanderson, M. R., and Steitz, T. A. unpublished results). (Stereopair for parallel viewing).

stranded B-DNA). Because the polymerase is stalled, the primer and template strands become frayed and dissociate from the polymerase site so that the end of the primer strand can reach over to the exonuclease site (Fig. 9.15) where the last, mismatched, base is removed.

Summary

The Klenow fragment of DNA polymerase I from *E. coli* has two widely separated domains, one carrying the polymerase activity, and the other the 3'-to-5' proof-reading exonuclease activity. The exonuclease is a typical six-stranded nucleotide-binding domain. The polymerase domain has a large cleft, presumed to bind DNA. Two helices protrude into the major groove of the model-built DNA and may help Pol I follow the DNA primer terminus. When the polymerase adds an incorrect base, the two DNA strands probably separate, which prevents the polymerase adding any more bases. The daughter strand then can reach over to the proof-reading exonuclease active site where the incorrectly added base is excised.

9.3 Specific interactions

9.3.1 The need for specificity

For an organism to control its development effectively, it must be able to regulate gene expression very precisely, so that it can control the timing of DNA replication precisely. Rather short sequences of nucleic acids can act as control sequences which specific proteins recognize and to which they respond. Many such control elements have been sequenced, but what protein or proteins they interact with is less well known, especially in eukaryotes. Furthermore, how the short DNA sequences interact in a sequence-specific manner with appropriate proteins is only understood in certain systems for which there are high resolution X-ray crystallographic structures.

9.3.2 Repressors and activators

Enormous progess in understanding the basis of specific sequence recognition was made when the structures were determined of several proteins that regulate transcription. First to be solved were the catabolite gene activator protein (**CAP**) from *E. coli* and the cro repressor (**cro**) from phage lambda, followed by the DNA-binding domain of cI repressor from phage lambda (**lambda repressor**) and *trp* **repressor** from *E. coli*. More recently, the high-resolution structures of a number of repressors complexed with their operator DNA have become available and this has helped to consolidate our ideas of how repressors interact with DNA.

   These proteins have quite different overall structures. The λ-cro protein, whose sole function is to bind tightly to a specific DNA sequence, is a dimer whose polypeptide chains fold as single domains. CAP and lambda repressor, whose mode of action is more complex, are also dimers, but each monomer is folded into two domains, one of which binds DNA while the other is involved in regulating the DNA binding. Such a division of functionality is found in other repressors and activators, such as *E. coli* **lac repressor** and yeast **gal activator**. The **trp** and **met repressors** from *E. coli* have very different structures; both are intertwined monomers which bind two molecules of their co-repressor, which are L-tryptophan and *S*-adenosylmethionine respectively. Repressors inhibit DNA transcription by preventing the RNA polymerase from forming an open complex with the promoter, whereas activators facilitate such complex formation.

One structural feature shared by many of these proteins is in their DNA-binding region. In several cases a pair of α-helices stack to form a V-shaped structure where the angle between the two α-helices is the same for all of the proteins. The first α-helix determines the position of the second α-helix so that it can project into the major groove and make contacts which allow the protein to recognize specific sequences (the **recognition helix**). This structure, with its unique interhelical angle, has become known as the 'two-helix' motif or the '**helix–turn–helix**' motif (Fig. 9.16), and is found only in DNA-binding proteins.

The very high level of similarity in tertiary structure engenders a lower level of sequence similarity. The helix-turn-helix motif is about 20 residues long and bends at position-9, which is always a glycine. Close to and on either side of the turn, at positions-5 and -15, two small hydrophobic residues in van der Waals contact help to maintain the correct interhelical angle which is close to 40°. Further from the turn, the bulky hydrophobic side-chain of residue-18 fits between the two helices and also helps maintain the correct angle. These residues comprise the major sequence similarities found in other repressor molecules. The parts of the helix which bind to the DNA have substantially different amino acid sequences because they recognize different DNA base-sequences. The *met* repressor from *E. coli*, however, while showing some peptide sequence similarity to the helix–turn–helix motif, actually binds to DNA via β-strands rather than via α-helices.

This helix–turn–helix motif is found in other DNA-binding proteins as well as in repressors. The homeodomain polypeptide from the *Antennapedia (Antp)* gene of Drosophila consists of about 60 residues which bind specifically to certain DNA sequences and protect them from DNAase I. Wüthrich has solved the solution structure for this homeodomain by NMR methods and compared it with the structures for the phage 434 repressor and the *trp* repressor from *E. coli*. It emerges that residues 30–50 form a helix–turn–helix motif virtually identical with those oberved in various prokaryotic repressors, while there are also significant differences in their model architectures.

Model complexes

Is it possible to predict the overall structure of a complex between DNA and repressor or activator in the absence of experimental data? Since all model complexes have to meet three major criteria, one can get a sense of the overall topology of the complex. First, the protein and DNA must dock so as to avoid bad contacts. Secondly, the electrostatic field of the protein should complement the electrostatic field of the DNA. Finally, the repressors and activators studied so far are all approximately symmetrical dimers of two identical polypeptide chains, which bind to approximately palindromic DNA operator sequences. If one assumes that the approximate two-fold axes of protein and DNA must be coincident, then the DNA-binding helix-turn-helix motif on one protein monomer fits into one major groove and the symmetry-related DNA-binding helix on the second protein monomer fits into the next major groove (Fig. 9.16). Using the additional constraints provided by DNase I footprinting studies and studies of where the protein protects the DNA from chemical modification, one can build protein-DNA models of remarkable subtlety. (However, much care has to be exercised in predictions of protein tertiary structures in cases where one cannot ascertain independently that the global molecular architecture is homologous to that of the protein selected as the basis for structure prediction.)

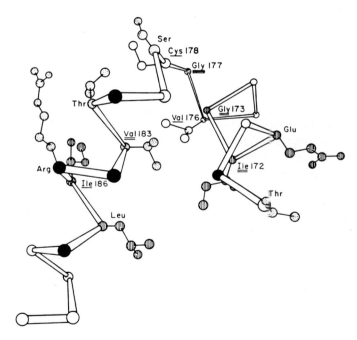

Fig. 9.16 Diagram of the helix-turn-helix motif. The α-carbon backbone is shown along with residues which are homologous in CAP, *lac* repressor, and *gal* repressor. The stippled side chains are identical in these three proteins. Side-chains similar in *lac* repressor and CAP are striped. Some of these seem to be involved in the helix–helix packing. Gly-177 is invariant in all helix-turn-helix proteins and the solid α-carbons represent residues which appear to interact with the DNA in the model-built CAP–DNA complex (from Weber, I. T., McKay, D. B., and Steitz, T. A. (1982). *Nucleic Acids Research*, **10**, 5085–101).

For instance, these models suggest that CAP, cro, and cI proteins interact with DNA in similar ways. CAP and cI direct the *N*-terminal end of an α-helix into the major groove so that the partial positive charge on the *N*-terminal end of an α-helix, due to the helix dipole, helps to balance the negative charge on the DNA. These model complexes make testable predictions about contacts between specific protein side-chains and specific DNA bases. For example, Ptashne and his colleagues predicted how to change the sequence specificity of 434 repressor (a close relative of lambda cI repressor) by altering the amino acid sequence of the DNA-binding helix. When they made the appropriate changes in the 434 repressor by site-directed mutagenesis (Section 10.3), they found the expected change in DNA sequence specificity.

Real complexes

Model complexes are not the complete answer, for both protein and DNA structure may change when a complex forms. Much new information has come from the crystal structures of appropriate operator DNA complexed with 434 repressor (Harrison and co-workers), with λ-repressor (Pabo and co-workers), with *trp* repressor (Sigler and co-workers), and with *met* repressor (Phillips and colleagues) proteins. In most cases, the observed mode of binding is similar to, but far more complicated than, that predicted by simple model-building using ideal B-DNA. In particular, the operator DNAs bend around the repressors resulting in the DNAs being overwound at the centre (the twist per base-pair is *more* than 36°) and underwound at the ends (Fig. 9.17b). This makes the major groove wider in the centre than it would otherwise be, presumably enabling more protein–DNA contacts to occur.

The observed binding of 434 repressor and λ-repressor to DNA is very similar to that predicted for the binding of cI repressor to DNA. For example, each monomer of 434 repressor is in contact with four phosphate groups (Fig. 9.17a), but there is an unexpectedly large number of main-chain amide groups that interact with the DNA backbone. This presumably creates a more precise orientation of the protein on the DNA than could be achieved by using side-chains alone, thus allowing more accurate positioning of the side-chains which interact with bases in the major groove. The overwinding of the DNA at the centre causes compression of the minor groove and there is a concomitant expansion of the minor groove at the ends compared to ideal B-DNA, which allows arginine-43 to form hydrogen-bonds to phosphate groups on both DNA strands.

Three glutamine residues on helix-3 interact with bases in the major groove (Fig. 9.18). As expected, most of the interactions which produce specificity are hydrogen-bonds, but one hydrophobic interaction appears to be important. The C_β and C_γ of glutamine-29 are in van der Waals contact with the 5-methyl group of the thymine in base-pair-3 in the operator site. If either the glutamine or the base-pair is changed, the repressor does not bind.

The details of the interaction of *trp* repressor with DNA appear to be somewhat different. It has been suggested that part of the specific binding of *trp* repressor to its operator DNA is due to the extreme flexibility of the particular DNA sequence. Some of the hydrogen-bond interactions between *trp* repressor and the DNA are mediated by water molecules. Finally, it appears that the interactions of the protein

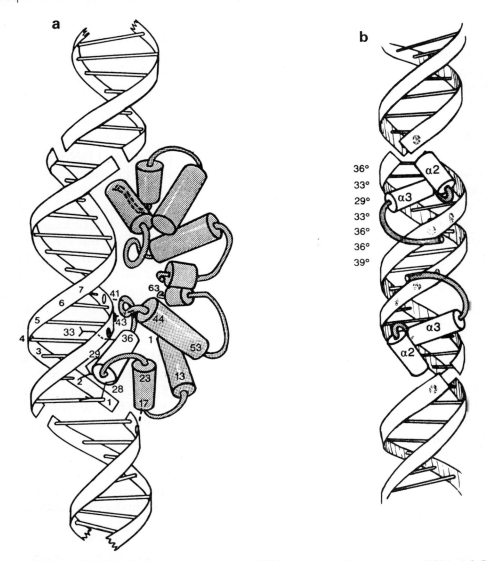

Fig. 9.17 Diagram showing the interaction of 434-repressor with its operator DNA. (a) Overall view of the complex. α-Helices are shown as cylinders. In the lower monomer, key residues in the DNA-repressor interface are shown. (b) Schematic representation of the interaction of the helix-turn-helix motif with DNA. The numbers shown are the twist in degrees between base-pairs, giving an indication of the local distortion of the DNA double-helix from the ideal value of 34.3°. The parts of the DNA polyphosphate backbone that are in contact with the protein are shown (from Anderson, J., Ptashne, M., and Harrison, S. C. (1987). *Nature*, **326**, 846–52. Copyright (1987) Macmillan Magazines Limited).

with the phosphate backbone of the DNA may not only stabilize the complex, but may also provide some of the specificity. This is in direct contrast to the models pro-

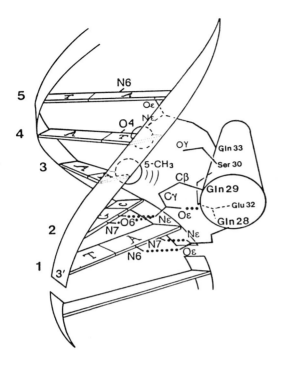

Fig. 9.18 Schematic representation of the interaction between α-helix-3 and an operator half-site. The α-helix is represented as a cylinder. The glutamine side-chains which are in contact with the DNA are shown and the hydrogen-bonds they make to the DNA are identified by dotted lines (from Anderson, J., Ptashne, M., and Harrison, S. C. (1987). *Nature*, **326**, 846–52. Copyright (1987) Macmillan Magazines Limited).

posed for 434 repressor and λ-repressor, where the contacts between the protein and phosphate backbone of the DNA are thought only to stablize the complex.

Phillips' crystal structure of the *met* repressor–operator complex shows two *met*-repressor dimers bound to a synthetic 19-mer oligonucleotide duplex containing two *met* boxes. The dimers do not have the helix-turn-helix motif characteristic of other bacterial repressor and activator structures. Rather, the dimers bind with two β-strands inserted into the major groove of the DNA. These dimers form a left-handed superhelix around the DNA, with protein–protein interactions between them forming the antiparallel pairing of β-strands which cover a buried DNA surface area of some 200 Å (see front cover).

While this discussion has focused attention on the results obtained from crystal structures, the role of NMR for the determination of solution structures clearly has an important future. A good example of the results which can now be achieved by high resolution NMR analysis of protein–DNA interactions in solution is Kaptein's work on complexes formed between the *lac* headpiece and various *lac* operator fragments. The headpiece binds with its recognition helix (the second helix of the helix-turn-helix motif) in the DNA major groove in an orientation opposite to that found for other repressors such as λcI, *trp*, and 434. There is support from results using genetic data on the *lac* system which is in agreement with these findings. It seems likely that NMR will also uncover other repressor interactions which, like the *met* repressor, are not based on the helix–turn–helix motif.

9.3.3 Motifs found in eukaryotic transcriptional proteins

The zinc finger and leucine zipper motifs were first identified in regulatory proteins from eukaryotic sources. Neither is as well-characterized structurally as the helix–turn–helix repressors and activators.

The **zinc finger** motif, so named because the side chains in the motif were thought to point outwards like fingers, has been seen in various eukaryotic transcriptional regulatory proteins. Zinc fingers were first observed in transcription factor TFIIIA from *Xenopus laevis* and they have since been identified in a variety of genes from species as diverse as yeast and man. The classic zinc finger is about 30 amino acids in length and contains two cysteine and two histidine residues which provide the ligands for one tetrahedrally co-ordinated zinc ion. Most zinc fingers have a spacing of cysteines and histidines that fits the consensus sequence: Cys-$(X)_{2-4}$-Cys-$(X)_{12-13}$-His-$(X)_{2-3}$-His where X corresponds to other amino acids. In some zinc fingers, one or more of the histidine residues is replaced by cysteine. Two-dimensional NMR data suggest that a single zinc finger domain folds up to form an α-helix in the second part of the structure, with a turn between the two cysteines. The data also support the existence of a hydrophobic core consisting of a phenylalanine and two leucine residues. A single zinc finger is not sufficient to bind DNA. For example, TFIIIA has about ten such domains on a single polypeptide chain.

The **leucine zipper** motif is also about 30 amino acids long and contains a leucine every seven residues. Proteins that include such motifs are the yeast GCN4 transcriptional activator as well as the *fos*, *jun*, and *myc* oncoproteins. These proteins bind as dimers where each monomer, containing a single leucine zipper motif, folds up to form a dimer of parallel α-helics in a coiled coil arrangement, like that of tropomyosin and keratin. However, it appears that the leucine zipper motif does *not* bind DNA. Instead it may well be a structural element that helps the protein dimerise and thus allows neighbouring parts of the protein to bind to DNA.

Summary

Repressors and activators are dimeric, often small proteins which bind to a specific DNA sequence and regulate transcription by their effect on RNA polymerase, either preventing or enhancing its binding to DNA. Both the repressor or activator and its specific DNA-binding site have approximate twofold symmetry. Some repressors and activators economically achieve sequence specificity by inserting an α-helix–turn–helix motif into the major groove of B-DNA, forming specific hydrogen-bond networks between the bases on the DNA and protein side-chains on the α-helix. Repressors like *met*, however, achieve recognition through the insertion of a β-pleated sheet into the major groove of the DNA. In both types of recognition, there can be distortion of the structures of DNA and protein when they bind to each other, which probably varies in extent from complex to complex.

9.3.4 Restriction endonucleases: *Eco*RI complexed with DNA

Bacteria have a system which effectively prevents any foreign DNA that might have entered the cell from being replicated. One or more restriction endonucleases recognize and bind to the DNA at specific sites and make a double-stranded cleavage. Each bacterial type has its own restriction enzyme or enzymes and each enzyme has its own specific recognition sequence. Some enzymes cleave symmetrically to leave a blunt (flush) end. Some cleave asymmetrically to leave a staggered (sticky) end. Their names consist of a three letter abbreviation for the organism followed by a Roman numeral (e.g. *Hae*III is from the bacterium *Haemophilus aegyptius*).

The bacterium protects its own DNA from being cut by specifically modifying the recognition site using its own methylase. There are three types of restriction and modification systems termed I, II, and III in order of their discovery. Type I consists of a large enzyme complex containing subunits encoding endonuclease, methylase, and several other activities. The recognition sequence comprises a trinucleotide and a tetranucleotide separated by about six non-specific base-pairs (Table 9.1), but the endonucleolytic cleavage site can be up to 7000 base-pairs distant. Type II systems have independent endonucleases and methylases that act on the same DNA sequence. These sequences are generally palindromic (i.e. they have a two-fold axis of symmetry) and the cleavage sites are usually within or very close to the recognition sites. A range of enzymes with different specificities has been isolated from a wide variety of organisms and type II restriction enzymes are now highly useful tools in recombinant DNA research (Section 10.2). The type III system shares features in common with both type I and type II. There are two independent polypeptides, one of which acts independently as a methylase, but both are required for specific endonucleolytic activity. In the case of *Eco*RI, for example, the recognition sequence is an asymmetric pentanucleotide and the cleavage site is 25 base-pairs downstream.

*Eco*RI is a restriction enzyme of type II from *E. coli*. It is a dimeric molecule with two identical subunits, each of molecular weight about 30 kDa. It binds to the self-complementary hexanucleotide duplex of sequence d(GAATTC) and hydrolyses the 3′-phosphodiester bond between guanine and adenine. The protein can also bind non-specifically to DNA: this may help it to locate its specific sequence by facilitated one-dimensional diffusion along the DNA. Magnesium is required for the cleavage of DNA and the correct conformation of the DNA:protein complex appears not to be formed until magnesium becomes bound. Even under optimal conditions, *Eco*RI cuts each strand of the DNA in a separate reaction so the hydrolysis of double-stranded DNA requires two consecutive reactions at the same site.

The *Eco*RI modification methylase uses *S*-adenosylmethionine to methylate the second dA residue in each strand of the recognition sequence at its N-6 position. Following such action, the endonuclease can bind to but does not cut the

Table 9.1. Some restriction endonucleases and their recognition sequences

Type	Enzyme	Recognition site
Type I	EcoK	A A C (N)$_6$ G T G C T T G (N)$_6$ C A C G
	EcoB	T G A (N)$_8$ T G C T A C T (N)$_8$ A C G A
Type II	EcoRI	G▾A A T T C C T T A A▴G
	SmaI	C C C▾G G G G G G▴C C C
	PstI	C T G C A▾G G▴A C G T C
	Sau3AI	▾G A T C C T A G▴
	NotI	G C▾G G C C G C C G C C G G▴C G
	MnlI	C C T C (N)$_7$▾ G G A G (N)$_7$▴
Type III	EcoPI	A G A C C T C T G G

Cleavage sites indicated ▾ ▴

recognition sequence. It appears that in the unlikely event of the endonuclease making a single-strand nick at a non-cognate site, for instance in the host chromosome, the nicked DNA would dissociate from the enzyme before cleavage of the second strand. *In vivo* repair of the nick by ligase action would then act as a proofreading mechanism to enhance the selectivity of restriction cleavage to the observed value of about 10^8.

The structure of the protein

A complex of *Eco*RI with a 13-mer oligodeoxyribonucleotide containing the specific RI binding site is stable in the absence of magnesium, and it is this complex which has been crystallized and its structure solved at 3 Å resolution. *Eco*RI can be categorized as a parallel-stranded alpha-beta protein. In such proteins a hydrophobic parallel-stranded β-sheet is sandwiched between α-helices, which form the surface of the protein. *Eco*RI has a five-stranded β-sheet (Fig. 9.19), where only strand-2 is antiparallel to the rest. Like the Klenow fragment, *Eco*RI also has an arm-like structure, composed primarily of a three-stranded antiparallel β-sheet which wraps around the DNA opposite the bond which is to be cleaved. This sheet has a left-handed supertwist which is complementary to the right-handed B-DNA.

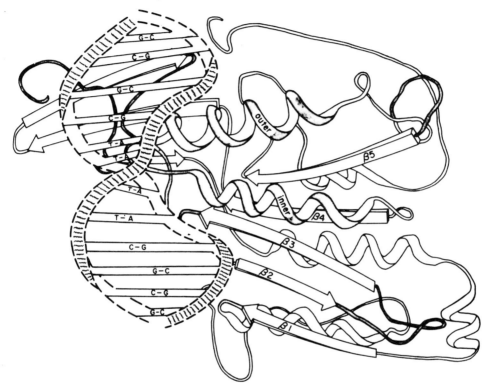

Fig. 9.19 Schematic view of the interaction of one subunit of *Eco*RI with its DNA substrate. β-Strands are shown as arrows, α-helices as coils, and the DNA backbone as ribbons. The inner and outer recognition helices (front) connect the third, fourth and fifth β-strands. The local kinking of the DNA in the centre of the recognition sequence can be seen (from McClarin, J. A., Frederick, C. A., Wang, B. C., Green, P., Boyer, H. W., Grable, J., and Rosenberg, J. M. (1986). *Science*, **234**, 1526–34. Copyright (1986) AAAS).

The structure of the DNA

The structure is known of a 12-mer oligodeoxyribonucleotide which differs by only one base-pair from the 13-mer used in the above crystals. So it is possible to examine how the DNA in the 13-mer is distorted by the protein in the complex. The DNA has become locally twisted in three places in the specific hexanucleotide. One such '**neokink**' is at the centre, and the other two are at symmetrical positions on either side (neokinks differ from ordinary DNA kinks in that they are induced by the protein). The central neokink widens the major groove by 3.5 Å to accommodate recognition elements from the protein. The two neokinks on the outside seem to distort the phosphodiester bond which is cleaved by *Eco*RI. The strain thus imposed may be important in increasing the efficiency of the enzyme.

The structure of the complex

The DNA is implanted into the protein so that only the major groove is buried. As expected, the two-fold axis which relates the two protein monomers of the *Eco*RI dimer also relates the symmetrical halves of the specific hexanucleotide sequence. Consequently, the binding of *Eco*RI to its specific sequence is symmetrical. Thus four α-helices, two from each polypeptide chain, cause the specific binding of *Eco*RI to DNA. The protein only recognizes the **purine** nucleotides in the specific sequence. The central adenines form bridged hydrogen-bonds to an arginine and a glutamate from the two 'inner' helices (Fig. 9.20). On each of the two 'outer' helices, there is an arginine which forms a bridged hydrogen-bond to the guanine.

In the five-stranded β-sheet, strands one, two, and three are all antiparallel, and carry the residues that may be involved in catalysis. Although the general location

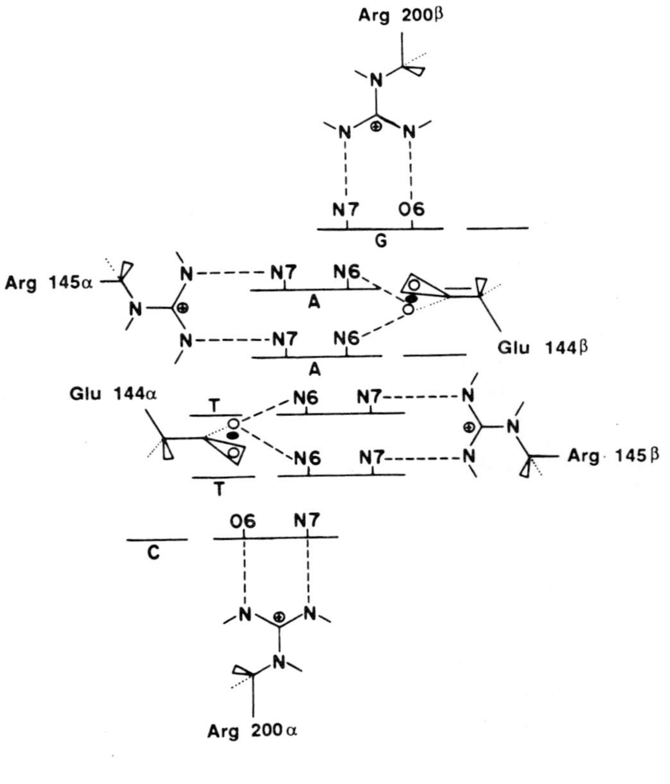

Fig. 9.20 The hydrogen-bond interactions that determine the specificity of *Eco*RI. All of the interactions shown involve bridging hydrogen-bonds, either to a pair of successive bases, like Arg-145, or to a single base, like Arg-200 (from McClarin, J. A., Frederick, C. A., Wang, B. C., Green, P., Boyer, H. W., Grable, J., and Rosenberg, J. M. (1986). *Science*, **234**, 1526–34. Copyright (1986) AAAS).

of the catalytic site is known, the particular residues responsible for catalysis and the mechanism of catalysis have still to be determined. When magnesium is added to the crystals, the enzyme becomes catalytically active and the 13-mer oligo-deoxyribonucleotide is specifically cleaved to give the enzyme-product complex. We shall soon understand precisely the mechanism of an enzyme that interacts with DNA.

Summary

*Eco*RI displays two methods of protein interaction with DNA. A β-strand with a left-handed supertwist carries the catalytic residues and fits into the minor groove of the right-handed B-DNA duplex. Four α-helices are located in adjacent major grooves in the centre of the specific hexanucleotide where they effect recognition of the specific sequence. To accommodate these four α-helices, the centre of the specific hexanuc-leotide is locally twisted as a result of binding of the protein to form a neokink.

9.3.5 The use of base modifications

Most models of sequence-specific recognition propose that a protein distinguishes specific base-pairs through the formation of multidentate hydrogen-bonds or by van der Waals contacts between amino acid side chains of the protein and the nucleo-bases of the DNA. This is because a protein can only recognize the bases edge-on via the surfaces of the major and minor grooves of the double-helix. Such processes are illustrated for direct readout of DNA sequences (Fig. 9.21).

Oligodeoxyribonucleotides can often substitute for the longer DNA polymers in studies of recognition phenomena. Consequently, the substitution of normal bases by changed or structurally-modified bases using chemically-synthesized oligomers has provided a useful means for investigating recognition. It formally permits the systematic modification of functional group patterns within a given sequence of bases. This approach has now been employed in work on repressor proteins, restriction endonucleases and modification enzymes, and on the interaction between RNA polymerase and its promoter site.

For example, substitution of 2'-deoxyuridine for 2'-deoxythymidine brings about deletion of the thymine methyl group but conserves the number of hydrogen-bonds in the base-pair. By contrast, the changes dA→dPu and dG→dI cause the loss of one hydrogen-bond while the change of an A:T base-pair for G:C^{Me} inserts an additional hydrogen-bond into the minor groove.

Caruthers has used such changes to identify some of the functional groups required for complexation of *E. coli* RNA polymerase and the λP$_R$ promoter using a variety of synthetic, 75-base-pair oligomers. Two adjacent thymines were identified in the −35 region whose methyl groups were crucial for binding. A strong require-ment for preservation of the dT:dA base-pair at position-7 was also established.

a

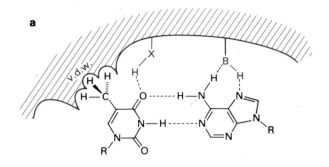

Fig. 9.21 (a) Protein (red) complementary recognition of a cognate T:A base-pair (black) by van der Waals and hydrogen-bond interactions. (b) Adverse interaction between the same protein (red) and an analogue base-substitution for a 4-thioU:N^6-MeA base-pair (black).

b

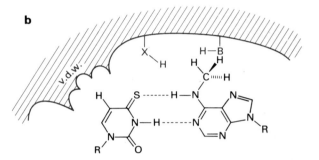

The recognition sequence dGAATTC for the *Eco*RI restriction endonuclease and methylase has been explored by Gumport through the use of 2-aminopurine (2AP), 2,6-diaminopurine, 5-bromocytosine (5-BrC), 5-bromouracil (BrU), hypoxanthine, N^6-methyladenine (N^6MeA), 5-methylcytosine (5-MeC), and uracil as single-substitutions for one of the normal bases in the octamer d(pGGAATTCC). All of the 11 base-changes studied show some effect on enzymatic activity and the results vary from complete inactivation of the endonuclease to enhanced activity for the methylase. Thus, for the endonuclease, the substitution of N^6MeA for A-3 or A-4, of U for T-6, or of 5-BrC or 5-MeC for C-7 results in complete loss of activity. Similarly, the restriction methylase activity is totally lost following substitution of N^6MeA for A-2, of 2AP for A-3, or of U for T-6. By contrast, the value of k_{cat}/K_m increases sevenfold for the substitution of BrU for C-6. It appears that:

(a) there is tight complementarity between the enzymes and the DNA;

(b) the endonuclease and methylase do not recognize an identical sequence in the same way.

The recognition site d(GATATC), for the *Eco*RV restriction endonuclease has been the subject of studies of the same sort in a number of laboratories. Generally, the results suggest that the enzyme interacts with the bases via contacts in the major groove of the B-DNA involving hydrogen-bonding to the adenine amino groups and van der Waals interactions with the thymine methyl groups. In a very thorough exploration, Connolly has determined apparent binding energies for a group of ten base-changes in the ATAT core (Table 9.2). Several other base-changes produced dodecamers that were not cleaved by the endonuclease and thus indicate direct readout of $A-5(N^7)$, $T-6(O^4)$, and $T-8(O^4)$. It appears that:

(i)   up to 20 sequence-specific site contacts exist between the enzyme and the central ATAT core of the recognition site;

(ii) they are located in both the major and minor grooves; and

(iii) they enhance the primary event of discrimination between the cognate (correct) base-sequence and non-cognate (incorrect) base-sequences for cleavage by *Eco*RV endonuclease.

Table 9.2. Steady-state parameters and apparent binding energies for the deletion of *Eco*RV interaction sites within the central ATAT residues of d(pGACGATATCGTC) (with permission from B. A. Connolly, results in press 1990)

Contact →modification	k_{cat} (min^{-1})	K_m (μM)	k_{cat}/K_m (s^{-1}M^{-1})	$\triangle G_{app}$ (kJ mol^{-1})	*Eco*RV/DNA interaction
Wild-type	6.9	3.8	30 200	0	normal
A-5: 6-NH$_2$→H	nd^a	nd	nd	nd	hydrogen-bond
A-5: 3-N→CH	0.057	4.0	238	−6.0	hydrogen-bond
T-6: 5-Me→H	0.013	3.8	56	−7.8	van der Waals
T-6: 4C=O→C=S	0.016	0.5	533	−5.0	hydrogen-bond
T-6: 2C=O→C=S	1.88	1.0	32 900	+0.1	none apparent
A-7: 6-NH$_2$→H	0.067	8.0	139	−6.7	hydrogen-bond
A-7: 7-N→CH	1.9	0.8	39 600	+0.3	none apparent
A-7: 3-N→CH	0.1	1.4	1 190	−4.8	hydrogen-bond
T-8: 5-Me→H	0.006	6.5	15	−9.4	van der Waals
T-8: 2C=O→C=S	0.1	4.0	417	−5.3	hydrogen-bond

$\triangle G_{app}$ is given for the loss of a single interaction in the duplex.
^{a} Too low for accurate determination.

It is possible that the picture deduced from such results is oversimplified. Work using base-changes in plasmids has shown that a single change, such as A:T → T:A, can result in a loss of 10^6 to 10^8 in the efficiency of specific cleavage. Moreover, the crystal structure of the *Eco*RV endonuclease (Fig. 9.19) shows contacts only in the major groove although base-modification studies suggest that changes in the minor groove can have just as much effect as ones in the major groove. Whilst it is evident that the use of base-changes can be a valuable probe for the study of protein–DNA interactions, the results must be interpreted with some caution. The future application of base-substitutions in DNA–protein recognition studies may well have to take account of changes in roll, tilt, and slide of the base-pairs which are capable of having indirect effects on the readout of DNA structure.

Summary

Multiple ionic interactions between a protein and the DNA phosphate backbone contribute strongly to non-specific, total, binding affinity whereas readout of bases allows protein discrimination between correct and incorrect sequences for binding. With **direct readout**, the protein recognizes, or interacts with, the surface of a unique functional group pattern that is characteristic of a given sequence. This seems to be the case especially for *Eco*RI endonculease and the repressor proteins.

With **indirect readout**, the functional group pattern of the recognition sequence contributes to the overall geometry of the site and is presumed to be sterically important for precise association between protein and nucleic acid. Such sequences are recognized primarily through changes in the geometry of the phosphate backbone. It follows that these changes must be typical of the specific sequence of nucleobases. Sigler's *Trp* repressor–operator complex appears to provide a prime example of such behaviour.

The use of **base-modifications** can provide a useful, if indirect, means for identification of the direct-readout of base-sequences and thereby pinpoint loci on the bases for hydrogen-bonding and van der Waals interactions between a protein and its cognate DNA recognition sequence.

9.4 Other nucleic acid-interacting proteins

While a very large number of proteins interact with nucleic acids, apart from those described in Sections 9.2 and 9.3, almost nothing is known about the structures of most other nucleic acid-interacting proteins. Until recently this was due to the lack of availability of multi-milligram quantities of such proteins for structural analysis. With the advent of gene cloning procedures the major problem is now one of growing crystals suitable for X-ray structural analysis. Either crystals cannot be obtained at all, or the crystals do not diffract well enough to permit high resolution analysis, or crystals of the protein complexed with nucleic acid cannot be obtained. The rapid progress being made using NMR analysis of protein–nucleic acid interactions in solution will, however, diminish the severity of this constraint.

We can now look at some other classes of enzymes that interact with nucleic acids and describe what we know about their structure. More detailed accounts of the enzymology and reaction mechanisms of these proteins are given in standard biochemical texts.

9.4.1 Introduction and removal of terminal phosphatase

A polynucleotide kinase isolated from bacteriophage T4 catalyses the transfer of the γ-phosphate of ATP to the 5′-hydroxyl terminus of DNA, RNA, or an oligonucleotide in a reaction that requires magnesium ions. The enzyme is particularly useful for introducing a radioactive label on to the end of a polynucleotide, where the phosphate donor is γ-[^{32}P]-ATP. Both single- and double-stranded polynucleotides can be phosphorylated although recessed 5'-hydroxyl groups in double-stranded DNA, such as those obtained by cleavage with certain restriction enzymes, are poorly phosphorylated. This sort of polynucleotide kinase activity, though not found in bacteria, has been found in some mammalian cells. The T4 enzyme is the only well characterized kinase which has polynucleotides as substrates. Its sole biological role seems to be in an RNA ligation pathway that is non-essential to infectivity of most T4 strains (Section 9.4.2). Little is known about its structure apart from its molecular weight (34 kDa) and amino acid sequence (determined recently from its gene sequence).

This T4 protein has a second, 3'-phosphatase activity, but it is not clear whether this is located on the same or a separate domain. Phosphatases catalyse the hydrolysis of phosphomonoesters to produce inorganic phosphate and the corresponding alcohol. The phosphatase activity of polynucleotide kinase is unusually specific for a 3'-phosphate of a nucleoside or polynucleotide (most phosphatases are non-specific). Alkaline phosphatases are found in bacteria, fungi, and higher animals (but not plants) and will remove terminal phosphates from polynucleotides, carbohydrates, and phospholipids. The *E. coli* enzyme is a dimer of molecular weight $\sim$ 89 kDa, requires a zinc(II) ion, and is allosterically activated by magnesium ions. During dephosphorylation of the substrate, its phosphate is transferred to a serine residue on the enzyme located in the sequence Asp.Ser.Ala. This same sequence is found in mammalian alkaline phosphatases (the calf intestinal enzyme is particularly well characterized) and it is similar to the active centre of serine proteases. Acidic phosphatases are also common, but these do not usually operate on polynucleotides as substrates.

9.4.2 Ligases

A ligase is an enzyme that catalyses the formation of a phosphodiester linkage between two polynucleotide chains. In the case of DNA ligases, a 5'-phosphate group is esterified by an adjacent 3'-hydroxyl group and there is concomitant hydrolysis of pyrophosphate in NAD$^+$ (bacterial enzymes) or ATP (phage and eukaryotic enzymes). Particularly efficient joining takes place when the phosphate and hydroxyl groups are held close together within a double-helix, typically where joining seals a 'nick' and creates a perfect duplex (Fig. 9.22). This situation occurs both in gene synthesis (Section 3.4.5) and in recombinant DNA technology in ligating identical 'sticky ends' formed by cleavage with a restriction endonuclease (Table 9.1 and Section 10.2.2). The enzyme will join blunt ends when used at high enzyme concentration.

E. coli and phage T4 DNA ligases are well characterized enzymes which have an important role in DNA replication (Section 5.4.3). In a common feature of the first

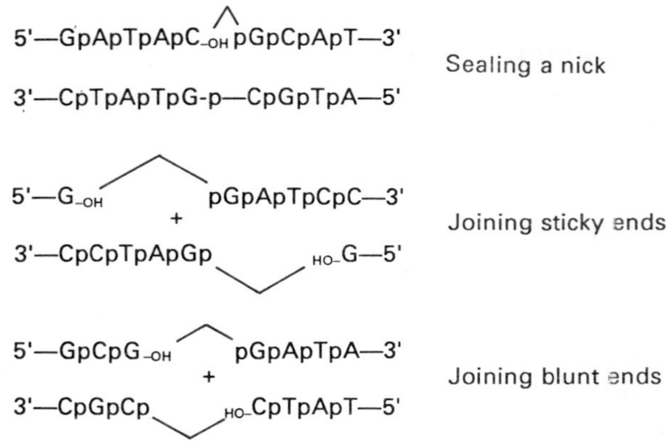

5'—GpApTpApC₋OH⌃pGpCpApT—3'

3'—CpTpApTpG-p—CpGpTpA—5'

Sealing a nick

Fig. 9.22 Joining reactions carried out by DNA ligase.

5'—G₋OH   ⌃   pGpApTpCpC—3'

+

3'—CpCpTpApGp   HO₋G—5'

Joining sticky ends

5'—GpCpG₋OH⌃pGpApTpA—3'

+

3'—CpGpCp⌄HO₋CpTpApT—5'

Joining blunt ends

step of the reaction for T4 DNA ligase and eukaryotic ligases, AMP becomes covalently attached to a lysine residue on the enzyme as a phosphoramidate (adenylylation) and pyrophosphate is released. A comparison of amino acid sequences of the T4 and yeast enzymes shows there is substantial homology in the region of the adenylylation site.

RNA ligases are ubiquitous and, unlike DNA ligases, catalyse the joining of single-stranded polynucleotides in the absence of a complementary template strand. They occur in both prokaryotic and eukaryotic cells and are involved in many aspects of RNA processing. Most of the RNA ligases isolated so far appear to be involved in processing tRNA. There are three types of RNA ligation pathway (Fig. 9.23). All begin with one substrate containing a 5'-hydroxyl group and the other containing a 2',3'-cyclic phosphate. *E. coli* RNA ligase catalyses the direct formation of a 2'-5'-phosphodiester, whereas in HeLa cells a 3'-5' linkage is formed. In yeast and wheat germ the cyclic phosphate is first opened to give a 2'- phosphate whilst the second substrate is phosphorylated by a kinase. The two substrates are now joined by the RNA ligase to form a 3'-5'-phosphodiester-2'-phosphate. Lastly, the 2'-phosphate is removed by a phosphatase.

The phage T4 RNA ligase is the only enzyme that catalyses the joining of a 5'- phosphate to a 3'-hydroxyl group *in vitro*, and it has therefore been valuable for ligating synthetic RNA (Section 3.5.3) and for 3'-labelling of RNA. The ligation pathway (Fig. 9.23) is not required for T4 infection of most strains of *E. coli* and is only utilized in rare cases where a mandatory host lysine tRNA has been cleaved endolytically in its anticodon during T4 infection and needs to be rejoined. Here the T4 polynucleotide kinase is believed to bring about transfer of phosphate from the 2',3'-cyclic phosphate of one half of the tRNA to the 5'-hydroxyl of the other half.

One unifying feature of RNA ligases so far is that they all appear to utilize ATP and become adenylylated at a lysine residue, just as for DNA ligases. However, there seems to be little sequence homology between RNA ligases and no X-ray structural data is available for any RNA or DNA ligase.

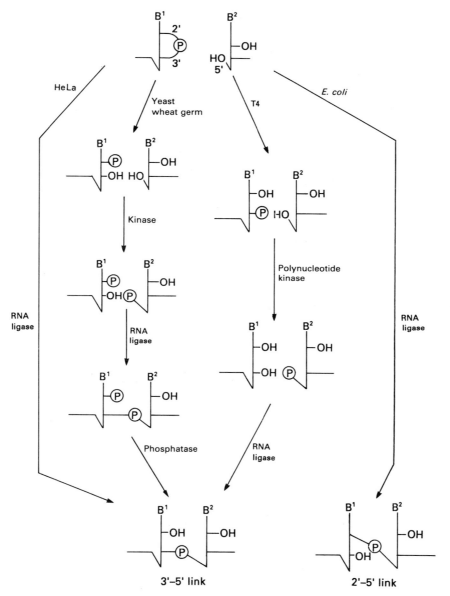

Fig. 9.23 Pathways of RNA ligation in prokaryotic and eukaryotic systems.

9.4.3 Nucleases

Almost every organism contains a wide variety of nucleases, of which some are involved in the salvage of nucleotides and some feature as intrinsic activities of proteins used in replication and repair processes. Restriction endonucleases (Section

9.3.3), non-specific endonucleases, such as DNAse I (Section 9.2.2), and ribonucleases (Section 6.3 and Section 9.2.3) have already been mentioned. There are a number of other nucleases which are used in manipulation of DNA and RNA and some of these are listed in Table 9.3.

Table 9.3. Some exonucleases and their activities

Nuclease	Origin	Activities
Exonuclease III	*E. coli*	(1) ss *exo*-cleavage from 3'-ends of dsDNA (2) *endo*-cleavage for apurinic DNA (3) RNase H (4) 3'-phosphatase
Exonuclease VII	*E. coli*	ss *exo*-cleavage from 5'- or 3'-end of ssDNA
Bal 31	*Alteromonas espejiana*	(1) ss *exo*- and *endo*-cleavage from 5'- or 3'-end of dsDNA (2) ssDNA *endo*-cleavage
S1	*Aspergillus oryzae*	ssDNA or RNA *exo*- and *endo*-cleavage
Lambda exonuclease	Infected *E. coli*	SS *exo*-cleavage from 5'-end of dsDNA
Phosphodiesterase I	Bovine spleen	ss *exo*-cleavage from 5'-end of ssDNA or RNA
Phosphodiesterase II	*Crotalus adamanteus* (or other snakes)	ss *exo*-cleavage from 3'-end of ssDNA or RNA

Summary

T4 polynucleotide kinase catalyses the transfer of the γ-phosphate of ATP to the 5'-end of DNA, RNA, or an oligonucleotide. Kinases that have polynucleotides as substrates exist in eukaryotes, but not in bacteria. Phosphates are generally non-specific and will remove terminal phosphates from nucleic acids, carbohydrates, and phospholipids.

DNA ligases seal nicks in double-stranded DNA. T4 DNA ligase will join complementary sticky ends and at high concentration will also join blunt ends of DNA. RNA ligases join single-stranded polyribonucleotides in the absence of a complementary strand.

Several exo- and endonucleases are useful in manipulation of DNA and RNA.

Further reading

9.1

Brennan, R. G. and Matthews, B. W. (1989). Structural basis of DNA–protein recognition. *Trends Biochem. Sci.*, **14**, 286–90.

Gabler, R. (1978). *Electrical interactions in molecular biophysics.* Academic Press, New York.

Hélène, C. and Lancelot, G. (1982). Interactions between functional groups in protein–nucleic acid associations. *Progr. Biophys. Molec. Biol.*, **39**, 1–68.

Honig, B., Hubbell, W., and Flewelling, R. (1986). Electrostatic interactions in membranes and proteins. *Ann. Rev. Biophys. Biophys. Chem.*, **15**, 163–93.

Ollis, D. L. and White, S. W. (1987). Structural basis of protein–nucleic acid interactions. *Chem. Revs.*, **87**, 981–96.

Saenger, W. and Heinemann, U. (ed.) (1989). Protein-nucleic acid interaction. Macmillan Press, London.

Tanford, C. (1973). *The hydrophobic effect—formation of micelles and biological membranes.* John Wiley and Sons, New York.

Travers, A. A. (1989). DNA conformation and protein binding. *Ann. Rev. Biochem.*, **58**, 427–52.

Tullius, T. D. (1989). Physical studies of protein-DNA complexes by footprinting. *Ann. Rev. Biophys. Biophys. Chem.*, **18**, 213–37.

9.2

Allen, D. J., Darke, P. L., and Benkovic, S. J. (1989). Fluorescent oligonucleotides and deoxynucleotide triphosphates: preparation and their interaction with the large (Klenow) fragment of *Escherichia coli* DNA Polymerase I. *Biochemistry*, **28**, 4601–7.

Brayer, G. D and McPherson, A. (1984). Mechanism of DNA binding to the gene 5 protein of bacteriophage fd. *Biochemistry*, **23**, 340–9.

Draper, D. E. (1989). How do proteins recognise specific RNA sites? New clues from autogenously regulated ribosomal proteins. *Trends Biochem. Sci.*, **14**, 335–8.

Joyce, C. M. and Steitz, T. A. (1987). DNA polymerase I: from crystal structure to function *via* genetics. *Trends Biochem. Sci.*, **12**, 288–92.

Kornberg, R. and Klug, A. (1981). The nucleosome. *Sci. Amer.*, **244**, 52–64.

Lohmann, T. M., Bujalowski, W., and Overman, C. B. (1988). *E. coli* single-stranded binding protein: a new look at helix-destabilising proteins. *Trends Biochem. Sci.*, **13**, 250–5.

McHenry, C. S. (1988). DNA polymerase III holoenzyme of *E. coli*. *Ann. Rev. Biochem.*, **57**, 519–50.

McPherson, A., Brayer, G. D., Cascio, D., and Williams, R. (1986). The mechanism of binding of a polynucleotide chain to pancreatic ribonucleases. *Science*, **232**, 765–8.

Oefner, C. and Suck, D. (1986). Crystallographic refinement and structure of DNAse I at 2 Å resolution. *J. Mol. Biol.*, **192**, 605–32.

Ollis, D. L., Brick, P., Hamlin, R., Xuong, N., and Steitz, T. A. (1985). Structure of large fragment of *Escherichia coli* DNA polymerase I complexed with dTMP. *Nature*, **313**, 762–6.

Schmid, M. B. (1988). Structure and function of the bacterial chromosome. *Trends Biochem. Sci.*, **13**, 131–5.

Tanaka, I., Appelt, K., Dijk, J., White, S. W., and Wilson, K. S. (1984). 3 Å Resolution structure of a protein with histone-like properties in prokaryotes. *Nature*, **310**, 376–81.

9.3

Anderson, J., Ptashne, M., and Harrison, S. C. (1987). The bacteriophage 434 repressor–operator complex at 3.2 Å resolution. *Nature*, **326**, 846–52.

Boelens, R., Scheek, R. M., Van Boom, J. H., and Kaptein, R. (1987). Complex of *lac* repressor headpiece with a 14 base-pair *lac* operator fragment studied by two-dimensional nuclear magnetic resonance. *J. Mol. Biol.*, **193**, 213–16.

Brennan, C. A., Van Cleve, M. D., and Gumport, R. I. (1986). The effects of base analogue substitutions on the cleavage by the *Eco*RI restriction endonuclease of octadeoxyribonucleotides containing modified *Eco*RI recognition sequences. *J. Biol. Chem.*, **261**, 7270–8; 7279–86.

Brennan, R. G. and Matthews, B. W. (1989). The helix-turn-helix DNA binding motif. *J. Biol. Chem.*, **264**, 1903–6.

Connolly, B. A. and Newman, P. C. (1989). Synthesis and properties of oligonucleotides containing 4-thiothymidine, 5-methyl-2-pyrimidinone-1-β-D-(2'-deoxyriboside), and 2-thiothymidine. *Nucleic Acids Res.*, **17**, 4969–74.

Dubendorff, J. W., deHaseth, P. L., Rosendahl, M. S., and Caruthers, M. H. (1987). DNA functional groups required for formation of open complexes between *E. coli* RNA polymerase and the λP$_R$ promoter. *J. Biol. Chem.*, **262**, 892–8.

Knight, K. L., Bowie, J. U., Vershon, A. K., Kelley, R. O., and Saver, R. T. (1989). A new class of sequence-specific DNA-binding protein. *J. Biol. Chem.*, **264**, 3639–42.

Mazzarelli, J., Scholtissek, S., and McLaughlin, L. W. (1989). Effects of functional group changes in the *Eco*RV recognition site on the cleavage reaction catalysed by the endonuclease. *Biochemistry*, **28**, 4616–22.

McClarin, J. A., Frederick, C. A., Wand, B. C,. Green, P., Boyer, H. W., Grable, J., and Rosenberg, J. M. (1986). Structure of the DNA–*Eco*R1 endonuclease recognition complex at 3.0 Å resolution. *Science*, **234**, 1526–34.

Otwinski, Z., Schevitz, R. W., Zhang, R.-G., Lawson, C. L., Joachimiak, A., Marmorstein, R. Q., Luisi, B. F., and Sigler, P. B. (1988). Crystal structure of *trp* repressor/operator complex at atomic resolution. *Nature*, **335**, 321–9.

Qian, Y. Q., Billeter, M., Otting, G., Muller, M., Gehring, W.J., and Wuthrich, K. (1989). The structure of the *Antennapedia* Homeodomain determined by NMR spectroscopy in solution; comparison with prokaryotic repressors. *Cell*, **59**, 573–80.

Rafferty, J. B., Somers, W. S., Saint-Girons, I., and Phillips, S. E. V. (1989). Three-dimensional crystal structures of *Escherichia coli* met repressor with and without corepressor. *Nature*, **341**, 705–10.

Roberts, R. J. (1989). Restriction enzymes and their isoschizomers. *Nucleic Acids Res.*, **17**, r347–73.

Struhl, K. (1989). Helix-turn-helix, zinc finger, and leucine zipper motifs for eukaryotic transcriptional regulatory proteins. *Trends Biochem. Sci.*, **14**, 137–40.

Thomas, G. A., Kubasek, W. L., Peticolas, W. L., Greene, P., Grable, J., and Rosenberg, J. M. (1989). Environmentally induced conformational changes in B-type DNA: comparison of the conformation of the oligonucleotides d(TCGCGAATTCGCG) in solution and in its crystalline complex with restriction nuclease *Eco*RI. *Biochemistry*, **28**, 2001–9.

9.4

Engler, M. J. and Richardson, C. C. (1981). DNA ligases. *The enzymes*, Vol. XIV. Academic Press, New York, pp. 3–29.

Gross, H. J. and Filipowicz, W. (1984). RNA ligation in eukaryotes. *Trends Biochem. Sci.*, **9**, 68–71.

Richardson, C. C. (1981). Bacteriophage T4 polynucleotide kinase. *The enzymes*, Vol. XIV. Academic Press, New York, pp. 299–314.

DNA sequence rearrangements and alterations

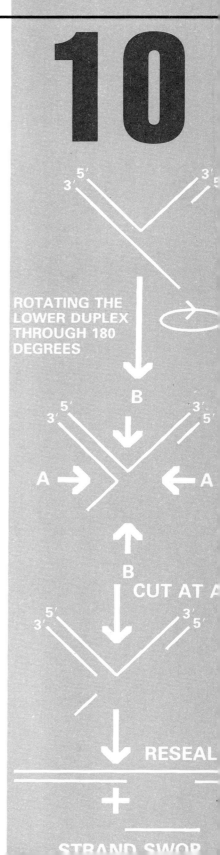

ROTATING THE LOWER DUPLEX THROUGH 180 DEGREES

B

A →

← A

B

CUT AT A

RESEAL

+

STRAND SWOP

10.1 Homologous and illegitimate recombination

One of the first discoveries in genetics was the observation that, before donation to the offspring, the different parental genes are shuffled. How do these 'beads on a string' become assorted? The answer is by **homologous recombination**. If a cell contains more than one copy of a given chromosome, then segments from one copy can recombine with corresponding segments of the other. As we shall see below, this recombination is ultimately dependent upon the DNA sequence homology between the two chromosomes (Fig. 10.1).

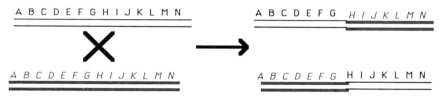

Fig. 10.1   Homologous recombination.

What advantages does homologous recombination provide? It enables the host organism to assort alleles (differing copies of the same gene) into novel groups. If a copy of a particular gene in a fruit fly has randomly acquired a mutation that yields a more efficient enzyme, then Darwinian natural selection can operate upon that gene provided it is not shackled to all the other genes on the chromosome. Thus, favourable and unfavourable alleles can be randomly shuffled and then the many combinations in the population can be tested by natural selection. Another advantage which recombination provides is the ability to repair a damaged gene in an otherwise favourable chromosome. If no ability to assort different alleles existed, then a single unfavourable mutation in a chromosome would consign the whole of it to oblivion. Lastly, homologous recombination is used in many discrete instances to regulate the expression of genes. On a more subjective note, the recombination process is often used as a tool in the laboratory to aid the researcher.

10.1.1 The mechanism of homologous recombination

Homologous recombination is linked to DNA replication, but does not occur at replication forks. Rather, it takes place between intact double helices. Genetic and DNA sequence analysis has shown that recombination is precisely accurate to one base-pair. The inference from this is that base-pairing is involved during the process. Damage to DNA stimulates recombination, strongly suggesting that homologous recombination is initiated from broken DNA strands. Presumably the cell normally creates such breaks enzymatically in order to aid this process.

10.1.2 The Holliday junction

Numerous models have been proposed for the mechanism of homologous recombination. A key intermediate in many of these is the **Holliday junction** (named after the person who first proposed it, Fig. 10.2). One of the major properties of this structure is its ability to move along the DNA helices. This movement is called **branch migration** (Fig. 10.3). A reversal of the Holliday junction formation results in the exchange of a segment of one strand of the DNA duplex with the corresponding segment of the other duplex (Fig. 10.4). If one DNA strand (colour) has a slightly different sequence from the other strand (black), the mismatch thus created will be repaired. The effect of this repair is either to 'fix' the recombination event or prevent it (Fig. 10.5).

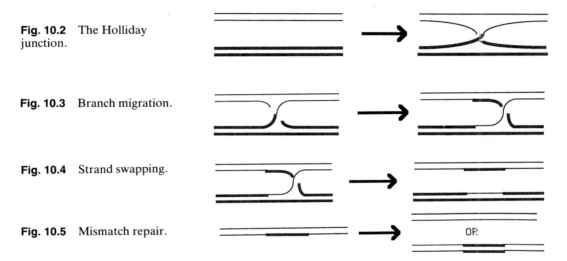

Fig. 10.2   The Holliday junction.

Fig. 10.3   Branch migration.

Fig. 10.4   Strand swapping.

Fig. 10.5   Mismatch repair.

The Holliday junction is a topologically symmetrical structure. A few simple manipulations of it result in another possible fate (Fig. 10.6). Thus rotation, cleavage at points B, and repair of nicks by ligation results in a recombination event (see Fig. 10.1). This mechanism requires enzymes which recognize and cleave Holliday junctions. Such enzymes are known, for instance bacteriophage T_4 endonuclease 7. So if a Holliday junction is formed, it is a plausible substrate for recombination.

The choice between which bonds are cleaved (Fig. 10.6) determines whether a strand swap or a recombination event occurs. This choice may be influenced by the DNA sequence at the junction, firstly because the enzymes may display sequence specificity for cleavage and secondly because the three-dimensional structure of the junction is not a simple tetrahedron but is distorted by the sequence. Choice of cleavage might be influenced by this geometry.

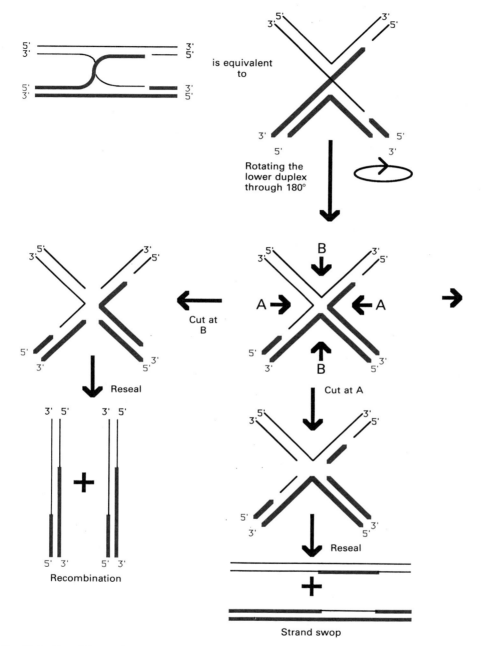

Fig. 10.6 Possible products of cleavage of the Holliday junction.

But how could such a structure be created in the first place? One possibility is as a consequence of the nicking of single strands of DNA (Fig. 10.7). We know of enzymes in *E. coli* which can catalyse this single-strand nicking and strand invasion process. The former is achieved by the RecBCD enzyme complex, a large protein complex of about 300 kDa molecular weight. RecBCD can only bind to a free DNA duplex end but, once bound, it moves along the duplex, unwinding the helix as it goes and rewinding the DNA behind it (Fig. 10.8). If RecBCD encounters a specific sequence termed a **Chi site** as it moves along the DNA, it cuts close to this site. RecBCD will continue unwinding the DNA, but the rewinding is prevented by the nick. This leaves a single-stranded region (Fig. 10.9).

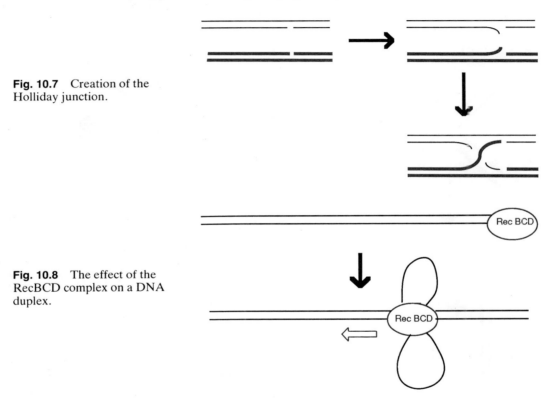

Fig. 10.7 Creation of the Holliday junction.

Fig. 10.8 The effect of the RecBCD complex on a DNA duplex.

The second phase of the process, namely **strand invasion**, is catalysed by the RecA protein. RecA binds to the single-stranded region and inserts it into a DNA duplex with which it is homologous (Fig. 10.10). Thus a combination of RecA and RecBCD proteins can catalyse the formation of a Holliday junction.

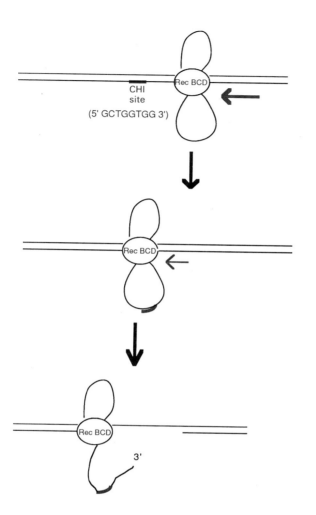

CHI
site

(5' GCTGGTGG 3')

Fig. 10.9 Nicking of DNA by RecBCD at a *chi* site.

10.1.3 The implications of recombination

We have already seen that recombination offers advantages to the organism by assorting alleles and repairing genes. What other advantages does it give? One major effect is the potential for expanding or contracting the number of copies of genes. If a chromosome contains a sequence which is repeated several times along its length, then different recombination products are possible. For example, the sequence shown below (Fig. 10.11) contains two copies of the segment CD on the upper chromosome. Recombination between either of these two segments and the single copy on the lower chromosome yields either the normal products or a duplication of the DEFGHC segment on one chromosome plus a corresponding deletion on the other.

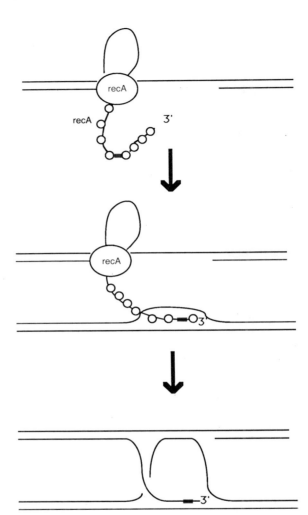

Fig. 10.10 Strand invasion catalysed by RecA protein.

The loss of such a sequence would probably be lethal in haploid organisms since they contain only one copy of each gene per cell. In diploid cells, however, the story would be different since one of the two chromosomes would retain the sequence. Each combination (addition and subtraction of DEFGHC) could then be tested by natural selection. This process, which is called **unequal crossing-over** is extremely important in the evolution of genomes and has led to the multiplication of single genes into gene families and superfamilies (see Section 5.5).

10.1.4 Gene conversion

The normal product of homologous recombination is the swapping of DNA between two duplexes (see Fig. 10.1). Sometimes, however, two copies of one

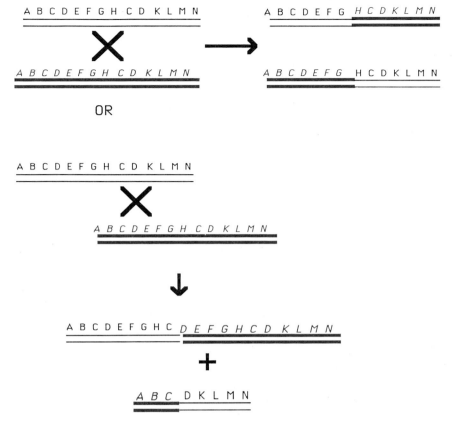

Fig. 10.11 Duplication and deletion of a sequence by recombination between repetitive DNA elements.

duplex are produced while the other is destroyed. This process is termed **gene conversion** (Fig. 10.12). It provides a mechanism for genes to edit and correct one another while remaining unchanged themselves. It is also used in some specific instances to control gene expression. We will see in the next section how the yeast *Saccharomyces cerevisiae* uses this process to control its mating type.

Fig. 10.12 Gene conversion.

10.1.5 Examples of recombination controlling gene expression

Recombination is used in many ways to control gene expression. Several very well characterized examples are described below.

Bacteriophage λ lysogeny

When bacteriophage λ virus infects *E. coli*, two outcomes are possible. Either a lytic growth of the virus results in the destruction of the host cell and release of many virus particles or the bacteriophage remains dormant in the cell (lysogeny). In the latter case, the viral DNA becomes integrated into the *E. coli* chromosome by a specific recombination event. The recombination sites on λ and *E. coli* are unique and the process is catalysed by a λ-encoded enzyme called an **integrase** (Fig. 10.13). The exact sequences (which are called *att* sites) are:

5′CTGGTTCA<u>GCTTTTTTATACTAA</u>GTTGGCAT3′   λ

5′TGAAGCCT<u>GCTTTTTTATACTAA</u>CTTGAGCG3′   *E. coli*

This process can be made to occur *in vitro* using only the two DNAs, Mg^{2+}, integrase, and an *E. coli* protein called IHF (integration host factor). In this way it has been shown that the integration event, which is virtually a one-step process, involves breakage, and re-ligation of eight phosphodiester bonds in a synchronous manner. Thus there is no net enthalpy loss. All that happens is that eight phosphodiester bonds are replaced by eight new bonds. In this way it resembles the type II

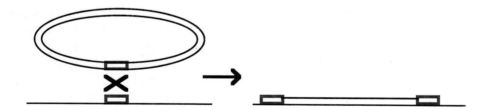

Fig. 10.13 Bacteriophage λ lysogeny.

topoisomerase-catalysed reactions and even self-splicing RNA introns (see Section 6.5). Lastly, this process can be reversed with the aid of another λ-encoded protein called an **excisionase** in addition to integrase.

Control of gene expression in *Salmonella* by inversion of a DNA segment

Salmonella typhimurium is closely related to *E. coli*. It controls the expression of its two flagellar genes (which are involved in motility) by a recombination event. Recombination between the two inverted repeat segments (Fig. 10.14, colour) causes inversion of the region between them. Note that recombination between direct repeats leads to excision of the intervening sequence. This is exactly what happens when the λ prophage DNA excises from the *E. coli* genome (an exact reversal of Fig. 10.13).

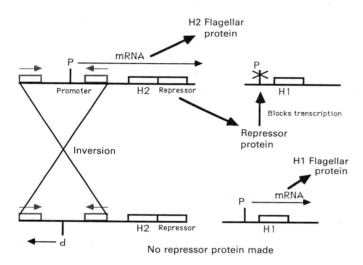

Fig. 10.14   Control of gene expression in *Salmonella* by inversion of a DNA segment.

Yeast mating type

The yeast *Saccharomyces cerevisiae* is a eukaryote with a primitive sex life. It can either exist as single haploid cells (containing one copy of each chromosome) or as a diploid cell. The switch controlling this interconversion resides at the *MAT* locus which can contain one of two 'cassettes' termed **a** and α (Fig. 10.15). These are short DNA stretches of about 700 bases. If an **a** cassette is present at *MAT*, the cell is of **a** mating type and can mate only with an α cell. If an α cassette is present the converse is true. The donor **a** and α cassettes lie at distant locations on the same chromosome as *MAT* (Fig. 10.15). These distant loci, termed *HML***a** and *HMR*α, are unaffected by the switching process. There is therefore always a silent copy of **a** and α in the yeast cell in addition to the expressed copy at *MAT*. If we look at the switch we find this process to be a specialized form of gene conversion since one copy at *HML* or *HMR* alters the sequence at *MAT* without being affected itself.

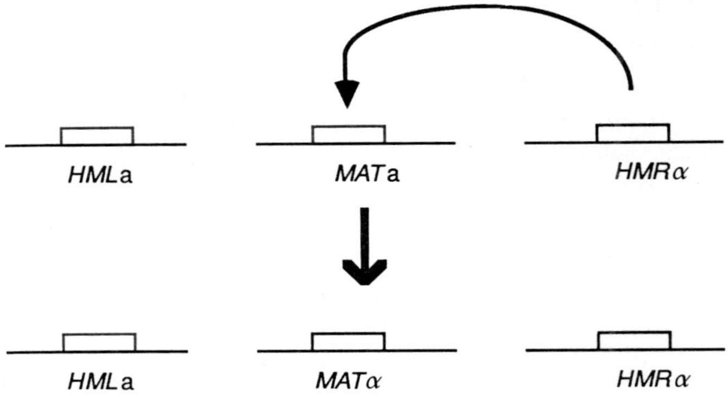

Fig. 10.15 Control of gene expression at the *MAT* locus in yeast by gene conversion.

The mechanism of this process has not yet been elucidated fully, but it is initiated by a double-stranded cut at a specific site at *MAT* (Fig. 10.16, upper). The enzyme that catalyses this cut is called the **HO endonuclease** and this is the only site recognized by this enzyme in the whole yeast genome. The ends thus generated probably engage in strand invasion into the *HML* or *HMR* donor cassettes (Fig. 10.16, middle and lower).

Another consequence of this process (and other gene conversion events) is the conversion of sequences near to the primary conversion site. The probability of any nucleotide being converted is inversely proportional to the distance from the primary site. In the example shown (Fig. 10.17), two out of three small, polymorphic nucleotide substitutions are co-converted during the mating-type switch.

The omega intron of *Saccharomyces cerevisiae*

The large rRNA gene of mitochondria in the yeast *Saccharomyces cerevisiae* is found in two configurations. The rRNA gene of some strains (termed Ω^+) contains an intron, but other strains lack this insertion. If Ω^+ and Ω^- strains are mated, the resultant yeast strains contain far more mitochondrial DNAs with introns than would be expected from simple mixing. Such biased transmission is another example of gene conversion (Fig. 10.18) and flanking polymorphisms may also be co-converted in an analogous manner to the mating type conversion. This event is also initiated by a double-stranded cut. In this case, the cleavage occurs at the intron-insertion site in the target rRNA gene and the endonuclease that carries out this cut is encoded by the omega intron itself. Thus an 'infectious' intron potentiates its own proliferation by use of a gene contained in it.

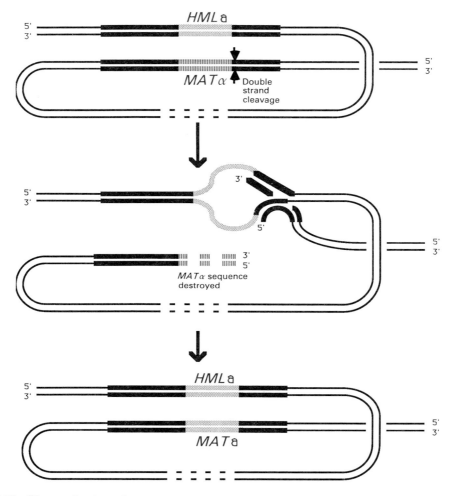

Fig. 10.16 The mechanism of gene conversion at the *MAT* locus.

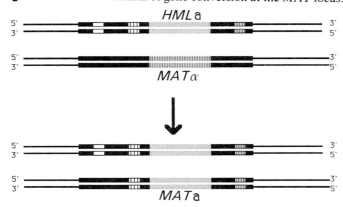

Fig. 10.17 Co-conversion of flanking markers during the mating type switch in *Saccharomyces cerevisiae*.

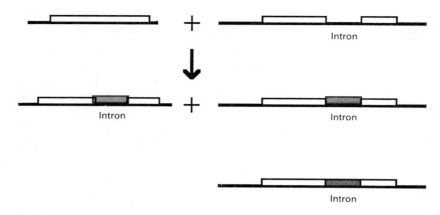

Fig. 10.18 The omega intron of *Saccharomyces cerevisiae*.

Rearrangements of antibody genes

All vertebrates possess immune systems. The major problem for these is how to recognize and respond to an almost infinite variety of invading entities. Antibody genes in most, if not all, vertebrates are multiple and varied. This solves part, but by no means all, of the problem. How is the additional diversity created?

Each immunoglobulin gene is a composite of a number of segments. During the process of development from a germ cell into one which expresses an immunoglobulin gene these segments are brought together. There are at least several (and sometimes many more) segments to choose from in the construction of an expressing immunoglobulin gene. For example, in the mouse ϰ light chain immunoglobulin locus each developing B lymphocyte chooses one V and one J DNA segment, say V3 and J4, and joins them together (Fig. 10.19). Transcription from a promoter a

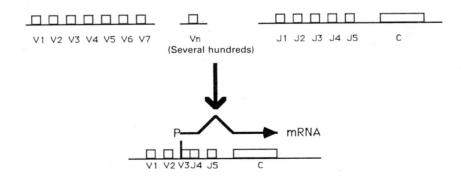

Fig. 10.19 The generation of antibody diversity by DNA rearrangement.

short distance upstream of V3 produces the V3/J4 hybrid spliced to the C region. There are thus 5 times *n* possible combinations of V and J generating 5*n* different antibody molecules. It should also be noted that random mutations in the rearranged genes arise at an abnormally high rate during this process and this increases the variety of possible combinations still further to the order of 10^8. The importance of these discoveries was recognized by the award of the 1987 Nobel Prize for Medicine to Susumu Tonegawa.

The mechanism of this DNA rearrangement is still under active study but it is known to involve partially homologous sequences (H and H′) which bracket regions to be joined (Fig. 10.20).

Chromosomal rearrangements and cancer

In several types of human leukaemia, the cancerous cells contain specific chromosomal rearrangements. For instance, most cases of Burkitt's lymphoma involve a

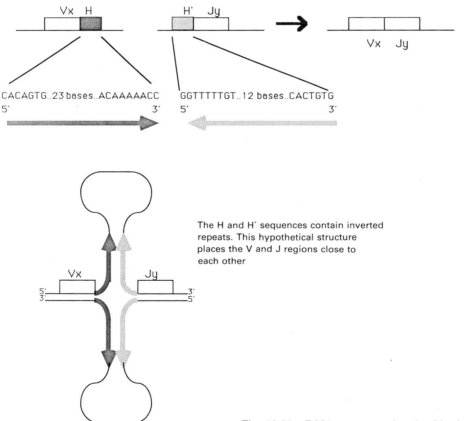

The H and H′ sequences contain inverted repeats. This hypothetical structure places the V and J regions close to each other

Fig. 10.20 DNA sequences involved in the recombination events at the immunoglobulin loci.

reciprocal translocation between chromosomes 8 and 14 such that a part of chromosome 8 is moved to chromosome 14. It turns out that one of the break points for this translocation lies at an immunoglobulin locus and presumably the translocation is due to an aberrant antibody-generating rearrangement. Other translocations also have break points lying at other immunoglobulin loci. The reason such rearrangements cause malignant transformation is not fully understood, but probably stems from aberrant expression of the genes at the other break point. In the case of Burkitt's lymphoma, this gene is the oncogene *myc* (see 10.1.9).

10.1.6 Illegitimate recombination

So far we have only considered homologous recombination. However, there is another major class of DNA rearrangement which is not dependent upon sequence homology. This is the movement of a piece of DNA called an **insertion sequence** (or IS element) into, or out of, the bacterial chromosome (Fig. 10.21).

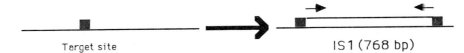

Target site   IS1 (768 bp)

Fig. 10.21 The IS element.

A small piece of DNA (typically 4–12 nucleotide-pairs, but constant for a given IS element) is duplicated as a result of IS insertion. Other features important to the transposition process are small inverted repeats at each end of the IS element (23 bases for IS1) and a **transposase** gene inside the IS element which encodes the enzyme that catalyses the process. Transposition depends upon endonucleolytic cleavage by the transposase enzyme at both the ends of the IS element and at the target site (Fig. 10.22). It is postulated that the small target site duplication arises as a result of staggered nicking of the target DNA. For a given IS element, the target sequence may be effectively random. More frequently, however, there are preferred 'hotspots' for insertion.

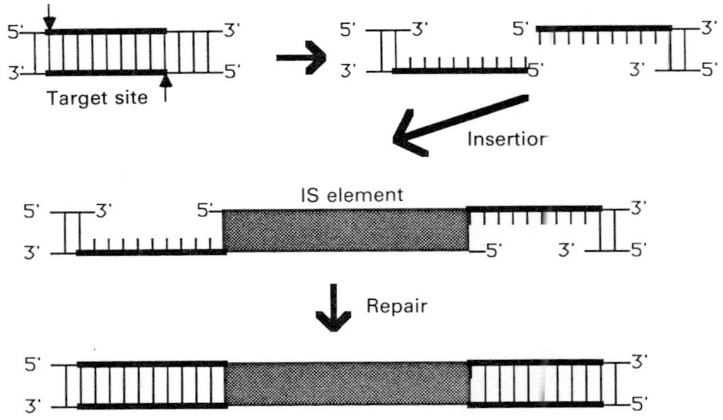

Fig. 10.22 Generation of flanking duplications during transposition of an IS element.

Transposon is the general term for a genetic element which moves by non-homologous recombination (such as an IS element). More complicated transposons exist in bacteria. These comprise a central region flanked by long repeats which are either direct or inverted with respect to each other. These long repeats are IS elements or derivatives thereof. An example of the former is Tn9 and of the latter, Tn10 (Fig. 10.23).

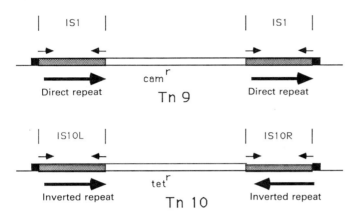

Fig. 10.23 Complex transposons in *E. coli.*

The central region of transposons contain antibiotic resistance genes and it is the transposition of these mobile DNAs that cause the rapid dissemination of antibiotic resistance between different strains and species of bacteria. They are therefore of considerable medical importance. It is presumed that complex transposons evolved from the fortuitous juxtaposition of two IS elements. In the laboratory, the IS elements which comprise a complex transposon can sometimes be mobilized independently. Such events, however, are much less common than transposition of the entire genetic element. Transposition in bacteria can result in two fundamentally different outcomes. In the first case, a transposon simply leaves one site on the one chromosome and enters another. In the second case, the original transposon is not lost, but a new mobile element appears at a distant site. Transposition in this instance is therefore replicative. Some transposons, including IS1, can use both mechanisms, suggesting that these two processes are mechanistically linked. One way in which this could be achieved is shown in Fig. 10.24.

The **Shapiro transposition intermediate** is as important to models of transposon movement as the Holliday junction is central to homologous recombination mechanisms. It was postulated in 1978 by Shapiro and its existence was recently proven during studies on transposition intermediates. It is important to note that replicative transposition results in the donor and recipient strands being joined to form a structure called a **cointegrate**. This can be seen during transposition from one plasmid to another and necessitates a homologous recombination step to separate the two duplexes (Fig. 10.25). This latter step is catalysed by a transposon-encoded **resolvase** enzyme. The resolvase stimulates recombination at a specific site (called

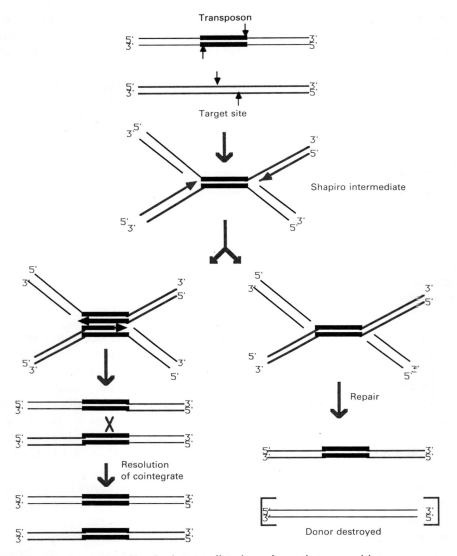

Fig. 10.24 Involvement of the Shapiro intermediate in prokaryotic transposition.

res) inside the transposon by endonucleolytic cleavage in a similar manner to that used by λ integrase (Section 10.1.5). Furthermore, the sequence of *res* is similar to that of the core sequence used by λ:

res   GATAATTTATAATAT

att   GCTTTTTTATACTAA

A final point to note from Fig. 10.24 is that during non-replicative transposition the donor chromosome is left broken and is usually lost.

In addition to simple transposition, transposons can catalyse rearrangements of the DNA surrounding them. Deletion or inversion of flanking DNA is often seen. It seems probable that such rearrangements are all abortive by-products of the normal transposition process.

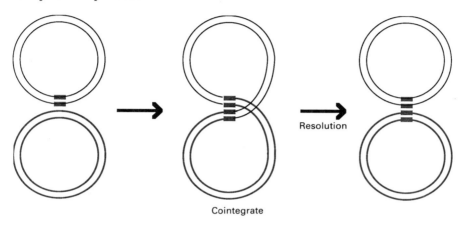

Fig. 10.25 Cointegrate formation and resolution.

10.1.7 Eukaryotic transposons

There exists a wide variety of different types of transposon in eukaryotes. Probably all eukaryotes contain at least one type and the variety of size and structure is bewildering. Virtually the only property shared by these elements is their ability to transpose, though most possess small inverted terminal repeats and generate small direct repeats of the target site during integration. We will only consider two types here. Others have been mentioned earlier (Section 5.5) because they form a significant part of the eukaryotic genome.

The P element transposon

The fruit fly *Drosophila melanogaster* is host to a wide variety of transposons which together comprise at least 10 per cent of its total genome size. One of the most interesting transposons is the **P element**. This is responsible for a complicated series of genetic phenomena which include lowered fertility, aberrant recombination, and increased mutation frequencies. These aberrations only occur in the progeny of a mating between a male whose genome contain P elements and a female lacking these transposons. No other combination of matings works. Furthermore, the damage seems to be confined only to the germline cells (those which give rise to the sperm or egg). All of the diverse genetic effects are believed to be the direct or indirect result of the transposition of the P element in the germ cells.

Two fundamental questions are raised by these observations. First, why does the P element only transpose when introduced via a sperm into an egg which lacks such sequences? Secondly, why is transposition restricted to germ cells? We now have a good idea of the answers to both questions, thanks to the studies of Gerald Rubin. We must first consider the structure of the P element and the nature of the proteins encoded by it. Functional P elements are 2907 base-pairs long and contain the usual features of inverted repeats and duplications of flanking sequence at their ends (Fig. 10.26). One strand of the P element contains four open reading frames. Mutations in any of these frames abolish the ability to transpose, suggesting strongly that they are all spliced together during RNA maturation to form a single mRNA which encodes a transposase enzyme. In support of this notion, the only RNAs encoded by P which are seen in fruit flies are those which contain all four regions. P RNA is synthesized in all flies containing these transposons and is made in all their tissues. If this RNA encodes the transposase, what is stopping transposition in most tissues of progeny flies and the whole body of the donor organism? We know that it is not due to the inability of the message to be translated, so the answer lies elsewhere.

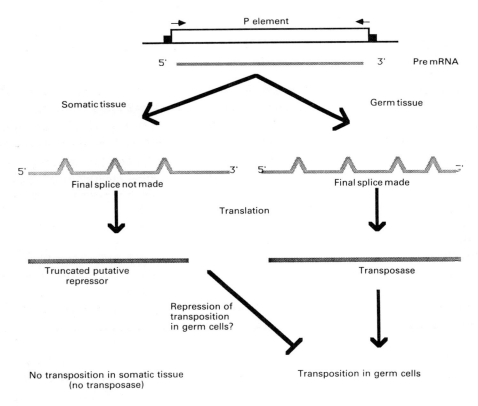

Fig. 10.26 The P element transposon of *Drosophila*.

It turns out that the protein made in somatic tissue (the entire body minus the germ cells) of the fly is not a transposase and it only contains sequence information from the first three introns of the P element (see Fig. 10.26). This truncated protein is ubiquitous in those cells of the fly where transposition is not occurring, which suggests that it may be a repressor of the transposase. What is the nature of the transposase? This was answered elegantly by the creation of a genetically engineered P element which lacked the intron separating the last two exons of the P element (Fig. 10.26). This mutated P transposon gained the ability to transpose in all tissues of the fly, thus proving that the four open reading frames together encode transposase and that the block to production of this protein is due to the removal of the third intron during mRNA splicing.

Now we have a clear picture of events. If an egg laid by a fly which lacks P elements is fertilized by a sperm whose DNA contains these sequences, transcription of P proceeds in all tissues, but complete RNA maturation is blocked at the level of splicing in all tissues except germ cells. In germ cells, active transposase is made and the result is massive transposition of the P elements and associated genetic lesions. In the rest of the fly, no active transposase is made and the P elements are immobile. When the embryo reaches adulthood, if it is female it lays eggs with a certain amount of the truncated putative repressor protein in them. This may have the effect of inhibiting transposition of P elements even in the germ cells. This resistance takes several generations to build up for reasons which are unclear.

Retroviruses as transposable elements

Retroviruses were first identified as agents involved in the onset of cancer about eighty years ago. More recently the AIDS epidemic has been shown to be due to the HIV retrovirus (see Section 4.7.2). In the early 1970s, it was discovered that retroviruses had the astonishing ability to replicate their RNA genomes via conversion into DNA which became stably integrated in the DNA of the host cell. It is only comparatively recently that retroviruses have been recognized as particularly specialized forms of eukaryotic transposons. In effect they are transposons which move via RNA intermediates that usually can leave the host cells and infect other cells. The integrated DNA form (or **provirus**) of the retrovirus bears a marked similarity to a transposon (Fig. 10.27).

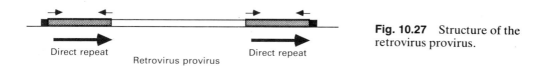

Fig. 10.27 Structure of the retrovirus provirus.

Comparison of the general structure of the provirus with that of Tn9 (Fig. 10.23) shows few differences. The long direct repeats of retroviruses rarely encode proteins and, if they do, these are not transposases. The central region does not contain antibiotic resistance, but it may contain cellular genes. Retroviruses which contain cellular genes are usually acutely cancer-inducing (oncogenic) because of these genes (called **oncogenes**). The normal counterparts of viral oncogenes (which are called **proto-oncogenes**) are genes vital for the correct functioning of the organisms they inhabit. It is the perturbation of their functioning as a result of being captured inside the retrovirus that causes cell transformation leading to cancer. One such oncogene, *myc*, is found at the other end of the translocation which causes Burkitt's lymphoma in man (see Section 10.1.5). Aberrant *myc* expression can therefore be caused by chromosomal rearrangement and retrovirus capture with both processes leading to tumourigenesis (Fig. 10.28). Another way in which retroviruses can disturb *myc* function and cause cancer is found in the infection of chickens by avian leukosis retrovirus (ALV). Infected chickens often develop a type of leukaemia after several months. An analysis of the tumour DNA shows that ALV DNA has become inserted near to the chicken *myc* gene. Again, it is the disturbance in *myc* gene expression resulting from the insertion of the retroviral genome that is presumably involved in the generation of this cancer.

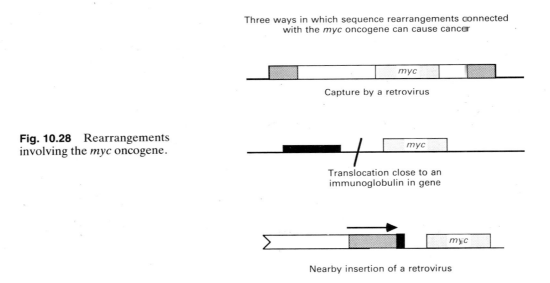

Three ways in which sequence rearrangements connected with the *myc* oncogene can cause cancer

Capture by a retrovirus

Fig. 10.28 Rearrangements involving the *myc* oncogene.

Translocation close to an immunoglobulin in gene

Nearby insertion of a retrovirus

Other retrovirus-related mobile elements inhabit organisms as diverse as yeast, flies, and plants. Many of these do not possess the ability to leave their host cell but still use an intracellular virus-like particle as a transposition intermediate.

The transposition cycle of retroviruses has other similarities to prokaryotic transposons, which suggest a distant familial relationship between these two types of transposon. Crucial intermediates in retrovirus transposition are extrachromosomal DNA molecules (Fig. 10.29). These are generated by copying (or reverse transcribing) the RNA of the virus particle into DNA by a retrovirus-encoded polymerase called **reverse transcriptase**. It is still unclear which of two molecules, a linear or a circle, is the direct precursor of the integrated element but, whichever is used, the insertion mechanism bears a strong similarity to prokaryotic transposition.

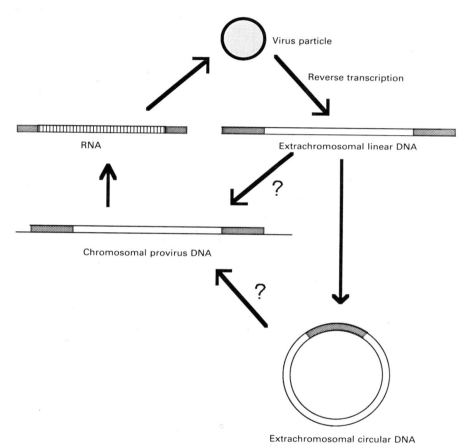

Fig. 10.29 Retrovirus transposition intermediates.

Summary

Two major types of recombination occur in all living organisms. Homologous recombination depends upon sequence complementarity and enables DNAs to exchange homologous sequences. The Holliday junction is a crucial intermediate in this process. Gene conversion can be associated with homologous recombination. In prokaryotes, homologous recombination is used as one way to control gene expression.

The second major class of recombination found in all living organisms is illegitimate recombination. It is practised by a wide variety of genetic elements which are collectively termed transposons. Transposons move into or out of stretches of DNA with which they share no direct sequence homology. They usually encode specific enzymes which catalyse this movement. A key intermediate in the mechanism of transposition of many of these sequences is the Shapiro intermediate. Two well studied classes of eukaryotic transposon are the P element of *Drosophila* and the retroviruses.

10.2 Gene cloning and recombinant DNA technology

It is not the object of the chapter to teach cloning and recombinant DNA methodology, which are covered in depth in other reference works (see Further reading; page 419). Instead, we shall survey the sorts of manipulation that are possible using cloning and recombinant DNA techniques.

Cloning is the technique of growing large quantities of genetically identical cells or organisms which are derived from a single ancestor (clones). Gene cloning is an extension of this whereby a particular gene of interest (or group of genes) is inserted into a carrier or **vector** DNA and introduced into cells by transformation. The cells containing vector DNA are propagated and each cell in a clump contains an exact copy (or copies) of the gene 'cloned' in the vector.

Gene cloning is carried out by many people in different areas of chemistry and biology. A chemist may be interested in structural studies of defined DNA sequences which are too large to be synthesized economically by chemical means. For biochemists, who spend much time trying to isolate tiny amounts of a protein from an inaccessible or expensive source, cloning can allow production of far greater amounts of a protein with greater convenience. A molecular biologist is interested in the sequence of a gene, its transcription, and the factors controlling its expression, all of which are extremely arduous without cloning. A molecular geneticist may be interested in a gene which causes an important effect, such as cancer or a developmental or neurological fault, and cloning enables the elucidation of what that gene encodes. Lastly, a population geneticist is interested in the sequence and evolution of DNA itself and the cloning of equivalent genes from different organisms allows a comparison of their sequences.

10.2.1 Vectors

Several classes of vector exist into which foreign DNA can be inserted and amplified. The three major classes are **plasmids, bacteriophage**, and **cosmids**. Prokaryotic plasmids are almost always circular double-stranded DNAs containing antibiotic resistance genes as markers and a variety of restriction sites which can be used for insertion of the foreign DNA. A large number of plasmids are available for use in *E. coli*. The most useful bacteriophage is λ. Bacteriophage λ has been engineered in many different ways to accept inserts of many different sizes and types. Because the transformation frequency is high, screening is easy, and it can accept large inserts, bacteriophage λ is often the vector of choice for an initial cloning step, especially for cloning of eukaryotic genes. Cosmids are large plasmids which contain the packaging site for bacteriophage λ DNA. They can therefore either be packaged into phage particles or they can be replicated as plasmids. Since the amount of DNA which can be packaged into a λ phage particle is 50 000 base-pairs, the potential size of cosmid inserts is very large. Cosmids are often used in **chromosome walking** (see Section 10.2.3), but they are somewhat more difficult to manipulate than is λ.

The choice of vector is dependent on four parameters (Table 10.1). Where the DNA of interest is a very small part of a mixture of DNAs that is being used for cloning, then the ease of screening for the correct clone is often the deciding factor. A second important parameter is transformation efficiency, especially if one needs to generate a large number of recombinants in order to ensure the presence of a correct one. However, if the average insert size is large, then less recombinants are needed in order to obtain a representative recombinant **library**. Finally, and least importantly, is the ease of making DNA from the recombinant. The use of a vector renders DNA isolation relatively simple and one can always subclone from the original vector into a more amenable one once the initial cloning is achieved.

Table 10.1. The choice of vector for gene cloning

	Plasmid	λ Phage	Cosmid
Ease of screening	++	++++	++
Transformation efficiency (recombinants per microgram)	10^7	10^8	10^8
Insert size (nucleotide-pairs)	≤15 000	≤30 000	≤40 000
Ease of making DNA	+++	+	+

10.2.2 Basic methodology

In virtually all cases, a fragment of DNA to be cloned is inserted as a duplex into a restriction site of the vector by use of the enzyme DNA ligase (see Section 9.4.2).

Usually the insert is a restriction fragment with termini compatible with the vector ends. Sometimes oligonucleotide **linkers** need to be joined to the insert. These are self-complementary synthetic duplex oligonucleotides that specify a recognition site for a restriction enzyme. Linkers are ligated to the fragment to be cloned and then treated with the restriction enzyme, thus generating new termini, which are now identical or compatible for joining to the cleaved vector. Often the insert is a cDNA copy of mRNA which has been generated using the enzyme reverse transcriptase (Fig. 10.30). The insert DNA can also be chemically synthesized (see Section 3.4.5).

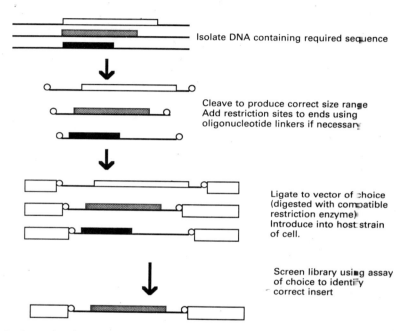

Fig. 10.30 Basic cloning procedure.

Another way of generating insert DNA is by means of the **polymerase chain reaction** (PCR). This method involves use of a pair of synthetic oligonucleotide primers flanking the region to be amplified, each complementary to a different DNA strand (see Fig 5.9, Chapter 5). The target duplex DNA is denatured by heat and annealed to the primers by a rapid lowering of the termperature to slightly below the melting temperature (Section 10.2.3). Use of a large excess of primers ensures that all the target DNA has bound the primers. DNA polymerase from *Thermus aquaticus* (a heat-stable polymerase) is used to extend the two primers in a 3'-direction such that each strand of the target DNA is copied once to give a two-fold amplification. Since the primers are derived from different DNA strands, each newly synthesized strand now contains a binding site for the primer used for copying of the *other* strand. A second round of denaturation by heat, annealing and primer extension results in a four-fold amplification. After twenty rounds of amplification, 2^{20} copies of the original target DNA a formed. Such a powerful technique can produce as much DNA as can be produced by cloning methods.

10.2.3 Assaying for the correct clone

The rate-limiting step in cloning is almost always the identification of the correct clone. Several approaches can be used, but there are some basic rules to be considered.

In some cases, nucleic acid **hybridization** (annealing of complementary strands, Section 2.5.1) is useful. This is particularly true if a related sequence, such as that from another species, has already been cloned. Alternatively, one can deduce the DNA sequence from the corresponding protein sequence (if available) and chemically synthesize oligonucleotide **probes** complementary to part of the target DNA for screening of clones. A complication here is that most amino acids are encoded by more than one nucleotide triplet. The result is that many different oligonucleotides need to be made to be sure of using the correct one. The number can be reduced by choosing a region of protein sequence containing the less ambiguous amino acids such as methionine and tryptophan which are specified by a single codon (TGG and ATG respectively). It is also possible to synthesize a mixture of oligonucleotides with two, three, or four bases at the points where ambiguity is present, since the first two bases are often invariant for a particular amino acid (Fig. 10.31; see Figure 6.20). Lastly, several different regions of a protein can be used to derive a battery of probes which can all be used to screen the library. In this way artefacts can be discounted.

The probe is labelled (either by radioactivity or by use of a non-radioactive reporter molecule) and a solution of the probe is brought in contact with the DNA of clones, which is usually immobilized on a filter. After careful washing of the filter, only that probe which is exactly complementary to the desired sequence is left attached to the filter and positive clones can be identified by autoradiography or by visualization of the reporter molecule.

```
Glu  Asp  Ile  Trp  Lys  Lys  Phe
GAG  GAT  ATT  TGG  AAA  AAA  TTC
 A    C    C         G    G    T
           A
```

Fig. 10.31 Mixed sequence oligonucleotides in gene cloning. A mixed oligonucleotide incorporating each alternative base can be used to probe for the gene encoding this oligopeptide.

The hybridization conditions used in such experiments are often crucial to a successful outcome. At this point it is therefore worthwhile to mention the major parameters affecting nucleic acid hybridization.

(a) *Temperature* (T). The rate of reassociation of single-stranded DNA into a duplex varies markedly with temperature (Fig. 10.32; Section 2.5.1). The shape of this curve is governed by two factors. At low temperatures, the reassociation rate is determined by the difference in free energy between the unassociated and the transition state:

$$k \propto e^{-E_a/RT}$$

Where k = reassociation rate constant, E_a = activation free energy, R = gas constant, T = temperature (in K). At higher temperatures, the stability of the duplex is markedly reduced until eventually it is unstable and the hybrid melts. Thus we see a fall off in reassociation rate as this point is approached.

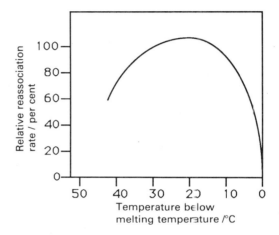

Fig. 10.32 Dependence of reassociation rate of DNA upon temperature.

(b) *Monovalent cation concentration (M)*. The **melting temperature** of a hybrid is reduced at lower ionic strength because cations stabilize the DNA duplex. Divalent cations such as Mg^{2+} are much more effective in stabilizing hybrids, but are rarely used in hybridization studies (Section 8.2.1).

(c) *Base composition (%G:C)*. G:C base-pairs are stronger than A:T, giving a higher melting temperature.

(d) *Duplex length (L)*. Hybrids shorter than a few hundred base-pairs have significantly lower melting temperatures.

In practice, these factors can be combined into an empirical equation giving the melting temperature T_m of a hybrid DNA:

$$T_m = 69.3 + 0.41(\%GC) + 18.5 \log_{10}M - 500L^{-1} \ /°C.$$

Use of hybridization temperatures from 10° to 20° below the calculated T_m of the hybrid is optimal in practice. For synthetic oligonucleotide probes of 15–20 residues and under standard hybridization conditions, the calculation of T_m is simplified to 2 K per dA:dT and 4 K per dG:dC base pair in approximately 1M sodium chloride solution. This is know as the 'Wallace Rule'.

Sometimes it is possible to enrich the source material in the required sequence. One way of doing this is to use cDNA generated from RNA which has been enriched by size-selection on gels or sedimentation velocity gradients. Alternatively, different tissues can be scrutinized for maximum content of the required RNA or protein. Several shortcuts in cloning can be considered. The polymerase chain reaction is an enrichment procedure which has already been mentioned (Section 10.2.2). A list of other methods includes the following.

(i) *Transposon tagging.* A transposon is used to create mutations in the required gene (Fig. 10.33) The transposon can then itself be used as a 'tag' to isolate the gene by hybridization (the transposon and its surrounding DNA must both be isolated by this method).

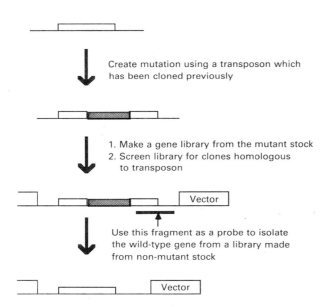

Create mutation using a transposon which has been cloned previously

1. Make a gene library from the mutant stock
2. Screen library for clones homologous to transposon

Fig. 10.33 Transposon tagging.

Vector

Use this fragment as a probe to isolate the wild-type gene from a library made from non-mutant stock

Vector

(ii) *Microdissection.* It is possible physically to dissect and clone the required part of the chromosome (provided the chromosomal location of the gene of interest is known). Chromosomes may be separated from one another by pulsed-field gel eletrophoresis or by fluorescence-activated sorting.

(iii) *Chromosome walking.* If an overlapping series of clones can be isolated it is possible to use one clone to isolate the next in line and thus 'walk' along the DNA to the required sequence (Fig. 10.34). This is a very time-consuming process, but nevertheless has been frequently used.

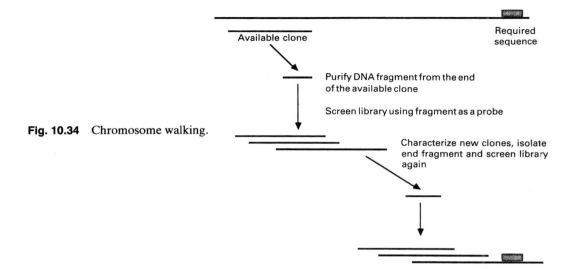

Fig. 10.34 Chromosome walking.

Summary

Gene cloning is the technique of isolation and amplification of a gene by insertion into a vector DNA followed by the transformation of cells in order to obtain cell clumps, each containing identical copies of the gene. Plasmids, bacteriophage, and cosmids are the main types of vector. Each has its own advantages and disadvantages depending on the circumstances. Often, assaying for the correct clone is the rate-limiting step in cloning. Transposon tagging, microdissection, chromosome walking, and the polymerase chain reaction are all possible shortcuts in cloning.

10.3 *In vitro* mutagenesis

The process of engineering specific changes in a DNA sequence in a test tube in order to assay the effect or function of that sequence is termed *in vitro* **mutagenesis**. It is an invaluable procedure, since it enables a region of interest in a DNA to be chosen and altered in a premeditated manner. In classical mutagenesis, alterations are created randomly and the effects of each mutation need to be screened in turn.

Before mutagenesis of a sequence, the DNA must first be cloned so that it can easily be manipulated. Three types of alteration can now be made: deletions, insertions, and replacements.

10.3.1 Deletions

These can be created at restriction sites by cleavage with the corresponding enzyme and then by treatment for a short period with the enzyme Bal 31 exonuclease, which removes both double- and single-stranded DNA from both ends (Fig. 10.35). Alternatively, if the restriction enzyme cut leaves overhanging, single-stranded ends, these may be trimmed prior to re-ligation by use of a nuclease that is specific for single-strands, such as S1 nuclease (Tables 9.1 and 9.2).

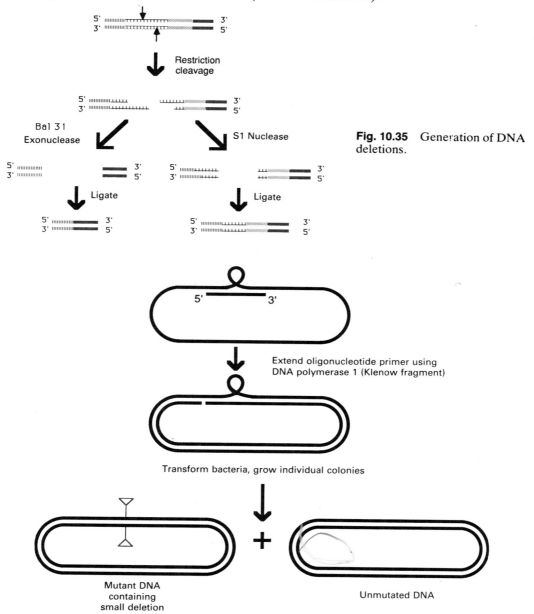

Fig. 10.35 Generation of DNA deletions.

Fig. 10.36 Oligonucleotide site-directed deletion mutagenesis.

Deletions of regions of DNA that are not near useful restriction sites may be achieved by use of synthetic oligonucleotides. In this procedure, an oligonucleotide which flanks the required deletion, but does not contain it, is used as a primer to create a complete complementary strand of the template DNA (Fig 10.36). In the process of cloning, mutant DNA segregates from wild type DNA and clones containing mutant DNA (i.e. carrying the deletion) can be selected.

One problem encountered with the above technique is that the bacteria often prefer to repair the mutagenized strand because the *in vivo*-generated DNA strand is methylated. This results in very low efficiencies of isolation of the mutated sequence. There are several ways around this problem. Eckstein has developed a

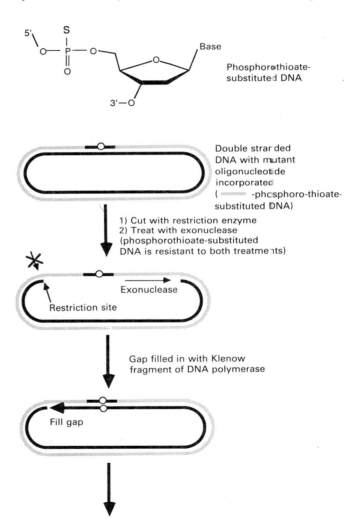

Fig. 10.37 The use of phosphorothioate-modified nucleotides for *in vitro* mutagenesis.

reliable method which incorporates **phosphorothioate**-modified nucleotides (Section 3.4.6) into the *in vitro*-generated strand. Such nucleotides are more resistant to nuclease degradation, with the result that the above procedure is possible (Fig. 10.37, page 414).

10.3.2 Insertions

Insertions at restriction sites may be made by ligation of oligonucleotide linkers into restriction sites after cutting with the appropriate enzyme. Sequence additions at other sites can be achieved by means of site-directed mutagenesis using oligonucleotides in an analogous way to that described for deletions (Fig. 10.36). Here, however, the oligonucleotide contains the DNA sequence specifying the insert and flanked by sequences exactly complementary to those in the target DNA on either side of the desired point of insertion.

10.3.3 Replacements

Sometimes it is desired to keep the same number of nucleotides in a sequence but to alter the sequence at a specific point. This can be achieved by making a small deletion at a restriction site and inserting into the gap an oligonucleotide linker of the same size but of different sequence.

A more general approach involves use of a synthetic oligonucleotide in an analogous way to the introduction of deletions and insertions, but with the same number of nucleotides in the mutant strand as wild type (Fig. 10.36). This procedure works particularly well for single-base alterations and is widely used to change the sequence of a cloned gene that codes for a protein. Expression of the mutated gene leads to the production of a protein with a single amino acid alteration (**protein engineering**).

10.4 Gene replacement in micro-organisms

Often one wishes to determine the effect of a mutation synthesized *in vitro* on gene function *in vivo*. For this purpose the normal gene must be replaced by the mutated gene. It is frequently impossible to remove the former and then add the latter because the first step is often a lethal event. Therefore, one needs to be able to substitute the mutated gene for the normal gene in two steps by keeping an extra copy of the original until the mutant gene is in place. This has been achieved in the yeast *Saccharomyces cerevisiae* by means of **homologous recombination** (Fig. 10.38). The mutated gene is ligated next to a marker gene which can be easily selected *for* under restrictive conditions (often a gene encoding a protein necessary for amino acid or nucleic acid biosynthesis). The construct is inserted into cells which lack the marker gene and these cells are placed in a medium lacking the specific metabolic precursor synthesized by the marker gene product. This procedure kills all cells except those in which the mutated gene has been exchanged for the normal one by a normal homologous recombination event. Subsequent removal of the normal gene is achieved by selection *against* the marker gene by use of a substrate analogue which is converted into a toxic metabolite by the marker gene product.

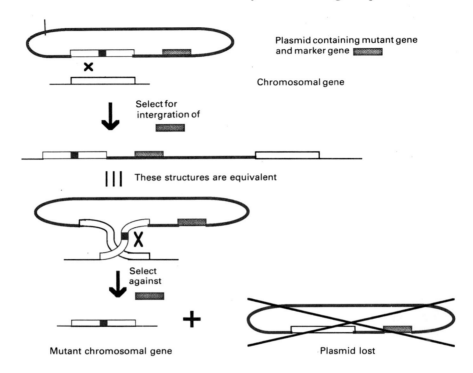

Fig. 10.38 Gene replacement in *Saccharomyces cerevisiae*.

Now the functioning of the mutated gene in its correct chromosomal location can be scrutinized. This method has been used to replace a normal yeast actin gene by a duplicate of the gene that lacks any intron (see Section 5.5.2). The mutated gene functioned perfectly well, showing that an intron can be dispensable to the gene carrying it.

10.5 Gene therapy and genetic engineering in mammals

The ability to introduce new or altered genes into a mammalian genome has tremendous implications. For example, it may prove possible to cure some genetic diseases by introducing a healthy copy of a gene into an afflicted individual. It is already possible to introduce into mammals genes which encode economically or medically important polypeptides such as insulin, growth hormones, and interferon, with the intention that the animal either grows faster or produces large amounts of protein which can be harvested. We will address the ways in which this can be carried out, leaving the ethical questions raised by this issue to others.

There are two major ways of introducing DNA into mammalian germ tissue such that the progeny of the recipient will carry the gene. The first involves microinjection of DNA solutions into the nucleus of an egg by means of an extremely fine capillary. Such a technique works very well with a mouse egg but is more difficult with other mammals, such as sheep, where it is extremely hard to see the nucleus. In this way **transgenic** animals have been created which carry functioning genes from another organism.

The second method involves the use of retrovirus-based vectors (Fig. 10.39). As described in Section 10.1.7, retroviruses can infect a cell and then insert their DNA into the chromosome of the host cell. The gene to be introduced into the host is

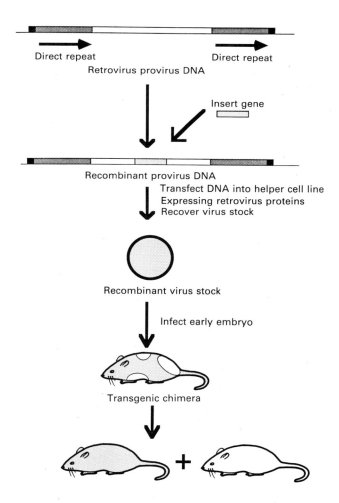

Fig. 10.39 Creation of transgenic animals by use of retroviral vectors.

ligated into the genome of the retrovirus. The retroviral DNA is then introduced into a cultured cell line which is capable of producing all of the components of a retrovirus except for the viral RNA (such a cell culture is called a **helper cell** line). This cell line will then package the recombinant virus stock into virus particles which can be harvested from the culture medium. Helper cells are necessary because the presence of the insert in the retroviral genome disrupts some of the normal retroviral genes needed for viral production. The harvested recombinant virus stock is then used to infect an early embryo which is then replaced into a donor mother. During growth, some cells of the embryo become infected by the virus and the retroviral gene, including the gene insert, becomes stably inserted into the DNA of these cells. Because not all cells become infected, the animal is a **chimera**. However, if the germ cells of this animal contain proviral DNA then its offspring will retain the recombinant in every cell of its body.

10.6 DNA sequence polymorphisms

Homologous recombination can occur more or less randomly along a chromosome. One important consequence of this is that the probability of a recombination event occurring between two points on a chromosome is inversely proportional to the distance between them. Thus, if two genes are very close together, there is little chance of a random recombination event happening between them. It is much more likely to occur somewhere else. This phenomenon is called **linkage** and was used as one of the first ways to map genes on chromosomes. In recent times, this phenomenon has been used to aid the diagnosis of genetic disease. Often the nature of the damaged gene causing a particular disease is unknown. However, if one can find a genetic marker of some kind which maps very close to the gene, then it can be used as a diagnostic tool. There is approximately a 1 per cent chance that any base on a chromosome in a particular individual will be different from the corresponding base on the same chromosome from another individual. Therefore, every chromosome has its own 'fingerprint' and if one of these so-called **sequence polymorphisms** can be identified close to the diseased gene, then it can be used to trace that particular gene through the relatives of the bearer. Of course, if the polymorphism creates or removes a restriction enzyme cleavage site, then it can be readily detected. In this way, markers close to several genetic diseases, such as muscular dystrophy and Huntington's disease, have already been isolated. From there it is an arduous, but feasible, task to clone such genes by chromosome walking.

One particular type of polymorphism in human DNA has leapt into the limelight recently. The human genome contains many copies of so-called **minisatellite** sequences. Each minisatellite region is comprised of tandem repeats. Individual repeats are very similar to one another in a particular cluster, but the consensus sequence of the repeat from one cluster is very different from that in another (Fig. 10.40). A radioactive copy of a minisatellite sequence is used to probe a restriction

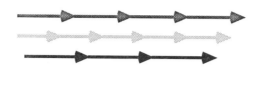

All repeats, however, share a 'core' sequence which resembles the *chi* site of *E. coli*

5' GGGCAGGAXG 3'

Fig. 10.40 Hypervariable minisatellite sequences.

digest of total human DNA which has been subjected to electrophoresis on an agarose gel and then transferred to a filter (this very important technique is called **Southern blot analysis**). A unique fingerprint pattern of bands corresponding to restriction fragments containing minisatellite sequences is generated. Since these repeats are inherited in a normal Mendelian manner, it is possible to test the genetic relationship between people by following this procedure. It is also possible to generate this fingerprint from relatively small amounts of tissue or body fluid. Alec Jeffreys has developed this procedure to make a powerful impact on forensic medicine.

Summary

In vitro mutagenesis is the process of engineering specific changes in a DNA sequence in a test tube and is now a major way of probing gene function. Complete genes can be replaced in microorganisms or introduced into plants and mammals.

DNA sequence polymorphisms are powerful tools in the diagnosis and mapping of genetic disease in humans and in forensic medicine.

Further reading

10.1

Berg, D. E. and Howe, M. M. (ed.) (1989). *Mobile DNA*. American Society for Microbiology, Washington, DC.

Bradshaw, R. A. and Prentis, S. (ed.) (1987). *Oncogenes and growth factors*. Elsevier, Amsterdam.

Lewin, B. (1987). *Genes III*. Wiley, New York, pp. 573–645.

Old, R. W. and Primrose, S. B. (1985). *Principles of gene manipulation*. Blackwell, Oxford.

Pai, E. F., Kabasch, W., Krengel, U., Holmes, K. C., John, J., and Wittinghofer, A. (1989). Structure of the guanine-nucleotide-binding domain of the Ha-*ras* oncogene product p21 in the triphosphate conformation. *Nature*, **341**, 209–14.

Watson, J. D., Hopkins, N. H., Roberts, J. W., Steitz, J., and Weiner, A. M. (ed.) (1987). *The molecular biology of the gene*, Vol. 1. Benjamin Cummings, Menlo Park, California, pp. 313–38.

10.2

Derbyshire, K. M. and Grindley, N. D. F. (1986). Replicative and conservative transpositions in bacteria. *Cell*, **47**, 325–27.

Finnegan, D. J. (1985). Transposable elements in eukaryotes. *Int. Rev. Cytol.*, **93**, 281–325

Glover, D. M. (ed.) (1985 and 1987). *DNA cloning: a practical approach*, Vols I–III. IRL Press, Oxford.

Sambrook, J., Fritsch, E. F., and Maniatis, T. (ed.) (1989). *Molecular cloning, a laboratory manual*, 2nd ed. Cold Spring Harbor Laboratory Press, New York.

10.3

Lathe, R. F., Lecocq, J. P., and Everett, R. (1983). DNA engineering; the use of enzymes, chemicals and oligonucleotides to restructure DNA sequences *in vitro*. in *Genetic engineering* (ed. R. Williamson), Vol. **4**. Academic Press, London, pp. 2–57.

Sayers, J. R. and Eckstein, F. (1988). Phosphorothioate-based oligonucleotide-directed mutagenesis. In *Genetic engineering*: *principles and methods*, Vol. 10. Plenum Press, New York, pp. 109–22.

10.4

Carter, B. L. A., Irani, M., Mackay, V., Searle, R. L., Sledziewski, A., and Smith, R. (1987). Expression and secretion of foreign genes in yeast. In *DNA cloning* (ed. D. M. Glover),Vol. III. IRL Press, Oxford, pp. 141–62.

10.5

Murphy, D. and Hansen, J. (1987). The production of transgenic mice by the microinjection of cloned DNA into fertilised one-cell eggs. In *DNA cloning* (ed. D. M. Glover), Vol. III. IRL Press, Oxford, pp. 213–48.

10.6

Jeffreys, A. J., Wilson, V., and Thein, S. L. (1985). Hypervariable 'minisatellite' regions in human DNA. *Nature*, **314**, 67–73.

Glossary

Agarose   A polysaccharide isolated from seaweed used as a matrix in gel electrophoresis.

Allosteric control   The ability of an interaction at one site of a protein to influence (positively or negatively) the activity at another site.

Amber codon   The nucleotide triplet UAG, one of three nonsense codons that cause termination of protein synthesis

Amber mutation   Any change in DNA that creates an amber codon in the corresponding mRNA at a site previously occupied by a codon specifying an amino acid.

Amplification   The production of extra copies of a chromosomal sequence found either as intra- or extrachromosomal DNA. With respect to plasmids it refers to the increase in the number of plasmid copies per cell induced by certain treatments of transformed cells.

Anneal (re-anneal)   The (re)establishment of base-pairing between complementary strands of DNA or a DNA and an RNA strand.

Anticoding strand   Duplex DNA used as template to direct the synthesis of RNA and is complementary to it.

Anticodon   A triplet of nucleotides in a constant position in the structure of tRNA that is complementary to the codon(s) in mRNA to which the tRNA responds.

Antiparallel   The orientation of the two strands in a DNA duplex where as seen from either end, one strand runs in the 5′→3′ and the other in the 3′→5′ direction.

Antisense   A strand of DNA that has the sequence complementary to mRNA (also non-coding strand).

Anti-terminator proteins   Allow RNA polymerase to transcribe through certain terminator sites.

Attenuation   The regulation of premature termination of transcription involved in controlling expression of some bacterial operons.

Autoradiography   The detection of radioactively labelled molecules present, for example, in a gel or on a filter by exposing an X-ray film to it.

Bacteriophages   Viruses that infect bacteria; often abbreviated as phages.

Base-pair (bp)   A partnership of A with T or of C with G in a DNA double helix; other pairs are possible in RNA under some circumstances.

B–DNA   Besides A–DNA one class of right-handed Watson–Crick double helix.

Bidirectional replication   Is accomplished when two replication forks move away from the same origin in opposite directions.

Blunt-end ligation   The covalent attachment of two duplex DNA molecules having no single-stranded extensions (sticky ends).

Buoyant density   The density of a particle or molecule when suspended in an aqueous salt or sugar solution.

Capping   The posttranscriptional attachment of a cap to the 5′-terminus of most eukaryotic mRNAs.

cDNA   A single-stranded DNA complementary to an RNA, synthesized from it by *in vitro* reverse transcription.

cDNA clone   A duplex DNA sequence representing an RNA, carried in a cloning vector.

Cell cycle   The period from one division to the next.

Centromere   A constricted region of a chromosome that includes the site of attachment to the mitotic or meiotic spindle.

Chromatids   Copies of a chromosome produced by replication.

Chromatin   The complex of DNA and proteins (mostly histones) present in the nucleus of a eukaryotic cell.

Chromosome   A discrete unit of the genome carrying many genes, consisting of a very long molecule of DNA, complexed with a large number of different proteins (mostly histones). Chromosomes are visible as a morphological entity only during the act of cell division.

Chromosome walking   Sequential isolation of clones carrying overlapping sequences of DNA to span large regions of the chomosome (often in order to reach a particular locus).

Cistron   The genetic unit defined by the cis/trans test; equivalent to gene in comprising a unit of DNA representing a protein.

Clone   A large number of cells or molecules.

Coding strand (of DNA)   Has the same sequence as mRNA.

Codon   A triplet of nucleotides that represents an amino acid or a termination signal.

Cohesive ends   See **Sticky ends**.

Concatenated (circles of DNA)   Interlocked like rings on a chain.

Consensus sequence   An idealized sequence in which each position represents the base most often found when many actual sequences are compared.

Constitutive genes   Genes that are expressed without additional regulation.

Copy number   The average number of copies of a particular (recombinant) plasmid present in a single host cell. Also used for individual genes.

Cordycepin   3′-Deoxyadenosine, an inhibitor of polyadenylation of RNA.

Core DNA   The 146 bp of DNA contained on a core particle.

Core particle
A digestion product of the nucleosome that retains the histone octamer and has 146 bp of DNA; its structure appears similar to that of the nucleosome itself.

Cosmids
Plasmids into which phage lambda cos sites have been inserted; as a result, the plasmid DNA can be packaged *in vitro* into the phage coat.

Covalently-closed-circular (CCC)
A completely double-stranded circular DNA molecule without nicks or other discontinuities, usually in a supercoiled conformation.

Crossing-over
The reciprocal exchange of material between chromosomes that occurs during meiosis and is responsible for genetic recombination.

Cross-linking
Introduction of covalent intra- or intermolecular bonds between groups that are normally not covalently linked. Used to detect proximity of (parts of) (macro)molecules.

Cruciform
The structure produced at inverted repeats of DNA if the repeated sequence pairs with its complement on the same strand (instead of with its regular partner in the other strand of the duplex).

Cyclic AMP
A molecule of AMP in which the phosphate group is joined to both the 3'- and 5'-positions of the ribose; one of its functions is the activation of CAP, a positive regulator of prokaryotic transcription.

Deletions
Constitute the removal of a sequence of DNA, the regions on either side being joined together.

Denaturation (of DNA or RNA)
Conversion from the double-stranded to the single-stranded state; separation of the strands is most often accomplished by heating.

Density gradient centrifugation
The separation of particles or molecules on the basis of differences in their buoyant density, usually in concentrated solutions of sucrose or caesium salts.

Diploid (set of chromosomes)
Contains two copies of each autosome and two sex chromosomes.

Discontinuous replication
Refers to the synthesis of DNA in short (Okazaki) fragments that are later joined into a continuous strand.

DNA sequencing
The determination of the order of nucleotides in the strands of a DNA molecule.

Downstream
Identifies sequences proceeding farther in the direction of expression; for example, the coding region is downstream from the initiation codon.

Duplex
The double-stranded structure of DNA.

End labelling
The addition of a radioactively labelled group to one end (5' or 3') of a DNA or RNA strand.

Endonucleases
Cleave bonds within a nucleic acid chain; they may be specific for RNA or for single- or double-stranded DNA.

Enhancer element A DNA sequence that increases the utilization of (some) eukaryotic promoters in cis configuration, but can function in any location, upstream or downstream, relative to the promoter.

Ethidium bromide A chemical that, upon intercalation between base-pairs of single-stranded DNA or RNA, fluoresces under ultraviolet light. Used in the detection of nucleic acid in density gradients or on gels.

Eukaryotic Organisms containing a nucleus.

Excision-repair Systems remove a single-stranded sequence of DNA containing damaged or mispaired bases and replace it in the duplex by synthesizing a sequence complementary to the remaining strand.

Exon Any segment of an interrupted gene that is represented in the mature RNA product.

Exonucleases Cleave nucleotides one at a time from the end of a polynucleotide chain; they may be specific for either the 5'- or 3'-end of DNA or RNA.

Exonuclease III An enzyme from *E. coli* which catalyses the stepwise removal of mononucleotides from double-stranded DNA carrying a 3'-OH terminus in the 3'→5' direction.

Fingerprint The characteristic array of oligopeptides or oligonucleotides obtained upon two-dimensional electrophoresis of a protein digested with a specific endopeptidase or an RNA digested with a specific endonuclease.

Footprinting In this context is a technique for identifying the site on DNA bound by some protein by virtue of the protection of bonds in this region against attack by nucleases.

Gap in DNA The absence of one or more nucleotides in one strand of the duplex.

Gel electrophoresis Electrophoresis performed in a gel matrix (usually agarose or polyacrylamide) which allows separation of molecules of similar electric charge density on the basis of their difference in molecular weight.

Gene A DNA sequence involved in the production of an RNA or protein molecule as the final product. Includes both the transcribed region and any sequences upstream and/or downstream responsible for its correct and regulated expression (e.g. promotor and operator sequences).

Genetic code The complete set of codons specifying the various amino acids, including the nonsense codons. The code is usually written in the form in which it occurs in mRNA.

Genome The entire genetic material of a cell.

Genotype The genetic constitution of an organism.

GT–AG rule Describes the presence of these constant dinucleotides at the first two (GT) and the last two (AT) positions of introns of nuclear genes

Gyrase	A type II topoisomerase of *E. coli* with the ability to introduce negative supercoils into DNA.
Hairpin	The double-stranded region formed by base-pairing of adjacent complementary sequences in the same DNA or RNA strand.
Haploid (set of chromosomes)	Contains one copy of each autosome and one sex chromosome; the haploid number *n* is characteristic of gametes.
Heteroduplex (hybrid) DNA	DNA generated by base-pairing between partly non-complementary single strands derived from the different parental duplex molecules; it occurs during genetic recombination.
Highly repetitive DNA	DNA sequences represented more than 100 000 times per genome; satellite DNA is one family thereof and also Alu-sequences.
Histones	Conserved DNA binding proteins of eukaryotes that form the nucleosome, the basic subunit of chromatin.
Homology	The degree of identity existing between the nucleotide sequences of two related but not complementary DNA or RNA molecules. 70 Per cent homology means that on the average 70 out of every 100 nucleotides are identical. The same term is used in comparing the amino acid sequences of related proteins.
Hybridization	The pairing of complementary RNA and DNA strands to give an RNA–DNA hybrid, and is also used to describe the pairing of two single-stranded DNA molecules.
Hyperchromicity	The increase in optical density that occurs when DNA is denatured.
Induced mutations	Result from the treatment of cells with a mutagen.
Induction	The expression of a gene in response to an external stimulus.
Initiation codon (AUG; sometimes GUG)	Codes for the first amino acid in protein sequences, which is formylmethionine in prokaryotes; fMet is often removed post-translationally.
Initiation factors (IF)	Proteins that associate with the small subunit of the ribosome specifically to render initiation of synthesis.
Insertions	Additional (sequences of) nucleotides in DNA.
Integration (of viral or another DNA sequence)	Insertion into a host genome as a region covalently linked on either side to the host sequences.
Intercistronic region	The region between the termination codon of one gene and the initiation codon of the next gene in a polycistronic transcription unit.
Intervening sequence	An intron.
Intron	A segment of DNA that is transcribed, but is removed from within the transcript by splicing together the sequences (exons) on either side of it.

Inverted repeats   Two copies of the same sequence of DNA repeated in opposite orientation on the same molecule. Adjacent inverted repeats constitute a palindrome.

***In vitro* (lit. 'in glass')**   Refers to any (biological) process occurring outside the living cell.

In vivo   Refers to any biological process occurring within the living cell.

Isoschizomers   Two restriction enzymes that recognize the same DNA sequence and cut it differently.

Klenow fragment   A piece obtained from DNA polymerase I by proteolytic cleavage; it lacks the 5′-to-3′ exonuclease activity.

Lariat   The branched intermediate of eukaryotic mRNA formed during splicing.

Library   A set of cloned fragments together representing the entire genome.

Ligase, (DNA ligase)   Catalyses the formation of a phosphodiester bond at the site of a single-strand break in duplex DNA. Some DNA ligases can also ligate blunt-end DNA molecules. RNA ligase covalently links separate RNA molecules.

Ligation   The formation of a phosphodiester bond to link two adjacent bases separated by a nick in one strand of a double helix of DNA. (The term can also be applied to blunt end ligation and to joining of RNA.)

Linkage   The tendency of genes to be inherited together as a result of their location on the same chromosome; measured by per-cent recombination between loci.

Linker number (linking number)   The number of time the two strands of a closed DNA duplex cross over each other.

Locus   The position of a chromosome at which the gene for a particular trait resides; locus may be occupied by any one of the alleles for the gene.

Loop   A single-stranded region at the end of a hairpin in RNA (or single-stranded DNA) or a non-paired segment in duplex DNA.

Map distance   Distance between genes (or rather mutations in genes) on the chromosome. Is measured as cM (centiMorgans) = per cent recombination (sometimes subject to adjustments).

Maxam–Gilbert sequencing   A DNA sequencing technique based on specific chemical modification of each of the four bases.

Melting (of DNA)   Results in denaturation.

Melting temperature (T_m)   The temperature where hyperchromicity is half-maximal.

Micrococcal nuclease   An endonuclease that requires Ca^{2+} ions for activity and thus can easily be inactivated by chelating these ions with EGTA. Used among other things in degrading the endogenous mRNA in cell-free translation systems. Chromatin DNA is cleaved preferentially between nucleosomes.

Missense mutation Any mutation that changes a codon specifying one amino acid to one coding for another amino acid.

Modification (of DNA or RNA) Includes all changes made to the nucleotides after their initial incorporation into the polynucleotide chain.

Modified bases All those except the usual four from which DNA (T,C,A,G) or RNA (U,C,A,G) are synthesized; they result from postsynthetic changes in the nucleic acid.

Multigene family A set of identical or related genes present in the same organism, usually coding for a family of related proteins.

Mutagens Increase the rate of mutation by causing changes in DNA.

Mutation Any change in the sequence of genomic DNA.

Negative supercoiling The twisting of a duplex of DNA in space in the opposite sense to the turns of the strands on the double helix.

Nick In duplex DNA is the absence of a phosphodiester bond between two adjacent nucleotides on one strand.

Nick translation The ability of *E. coli* DNA polymerase I to nick as a starting point from which one strand of a duplex DNA can be degraded and replaced by resynthesis of new material; is used to introduce radioactively labelled nucleotides into DNA *in vitro*.

Nitrocellulose A type of paper which binds nucleic acids. Used in Southern, Northern (and also Western) blotting as well as other filter hybridization techniques.

Nonsense codon Any one of three triplets (UAG, UAA, UGA) that cause termination of protein synthesis. (UAG is known as *amber*; UAA as *ochre*; UGA as *opal*.)

Nonsense mutation Any mutation that changes a normal codon into a nonsense codon.

Non-Watson–Crick base-pair Any base-pair other than the standard G:C and A:T (U in RNA) pairs. Occur mainly in intrastrand pairing of RNA.

Northern blotting A technique for transferring RNA from an agarose gel to a nitrocellulose filter on which it can be hybridized to a complementary DNA.

Nucleosome The basic structural subunit of chromatin, consisting of ca. 200 bp of DNA and an octamer of histone proteins.

Ochre codon The triplet UAA, one of three nonsense codons that cause termination of protein synthesis.

Ochre mutation Any change in DNA that creates an ochre codon in the corresponding mRNA at a site previously occupied by a codon specifying an amino acid.

Ochre suppressor A gene coding for a mutant tRNA able to respond to the UAA codon to allow continuation of protein synthesis.

Okazaki fragments The short stretches of 1000–2000 (in eukaryotes 100–200) nucleotides produced during replication of the lagging strand of the DNA duplex which are subsequently covalently linked.

Oncogene
A retroviral gene that causes transformation of the mammalian infected cell. Oncogenes are slightly changed equivalents of normal cellular genes called proto-oncogenes. The viral version is designated by the prefix **v**, the cellular version by the prefix **c**.

Open circular
The conformation taken up by supercoiled circular double-stranded DNA molecules when a nick is introduced into one of the strands (same as **Relaxed**).

Operator
The site on DNA at which a repressor protein binds to prevent transcription from initiating at the adjacent promoter.

Operon
A complete unit of bacterial gene expression and regulation, including structural genes, regulator gene(s), and control elements in DNA recognized by regulator gene product(s).

Origin (ori)
A sequence of DNA at which replication is initiated.

Overwinding (of DNA)
Caused by positive supercoiling (by applying further tension in the direction of winding of the two strands about each other in the duplex).

Packing ratio
The ratio between the length of the DNA double helix and the length of the fibre containing the DNA.

Palindrome
A sequence of DNA that is the same when one strand is read left to right or the other is read right to left; consists of adjacent inverted repeats.

PCR (polymerase chain reaction)
The use of oligonucleotide primers and a DNA polymerase to amplify a section of DNA.

Periodicity (of DNA)
The number of base-pairs per turn of the double helix.

Phenotype
In contrast to genotype, it is the appearance of an organism.

Plasmid
An autonomous self-replicating extrachomosomal circular DNA.

Polyacrylamide gel electrophoresis
See **Gel electrophoresis**.

Polyadenylation
The post-transcriptional attachment of up to 200 AMP residues to the 3'-terminus of most eukaryotic mRNAs.

Polycistronic mRNA
Includes coding regions representing more than one gene.

Polymerase
An enzyme that catalyses the assembly of nucleotides into RNA or of deoxynucleotides into DNA; usually the enzyme requires single-stranded DNA (sometimes RNA) as a template.

Polynucleotide kinase
An enzyme which transfers the γ-phosphate group from ATP on to the 5'-OH terminus of a DNA or RNA molecule. Used in end-labelling DNA and RNA for sequencing.

Polysome (polyribosome)
A mRNA associated with several ribosomes engaged in translation.

Positive supercoiling
When the double helix coils itself in the same direction as the winding of the two strands of the double helix itself.

Primary transcript
The original unmodified RNA product corresponding to a transcription unit.

Primer
A short sequence (of DNA or RNA) that is paired with one strand of DNA and provides a free 3'-OH end at which a DNA polymerase starts synthesis of a deoxyribonucleotide chain.

Probe (hybridization)
A labelled DNA or RNA molecule used to detect a complementary sequence by molecular hybridization.

Prokaryotic
Organisms lacking membrane-enclosed nuclei.

Promoter (in bacteria)
The region of the gene involved in binding of the RNA polymerase. (In eukaryotes) usually all regions of the gene required for maximum expression (excluding enhancer sequences).

Proofreading
Refers to any mechanism for correcting errors in protein or nucleic acid synthesis that involves scrutiny.

Pseudogene
A DNA sequence that shows a large degree of homology to the normal (expressed) gene but is itself not expressed.

Reading frame
One of three possible ways of reading a nucleotide sequence as a series of triplets.

RecA protein
The product of the *recA* locus of *E. coli* with dual activities, acting as a protease in the presence of single-stranded DNA (resulting from strong mutagenesis) and also able to exchange single strands of DNA molecules in co-operation with RecB and RecC proteins. The protease activity controls the SOS response; the nucleic acid handling facility is involved in general homologous recombination.

Recombinant DNA
Any DNA molecule created by ligating pieces of DNA that normally are not contiguous.

Relaxed
See **Open circular**.

Renaturation (of DNA or RNA)
The re-establishment of the DNA duplex or intrastrand hairpin structures in an RNA molecule after denaturation. (Of a protein): the conversion from an inactive into a biologically active conformation.

Replication eye
A region in which DNA has been replicated within a longer, unreplicated region.

Replication fork
The point at which strands of parental duplex DNA are separated so that replication can proceed.

Replicon
The regulatory unit of an origin and proteins necessary for initiation of replication (specific for this origin).

Repression	The blocking of the synthesis of certain enzymes when their products are present; more generally, refers to inhibition of transcription (or translation) by binding of repressor protein to specific site on DNA (or mRNA).
Restriction enzymes	Recognize specific short sequences of (usually) unmethylated DNA and cleave the respective DNA molecule (sometimes at target site, sometimes elsewhere, depending on type).
Retrovirus	A virus containing a single-stranded RNA genome that propagates via conversion into double-stranded DNA by reverse transcription.
Reverse transcriptase	RNA-dependent DNA polymerase. Originally detected in retroviruses. It is, however, also present in normal eukaryotic cells and even in *E. coli*.
Reverse transcription	Synthesis of DNA on a template of RNA; accomplished by reverse transcriptase enzyme.
Reversion (of mutation)	A change in DNA that either reverses the original alteration (true reversion) or compensates for it (second site reversion in the same gene).
Ribosomes	Subcellular particles consisting of several RNA and numerous protein molecules. Involved in translating the genetic code on mRNA into the amino acid sequence of the corresponding protein.
Ribosome binding site	See **Shine–Dalgarno sequence**.
Rifamycins (including rifampicin)	Antibiotic inhibiting transcription in bacteria.
rRNA	Ribosomal RNA. Forms part of the ribosome.
Sanger–Coulson sequencing	DNA sequencing technique based on transcription of single-stranded DNA by Klenow polymerase in the presence of dideoxynucleotides. The same technique can also be used for sequencing of RNA.
Satellite DNA	DNA consisting of many tandem repeats (identical or related) of a short basic repeating unit.
SDS (sodium dodecyl sulphate)	A detergent.
SDS gel electrophoresis	Gel electrophoresis of proteins in polyacrylamide gels in the presence of SDS. Molecules of SDS associate with the protein molecules giving them all a similar electric charge density and thus allowing separation on the basis of differences in molecular weight.
Semiconservative replication	Separation of the strands of parental duplex, each then acting as a template for synthesis of a complementary strand.
Semidiscontinuous replication	Mode in which one new strand is synthesized continuously while the other is synthesized discontinuously.

Shine–Dalgarno sequence	Part or all of the polypurine sequence AGGAGG located on bacterial mRNA just prior to an AUG initiation codon; is complementary to the sequence at the 3′-end of 16S rRNA; involved in binding of ribosome to mRNA.
Silent mutations	Do not change the product of a gene.
Site-directed mutagenesis	Introduction in the test tube of (a) specific mutation(s) into a DNA molecule at a predetermined site.
Site-specific recombination	Occurs between two specific (not necessarily homologous) sequences, as in phage integration/excision or resolution of cointegrate structures during transposition.
Southern blotting	A procedure for transferring denatured DNA from an agarose gel to a nitrocellulose filter where it can be hybridized with a complementary nucleic acid.
Spliceosome	The ribonucleoprotein particle containing precursor-mRNA and the various splicing factors on which splicing takes place.
Splicing	Describes the removal of introns and joining of exons in RNA; thus introns are spliced out, while exons are spliced together.
SSB (single-strand binding)	Protein of *E. coli*, a protein that binds to single-stranded DNA.
Sticky ends	Complementary single-stranded extensions at the ends of a DNA fragment or two different fragments resulting from a staggered cut (or introduced by tailing).
Stop codon	Same as termination codon.
Strand displacement	A mode of replication of mitochondrial and some viral DNA in which a new DNA strand grows by displacing the previous (homologous) strand of the duplex.
Structural gene	Gene coding for any RNA or protein product other than a regulator.
Supercoiling	The torsional twisting of a DNA duplex around its long axis causing the duplex to cross itself in space. Can be either negative (that is twisting in the direction opposite to the turns of the double helix) or positive (twisting in the same direction as the turns of the double helix).
Suppressor tRNA	A minor tRNA species that responds to a termination codon (e.g. a minor tyrosine tRNA with an AUC anticodon).
T_m	Is the abbreviation for melting temperature.
Template	Portion of single-stranded DNA or RNA used to direct the synthesis of a complementary polynucleotide.
Termination codon	One of three triplet sequences, UAG (amber), UAA (ochre), or UGA, that cause termination of protein synthesis; they are also called nonsense codons.
Terminator	A sequence of DNA that, after being transcribed, causes RNA polymerase to terminate transcription.

Topoisomerase	An enzyme that can change the linking number of DNA (in steps of 1 by type I; in steps of 2 by type II).
Topological isomers	Molecules of DNA that are identical except for a difference in linking number.
Transcription	Usually the synthesis of RNA on a DNA template. Also used to describe the synthesis of DNA on an RNA template by reverse transcriptase, the copying of a (primed) single-stranded DNA by DNA polymerase, and the copying of RNA by (viral) RNA polymerase.
Transformation	The acquisition by a cell of new genetic markers by incorporation of added DNA. In eukaryotic cells it also refers to conversion into a state of unrestrained growth in culture resembling or identical to the tumourigenic condition.
Transition	A mutation in which one pyrimidine is substituted by the other or in which one purine is substituted for the other.
Translation	Synthesis of protein on the mRNA template.
Translocation (of a chromosome)	A rearrangement in which part of a chromosome is detached by breakage and then becomes attached to some other chromosome.
Translocation (of the ribosome)	Moving one codon along mRNA after the addition of each amino acid to the polypeptide chain.
Transposition	The movement of part of the DNA to another location within the genome.
Transposon	A DNA sequence able to replicate and insert one copy at a new location in the genome, carrying genetic information additional to that necessary for transposition mechanism.
Transversion	A mutation in which a purine is replaced by a pyrimidine or vice versa.
Triplet	A sequence of three nucleotides in DNA or RNA. Usually means the same as codon.
Twisting number (T)	The number of turns in a DNA duplex (see linking number).
Underwinding (of DNA)	Produced by negative supercoiling (because the double helix is itself coiled in the opposite sense from the intertwining of the strands).
Upstream	Identifies sequences proceeding in the opposite direction from expression; for example, the bacterial promoter is upstream from the transcription unit, the initiation codon is upstream from the coding region.
Vector (cloning)	Any plasmid or phage into which a foreign DNA may be inserted to be cloned.
Watson–Crick rules	The base-pairing rules that underly gene structure and expression. G pairs with C and A with T (U in RNA).
Wild type	The genotype or phenotype commonly encountered in the natural population or laboratory stock of a given organism.

Wobble hypothesis   Accounts for the ability of a tRNA to recognize more than one codon by unusual (non-G:C, A:T) pairing with the third base of a codon.

Writhing number (W)   The amount of supercoiling of a DNA molecule (see linking number).

Z–DNA   A left handed form of Watson–Crick double helix.

Subject Index

Name Index